THE STEEL INDUSTRY IN THE NEW MILLENNIUM VOL 1:

Technology and the Market

THE STEEL INDUSTRY IN THE NEW MILLENNIUM VOL 1:

Technology and the Market

Edited by

R. Ranieri and J. Aylen

Papers presented at a European Conference
sponsored by
ICSIM, Istituto per la Cultura e la Storia d'Impresa,
'Franco Momigliano'

Volume published thanks to the contribution of the
EU Commission DG III Industry

Book 666
Published in 1998 by
IOM Communications Ltd
1 Carlton House Terrace
London SW1Y 5DB

IOM Communications Ltd
is a wholly-owned subsidiary of
The Institute of Materials

ISBN 1 86125 019 3

Typeset by
Fakenham Photosetting Ltd
Fakenham, UK

Printed and bound by
The University Press
Cambridge, UK

Contents

NOTES ON CONTRIBUTORS

FRANCO GIUSTINELLI, President of ICSIM (Istituto per la Cultura e la Storia d'Impresa, 'Franco Momigliano').

RUGGERO RANIERI, Research Associate and Jean Monnet Lecturer, University of Manchester. A modern economic historian, he sits on the Advisory Board of ICSIM and organised and chaired the Conference 'The Steel Industry in the New Millennium'.

ATTILIO ANGELINI is Chief Executive Officer of Acciai Speciali Terni S.p.A. He entered the steel sector in 1964 and after a short work experience in the U.S he was employed at the Taranto steel plant, where he progressed to Director of Plant and then Director General. From 1988 he has been CEO at Terni, and between 1989 and 1993 he was also Deputy Director and Director General of ILVA, with responsibility for Special Steel Flat Products.

PETER MARCUS is Managing Director of Paine Webber in New York. His much awaited, authoritative Steel Dynamics newsletter provides detailed cost comparisons across the world steel industry.

JONATHAN AYLEN was educated at the University of Sussex and Worcester College, Oxford. He is now Senior Lecturer in Economics at the University of Salford, where he has taught since 1974. His published papers in economics cover cost-benefit analysis, international comparisons of performance, privatisation and forecasting. He has travelled widely through the world steel industry and advised various national and international organisation on steel issues.

HANS BODO LÜNGEN is Deputy Senior Manager of the Raw Materials and Metallurgy Department of VDEH (Verein Deutscher Eisenhüttenleute), Düsseldorf. He graduated in metallurgy at the Rheinische-Westfälische Technische Hochschule (RWTH), Aachen, where he also earned a doctorate in 1990. He held posts, including Manager of Ironmaking, in VDEh since 1985.

KEVIN ALBERTSON is Lecturer in Econometrics at the Department of Economics of the University of Salford. He took his first degree in maths and operations research at the University of Canterbury in Christchurch, New Zealand before joining the New Zealand Department of Statistics for two years. He subsequently took a first class Masters Degree in Economics and completed a Doctorate on econometric theory.

PATRICK GENEVAZ is Director, Steel and Materials of Euroconsult (Paris). He started his career by teaching economics. In 1971 he joined CERU, a consulting company specialised in market research and project analysis. In 1973 he joined Eurofinance, an international consulting and financial organisation based in Paris, as senior economist in charge of the rubber, steel, nuclear, petroleum and electrical equipment industries. Between 1982 and 1995 he was Director of La Chambre des Cartes, a consulting firm specialised in the economics of steel and related industries. Among his publications: *Structure of the Steel Scrap Market* (European Commission 1994); *Evolution du marché mondiale des ferrailles et possibilités de substituts* (United Nations, Düsseldorf, 1995); *Steel Industries and Markets in Asia: Prospects to 2005* (Euroconsult 1997).

E. BIAMONTE is Managing Director of ITM – Innovation and Technology Management S.A. (the Battelle affiliate consultancy based in Geneva, Switzerland) and he supervises the Battelle Steel Programme. He graduated as an Industrial Chemist from the University of Padua, Italy, and covered several management positions at Battelle-Europe, where he specialised in strategic technologies consulting.

ANTHONY B. HINDER is a senior partner at ITM – Innovation and Technology Management S.A., where he has conducted several assignments related to the iron and steel industries and coordinates the Battelle Steel Programme. An economist by training with degrees at the University of Leicester and the Institute of International Studies (HEI), Geneva before specialising in strategic consulting at Battelle-Europe he worked in the Corporate Planning of Esso, Africa.

AZUSA TOMIURA is an Executive Advisor of Nippon Steel Corporation. After earning degrees from Kyushu and Tokyo University he has had a distinguished career inside the Corporation. He first joined the Yawata Steel Works in 1956, and has held a variety of posts, including Rolling Mill Engineer, Project Manager of the Construction of Kimutsu Works, Manager, General Manager, Director and Managing Director of Research, Development and Engineering. He is a Member of the Engineering Academy of Japan, of the Science Council of Japan and of the Loyal Swedish Academy of Engineering Science. Government service includes advisory roles for the Ministry of Education, The Ministry of International Trade and Industry and the Ministry of Science and Technology. In the latter capacity since 1994 he is a Member of the Strategic Committee of the Science and Technology Council. He has earned a number of prestigious awards, including ones from the Iron and Steel Institute of Japan and the Italian Society for Metals and he has served as Chairman of the Technical Committee of the IISI. He is also Vice President of the Society of Engineering Education in Japan. From 1996 he is a Visiting Professor at Keio University.

FRANCO BONOMO is an industrial consultant. After graduating in mechanical engineering at Padua University he worked at CSM from 1962 to 1997. He was, amongst others, responsible for technology, engineering and systems and processing programmes and Assistant Director for Innovation and Strategic Controls.

ANTHON F. EBERLE is Senior Vice President of the Division of Mineral and Reduction Technologies at VAI (Voest Alpine). He graduated at the Montanuniversity of Leoben, in Austria, in Metallurgical Engineering and was Project Manager and Vice President of the Department of Integrated Metallurgical Plants at VAI.

W. KEPPLINGER is a Professor of Environmental Process Engineering at the Montanuniversity of Leoben. After Graduation in Metallurgical Engineering he was employed by Chemie Linz, in Austria and joined the COREX® development at Voest Alpine in 1980. In 1986 he was appointed Division Manager for the COREX® plants.

SIEGFRIED ZELLER is Division Manager for FINMET plants at VAI. After graduation at the Montanuniversity of Leoben in Metallurgical Engineering he was Assistant to the Head of the Institute for Maschinenelemente und Allgemeine Maschinenkunde, at the same University. Since 1972 he has held posts within VAI, including Division Manager for Direct Reduction Plants.

EUGENIO REPETTO is Project Manager at CSM for the development of EAF technologies and the treatment of related materials. He has worked with CSM since 1966 as a researcher and has been involved in assisting production at various Italian electrical and BOF steel plants. He is currently in charge of ECSC multinational programmes in the field.

GUSTAVO BRASCUGLI, responsible for stainless steel production, raw materials testing and programming repairs at Acciai Speciali Terni, S.p.a. He graduated in Mechanical Engineering at the University 'La Sapienza' in Rome and since 1979 has held various managing jobs at AST's Steel works, including responsibility for the production of electric and construction steel.

G. PERNI is a Manager at Acciai Speciali Terni, S.p.a. in charge of the steelmaking plant for silicon, carbon and forging steels. He graduated at Perugia University in Chemistry of Inorganic Materials and worked for two years as a Research Assistant at the Chemical Science Faculty of Perugia. He joined AST in 1977 and worked for ten years in Quality Control, Chemical and Metallurgical Laboratories. He is currently a member of ECSC committees for steel matters.

WOLFGANG HENNIG is General Manager Sales and Coordination Bid, Preparation CSP, at SMS Schloemann-Siemag AG, Continuous Caster Division.

F. Hoffman, Vice President Sales, SMS Schloemann-Siemag AG, Düsseldorf.

B. Krüger is Senior Group Leader Sales and Coordination Bid, Preparation CSP, at SMS Schloemann-Siemag AG, Continuous Caster Division.

Federico Mazzolari is currently a consultant for Finarvedi Spa. After graduating from Milan Politechnic in mechanical engineering, he was a manager with SAVA SpA, an integrated manufacturer of aluminium, part of the Alusuisse group. In 1974 he moved to the Arvedi Group, first as Tube production Manager then as Marketing Manager and subsequently as Managing Director of one of the Group's subsidiaries. Since 1990 he has been at Finarvedi SpA, the holding company of the Group, and has held the position of Marketing and Planning Manager.

Ulrich Wengenroth is a professor at the Zentraalinstitut für Geschichte der Technik in the Technische Universität of Munich, at the Deutsche Museum. Among his works, Enterprise and Technology. *The German and British steel industries 1865–1895*, Cambridge University Press, 1994.

Matthias Kipping is a Reader in European Management in the Department of Economics at the University of Reading. He holds Masters Degrees from the Sorbonne in Paris and from Harvard University as well as a Doctorate from the University of Münich. Among his publications *Zwischen Kartell and Konkurrenz* (Berlin 1996) and a number of articles in English and French on industrial policy in the steel industry in historical comparative perspective. His current research interests include the transfer of management knowledge, namely the role of business education and consultancies.

Franco Audino has studied economics at the University of Genova and worked as a junior economist in various countries and as an economic planning analyst in Central Africa for the Office of Technical Co-operation of the United Nations. In the past twenty years he has worked in the steel industry in various development project in Italy and in South America. During a sabbatical stint at Paine Webber's World Steel Dynamics of New York he has contributed in drafting aggregate price cost models. As an economist in charge of the economics and market analysis department of a major Italian company in the past ten years he has analysed the development and prospects of the steel industry.

Pier Soulignac is now R & D Manager of special steel producer CLI in Le Creusot, France. He graduated in Materials Science at the Nancy School of Mines and worked in various fields of the stainless steel industry, including process, quality and applications development. In 1991 he joined Acciai Speciali Terni as technical marketing director.

ALAIN MOREAU is Marketing Director at the Direction of Strategy and Planning at Usinor. He has a degree in Economics.

CARLO PAGLIUCCI is Director of Finance, Marketing, and Technology Transfers at Centro Sviluppo Materiali, Rome. He graduated in Chemistry at Bologna University and specialised in Applied Economics at Naples Polytechnic. He has held posts within the IRI group since 1970 and within CSM since 1985. He is also President of Terni's Business and Technological Park and has published a number of articles on Management issues related to R & D.

JACQUES BROZZETTI is a Professor at the 'Ecole Nationale des Ponts et Chaussées' in Paris and Scientific Director at the French Institute of Steel Construction (CTICM). He graduated from the 'Ecole Nationale Supérieure des Arts et Métiers' (Paris) and from the 'Centre des Hautes Etudes de la Construction' (Paris). He received his M.S. at Lehigh University, USA. Being a specialist in fatigue problems, composite construction and fire protection of steel structures, he has acted as consulting engineer for projects in France and abroad. He is Chairman of the CEN/TC 250/SC3 in charge of drafting Eurocode 3 concerning steel construction.

MAURO PONTREMOLI is Deputy Director Manager for Products and Application at the CSM, Rome. He graduated in Physics at the University 'La Sapienza' in Rome and since 1980 has been a Researcher, Section and Project Manager at CSM. He is author of more than forty scientific articles on the metallurgy of flat rolled products, micromechanisms of fracture and environmentally induced fracture and fatigue. He is also active at the ECSC level as Italian expert for Flat Rolled Products and Structural Steel and as Chairman of the 'Corrosion Committee' of the European Pipeline Research Group, EPRG (since 1991).

LUCIANO FASSINA is a Consultant to NiDi (Nickel Development Institute of Toronto) a Board Member of The Milan Order of Engineers and of ADICO, Marketing and Commercial Director's Association, and the Chairman of CMM, Materials and Mechanical Committee at the Italian Metallurgical Association. He graduated in Chemical Engineering at Padua University and after a spell at Union Carbide-Edison as chemical plant designer he worked for Cofermet, Milan in the commercial area of flat rolled stainless steel, where he held a number of positions., including Marketing Director. He was a delegate for Italy's stainless steel at Eurofer in Brussels and at the Europe-Japan Working Group in charge of monitoring the world stainless steel market.

PHILIPPE DECLÉTY is an independent consultant to to NiDi (Nickel Development Institute of Toronto). His previous career was spent, for almost 30 years, with the French stainless steelmaker UGINE, where he was first development engineer and then marketing director. He has a specific expertise in the marketing and development of stainless steel products, particularly in the consumer goods, transport and food industries.

MASSIMO BREDA is Purchaser Reponsible for Public Areas Furniture and Arrangements on Cruise Vessels New Buildings at Fincantieri, Trieste. After graduating with distinction in Naval Architecture and Marine Engineering at the University 'Federico II' of Naples, he has been with Fincantieri since 1988, in charge, among other things, of information technology development and new products and engineering process methods. He has been involved in a number of international research activities including with the Working Group CESA (Committee of European Shipbuilders' Association) sponsored by the EU Commission, and in 1997 he was a member of the Standardisation technical committee sponsored by Euroyard.

CARLO VITTORIO FOLONARI holds a degree in industrial chemistry from the University of Turin. He started his career with Fiat in the Companies Central Laboratories in 1968. He joined the Fiat Research Center in 1973 and is currently Executive Vice-President for Processes and Materials.

FATHER WILLIAM T. HOGAN S.J. is Professor Emeritus in Economics at Fordham University. He has been engaged in economic studies of the steel industry and other basic, heavy industries for over 35 years. During this time he has authored a number of books, including his landmark five volume work, *Economic History of the Iron and Steel Industry in the United States*, (Lexington Books 1971), covering developments in the industry from 1860 to 1971. This was followed by a number of companion volumes, covering developments in the industry world wide, among which *Steel in the 21st Century. Competition Forges a New World Order* (Lexington 1994). In 1950 Father Hogan inaugurated the Industrial Economics Research Centre at Fordham University. He has appeared before legislative Committees of both the US Senate and House of Representatives on matters connected with capital investment and business taxation. In 1985 Father Hogan was awarded the Gary Memorial Medal by the American Irona and Steel Institute. He has also been honoured by the Republic of Korea with the prestigious Gold Tower of the Order of Industrial Merit. In 1993 he was awarded the honorary membership of IISI (The International Iron and Steel Institute).

PREFACE

In the Italian and European collective imagination of the 20th century, Terni has always represented industry and innovation in the iron and steel sector. Its steelworks are famous all over the world; and, although their staff has been greatly reduced, they excel, even today, in the production of special steels, those of the highest quality and technology.

Today Terni's economy benefits from the presence of chemical, textile, mechanical, as well as iron and steel companies. Historically, however, Terni is synonymous with 'steel'; seldom has a town identified itself so strongly with its main factory. The town almost became an appendix of the factory, thus creating a peculiar industrial 'mono-cultural' model. The origins of this lie in the second half of the 19th century, when, due to a number of strategic reasons ranging from the abundance of hydraulic power in the area to its distance from the sea considered to be a strategic asset, Terni was elected as the site of the first modern steel plant in Italy.

All this, however, was not sufficient to defend the factory and the town from recurrent crises linked to the instability of the domestic and international steel markets and to the restructuring processes, especially in the aftermath of the Second World War.

After having been part of the State shareholdings system for over fifty years, Terni's steelworks have recently been privatised, as a result of their acquisition by an Italian-German venture led by Krupp.

This was the background for the international Conference that took place in the summer of 1996: 'Steel in the new millennium: innovation, strategy and markets'.

The Conference was intended as part of the cultural activities launched in this town by ICSIM, whose full name is Istituto per la Cultura e la Storia d'Impresa, 'Franco Momigliano'. ICSIM was created recently by the major local government bodies belonging to the Umbria Region, by the ASSI Foundation for studies on business history, and by the Adriano Olivetti Foundation.

From its beginnings, the Momigliano Institute intended to devote a special, if not exclusive, attention to the new prospects of the world iron and steel industry. In the last two years this approach has led to the launching of the SteelMaster, an advanced training summer course aimed at technicians and managers. The SteelMaster was first held in 1997 at our headquarters in Villalago di Piediluco, an 18th century villa situated on the hills between Umbria and Lazio.

We intend to focus on the problems of the iron and steel industry in the new millennium, and each year, we are more and more committed to this issue.

These two volumes published in collaboration with the Institute of Materials in London testify our commitment. They contain the proceedings of the 1996

Conference. We owe a special thanks to Professor Ruggero Ranieri, of the University of Manchester, who co-ordinated the Conference and to Directorate General III of the European Commission; to the Region Umbria; to the Province and Commune of Terni; to the Banca Nazionale del Lavoro; to the National Council for Scientific Research (CNR), to Acciai Speciali Terni Ltd; to the Regional Association of Umbria's Chambers of Commerce; to the Terni's Employers' Association; and finally to the companies Arvedi and Dalmine. The financial contribution of all these bodies was invaluable.

Finally, I would like to mention all the scholars and experts who took part in the conference and are now included in these two volumes. We hope that their publication will be interesting for all those who operate in the steel industry.

Franco Giustinelli
Chairman of ICSIM
Terni, April 1998

FOREWORD

This is the first of two volumes bringing together the proceedings of the international Conference, 'The Steel Industry in the New Millennium. Innovation, strategy and markets' held in Terni on July 4 and 5, 1996. I would like to thank all the persons and institutions, who made that event possible. Organising a Conference for over 200 highly qualified participants from all over the world requires an important measure of both intellectual and organisational commitment and I am both proud and grateful of having been part of a large and dedicated team.

In the first place I wish to thank ICSIM, and particularly its president Franco Giustinelli, who not only sponsored and organised the Conference, but also encouraged its development from a very early stage, when few people were prepared to bet on its success. As a relatively new institution, committed to developing both research and teaching in business history and economics, while at the same time reinvigorating the links between the town of Terni and the wider international scholarly and business community, ICSIM provided a very valuable setting for the project.

I would then like to thank the Conference's many sponsors. In the first place the DG III of the European Commission, and particularly its Director General Pedro Ortun Silvan, who also enlivened the Conference's debate and proceedings with his intellectual contributions. The local government institutions all played an important role in allowing the Conference to take place: the Comune di Terni, and in particular Terni's Mayor, Gianfranco Ciaurro, the Provincia di Terni with its president Nicola Molè' and the Regione Umbria, whose president Bruno Bracalente enjoys the added advantage of being an academic economist, so that we were able to have a special dialogue with him. The Conference also benefited greatly from the support of AST, Acciai Speciali Terni, and we would like to thank its Chief Executive Attilio Angelini, whose introduction to the Conference opens this volume. A number of managers and employees of the Terni steel plant were also of great help in organising the Conference: above all I would like to thank dr. Francesco Menti. I would also like to thank the other sponsors of the conference: the CNR (Consiglio Nazionale delle Ricerche), the steel firms Arvedi and Dalmine, the Associazione Industriale of Terni and the Umbrian Unione Regionale of the Chambers of Commerce, the Banca Nazionale del Lavoro and the Azienda di Promozione Turistica del Ternano.

I am deeply indebted to all my colleagues with whom I worked for over fifteen months on the organising panel: Margherita Balconi, Roberto Bruno, Franco Bonomo, Gian Lupo Osti, Enrico Gibellieri and Carlo Pagliucci, Jonathan Aylen. Setting up the Conference's complex architecture was not an easy task, but we were able to overcome all the difficulties and disagreements by working together informally and constructively. In the context of the present volume, which deals with technology and markets, I am particularly grateful to the members of the Committee belonging or connected to the CSM (Centro Sviluppo Materiali) of Rome, and that is Roberto Bruno, Carlo Pagliucci and Franco Bonomo. They were able to contribute the extra-value added of the intellectual resources of a prestigious and well connected institution at the cutting edge of research. A special thanks to Franco Bonomo, who in his quiet and gentlemanly manner, provided some of the required mettle to get the whole project off the ground. He also competently chaired one of the most important Working Groups.

A great debt on my part goes to my co-editor Jonathan Aylen, whose input into this volume is of great value and importance, both as a contributor and as an editor. It is fair to say that I bear the whole responsibility of dragging him into this enterprise, well knowing of his burden of research, teaching and family commitments. I have no regrets, however, since I have treasured the value of his advice and scholarship. He is a first class expert on the steel industry, wonderfully opinionated and enormously wide-ranging. He is also a first class editor, when he sets his mind to it.

Warm thanks is also due to the Secretariat of the Conference, namely dr. Francesco Chiapparino, dr. Marco Marmottini and the personnel of Terni's Lingua Due, particularly Monica, Ambra, Isabella and Enza. All worked very hard and displayed great professionalism. The Conference would not have happened without them, and certainly not in the same way. The Administrators are the ones who have the most ungrateful task of all and very seldom do they get credit for the things that go right, so I would like to extend my thanks to Alberto Pileri of ICSIM, who helped overcome more than one obstacle.

Hopefully the contributors will find their final reward in the pages of this book. I will single out for a special mention, Peter Marcus, who lent his great authority to our Conference. I would also briefly like to mention those speakers, whose contributions, for a number of reasons, do not appear in print. In particular Luigi Agarini of Tad Fin Spa, who gave an important presentation to the Working Group on Markets, together with Guy Dollé, of Usinor Sacilor (whose piece on privatisation appears in volume II). Also a word of thanks to the French participants in the Technology section, Monsieurs Charbonnier and Birat of Irsid and Monsieur Quantin of Sollac. Particular thanks to Tim Smith of Steel Times, who did a wonderful job in chairing one session of the Conference as well as in publicising the event and summarising its contents in his lively and important journal. I am grateful to him also for his advice and support in making the necessary preliminary contacts for the Conference in the UK.

When it came to turning the Conference papers into a volume we found that there were new challenges and difficulties to face and inevitably we contracted new obligations. I would like to thank Mary Chim of the Institute of Materials, who was instrumental in offering ICSIM the opportunity to publish the entire Conference's proceeding in two volumes in the Institute's series. In fact my debt to Mary Chim is much greater, for she was a source of constant of advice and valuable contacts throughout the preparation of the Conference and beyond. We are also deeply indebted to Peter Danckwerts and to Simon Crump, of the Institute of Materials. The latter has been patient enough to work with us, e-mail by e-mail and chart by chart through the long and complicated editorial process. Finally ICSIM has supported my editorial task with a research grant at the University of Manchester.

The nature of this volume is such that it does not require any particular explanation or introduction. Following the works of the Conference, it intends to explore different aspects of the steel industry's present and future development. It is divided in four main sections, the first two of which correspond to the Conference's Working Group II on 'Technological Innovation', while the third contains the papers delivered during Working Group III, on the 'The Market for Steel'. The last section contains the final speech by Father William Hogan, written very much from the perspective of demand, whereas the introductory paper by Marcus (a much scaled down version of the large report he submitted to the Conference) looks at the steel market, mainly from the technological angle. Sub-sections are generally preceded by a brief introduction, laying out the main themes explored by the papers and offering a brief summary. The papers presented at the three remaining Working Groups of the conference – Working Group I on 'Privatisation and New Business Strategies', Working Group IV on the 'The Social Dimension' and Working Group V on the 'Steel industry in the European Union – are contained in a separate volume: 'The Steel Industry in the New Millennium: Institutions, Privatisation and Social Dimensions', also published by the Institute of Materials and edited by me and by Enrico Gibellieri. Since the Conference was held ICSIM's commitment to the steel industry has resulted in setting up the Steelmaster, an advanced training course for managers, technicians and trade unionists directed by Enrico Gibellieri.

Ruggero Ranieri

THE STEEL IN THE NEW MILLENNIUM CONVENTION

ATTILIO ANGELINI
Acciai Speciali Terui SpA, Viale B. Brin 218, I–05100 Terni, Italy

I would like to take this opportunity to thank the ICSIM Foundation and in particular its president, Senator Franco Giustineli for his invitation.

On behalf of Acciai Speciali Terni, which is the company I represent, it is a great honour for me to present the topics to be discussed at this Convention, which is dedicated to the steel industry in the year 2000. Of course the steel industry is an industry where major modernisation programmes are already currently leaving their mark and form the basis of far-reaching changes which are destined, over the next few decades, to alter its structure and performance.

The major issues which are currently being addressed with great vitality and a more market-oriented approach by the steel industry world-wide include:

- the privatisation of the steel industry within an international context and its consequent social problems;
- the major process/plant innovations which are having a substantial effect upon manufacturing technology as well as on the organisation of steel manufacturing sites;
- a technical and financial comparison between the large integrated sites and the more flexible mini mills which are largely dependent upon new technologies;
- the problem of the anticipated shortage of scrap iron for electric steel-works;
- the development of replacement products.

These topics form the basis of a discussion for the various different working groups for this Convention. The Convention is of particular significance and importance on account of the comparison between the different strategies implemented by the various major steel manufacturing companies world-wide, especially from the European Union.

TECHNOLOGICAL INNOVATION

The importance of recent innovations in manufacturing processes stands out among the various topics under consideration. This is a topic dealt with in detail by the 2nd Working Group of the Convention.

I would remind you that innovations of this type will affect steel manufacturing methods. In terms of value at least, steel is the most widely used material in the world. Indeed in terms of volume it is second only to cement with current production running at approximately 750 million tonnes per year (Mt/year), in an industry which globally employs some 3 million people and which currently accounts for approximately 4% of total industrial and civil energy consumption world-wide. This gives you some idea of the impact which innovations within the steel industry will have in the short and medium term. Steel is an industry which, I would reiterate, is in the full throes of technical development – thereby resoundingly putting pay to the gloomy predictions of decline which were being touted during the 80s.

Right at the end of the 80s two significant technological advances were made over conventional steel and hot coil manufacturing methods: on the one hand, in the integrated route, 'Reduction Smelting' methods, particularly the COREX process, on the other hand, in the electric furnace section the manufacture of the so-called 'Thin slab technologies' (Mannesmann/Arvedi ISP and SMS Compact Strip Production).

The first of these two methods was designed for implementation in the manufacture of molten iron in the absence of coke. The second set of processes were intended for use in the manufacture of coils and eliminated the need for conventional heating furnaces and a substantial amount of the hot rolling which was associated with this process.

Significant practical consequences of these technologies have been evident from the beginning of the 90s and it is anticipated that this will in turn have important implications and consequences before the end of the century and at the beginning of the new millennium, as the technology is slowly consolidated and perfected while, at the same time, similar technologies are developed by other operators within the steel engineering sector.

Both technologies are aimed at simplifying the manufacturing cycle with notable savings in operational and, above all, in investment costs.

This is opening the way for new operators to become involved in steel manufacturing. At traditional levels of investment, all such operators would have been excluded from getting involved in the industry. This trend would now appear to be irreversible; so much so in fact that it is thought that it will become even more firmly established, with the introduction of more advanced technologies such as second generation reduction smelting processes and finished shape direct forming processes, such as 'Strip Casting'.

Use of these new technologies will bring, in the case of the integrated route, savings for each tonne of coiled steel of up to over 30% of the investment costs and approximately 50% of the transformation costs. Similarly, in the case of the electric arc route, savings of almost 50% of investment costs and approximately 40% of transformation costs are expected. All of which holds true for small and medium sized manufacturing units (0.5–2.0 Mt/year), making the levels of investment required increasingly more accessible to new operators. This in turn will generate new and changed priorities for future manufacturing facilities where the trend towards 'Simple–Continuous–Flexible' will predominate.

It must be remembered that one of the most significant consequences of the anticipated simplification in the manufacturing cycles is the large energy saving which will occur together with the associated dramatic reduction in environmental pollution. The latter, in fact, has always made the location of steelworks a significant problem irrespective of where it was eventually located, but especially if this happened to be in the vicinity of an urban area.

The electric arc route would appear to be more compatible with this trend towards simplification, at least in the short to medium term. This is supported by recent increases in EAF (Electric Arc Furnace) production and by future forecasts which point to a gradual percentage increase in use of this process from the present level of 30% to 40% in 2005, and to 50% in 2010. An ever increasing share of flat products, traditionally manufactured in integrated works, will also be produced using the electric arc furnace route.

Over the last few years the growing importance of EAF production has led to significant innovative activity within the industrial sector in question. In turn, this has resulted in a dramatic reduction in energy consumption and an associated increase in productivity with development of direct current furnaces, continuous feeding and use of preheated scrap.

However, the increasing number of EAF plants has lead to questions emerging concerning the availability of scrap steel. Indeed for quite some time now a debate has been raging on this subject. The debate centres both on the quantity and quality of the scrap and obviously involves considering a range of ferrous materials which could potentially be used as a substitute for scrap. The overall view of process choice is therefore somewhat complicated by assumptions and forecasts in relation to the amounts of scrap and pre-reduced material which will be available for use in electric arc furnaces in the future. It therefore goes without saying that no single general view can be produced which will be valid for all manner of different geographic locations.

For instance, there is clearly a marked lack of scrap iron in the emerging Asian countries where it is expected that substitute products will be widely used. Yet, the situation appears to be less extreme in countries which have enjoyed a long period of industrialisation such as the USA, Japan and Europe. However, the widespread introduction of mini-mills for the manufacture of flat products, could even lead to a marked shortage of high quality scrap iron in these areas. It is therefore reasonable to predict that a certain amount of substitute ferrous material will also be used in areas which are, generally speaking, richer in scrap.

As a result particular attention will fall on 'Pre-reduced' material. In addition to on-going improvements in terms of yield within the context of existing manufacturing technologies, direct reduction has also seen intense experimental activity aimed at the development of new processes which are more competitive in terms of investment, energy consumption and diversification of raw materials which may be used. Examples include Iron Carbide, Cicored, Arex, Finmet etc. In this industrial sector it might be possible to reduce production costs to a level at which pre-reduced material is competitive with the current price of scrap, even in areas which currently lie outside geographical areas traditionally blessed with low energy costs.

In addition to this framework of active renewal and quasi-revolution within steel over the coming few years, the associated plant supply industry must also be carefully considered. This refers both to the modernisation of existing plant and to the installation of new 'Green Field' plant. Even if it is true to say that new technologies are likely to be introduced gradually, simply on the grounds that it is uneconomical to replace conventional plant machinery which is not yet inefficient, this business is nevertheless set to become significant in future years. To give you an idea of the volume of business involved, investment estimates exist for approximately 160 billion dollars worth of business up to 2010 solely in relation to the primary area of iron manufacture and basic oxygen steelmaking.

PRIVATISATION AND THE AST CASE

The choice of Terni as the location for the Convention, seems to strike a particular chord with the topics in question since it also happens to be the location of the main Italian manufacturing plant for special steels.

Acciai Speciali Terni is in fact one of the most notable examples of recent privatisation projects which have taken place in Italy, as a result of a programme which began back in 1993 and which was completed in 1994. The 'Acciai Speciali Terni' ILVA division was acquired by the Italo-German Consortium KAI which is made up of Krupp (50%) and FAR (50%). FAR itself was a group the shares of which were held by RIVA, FALCK and AGARINI. The subsequent withdrawal of RIVA and FALCK from the scene means that currently KRUPP owns 75% and AGARINI owns the remaining 25% of Acciai Speciali Terni. Currently Acciai Speciali Terni manufactures over 1 Mt steel per year. This breaks down into 0.65 Mt of stainless steel and 0.35 Mt of electrical and carbon steel. Approximately 3,400 employees are employed overall at the two sites in Terni and Turin.

As will be outlined in the papers delivered in the first Working Group of the conference, privatisation has often had significant negative economic and social effects on the areas affected by it, and has sometimes led to dramatic cutbacks at, and closures of, manufacturing units. In the case of Acciai Speciali Terni, privatisation has to be seen as part of the Rationalisation Programme which was initially implemented at the end of the 1980s. The introduction of the Rationalisation Programme was made necessary by the competitive market forces which were apparent and as a result of various restructuring and internationalisation requirements, and has lead to consistent reductions in staffing levels.

The negative social impact of this restructuring has been contained thanks to manufacturing growth which have been achieved by the company over the last few years and to the significant re-industrialisation efforts which have been undertaken by Acciai Speciali Terni. These efforts have led to the creation of jobs for more than 200 people (I am referring to Tubificio di Terni, Società Titania, ILSERV, and the Centro Servizio per l'Inossidabile [Stainless Steel Service Centre]).

Since privatisation, there is increased reticence, at least initially, to embark upon strategic type research which by its very nature requires both considerable investment on the one hand and the delivery of a return in the longer term on the other. As a consequence of this, public involvement is needed as much as ever in order to ensure that medium and long term research continues to be carried out, even in the case of a newly 'Privatised Industry'.

The USA and Japan are cases in point. The US government lends its support to the Direct Steel Program, while the Japanese government supports the D.I.O.S. Programme. Both programmes are aimed at developing integrated direct processes for the manufacture of molten iron and steel from iron ore mineral.

From this point of view it is possible to report that, in collaboration with CSM, Acciai Speciali Terni has succeeded in accessing and activating both state and European community related financial mechanisms in order to undertake important research projects as in the case of the new continuous casting technologies. This in turn has enabled the company to preserve and maintain its position as a leader in the field of innovation.

THE MARKET

It is clearly a requirement for further development that, in addition to preserving a competitive edge in relation to the competition in terms of quality and costs, there should also be a number of favourable market areas in existence at both a national and an international level.

It is anticipated that, while fluctuating in the short term, the demand for steel world-wide is set to continue to rise. Indeed some forecasts would tend to point to a gradual overall increase from the current level of 750 Mt/year to a level of 900 Mt/year in 2005, which would involve an average annual increase of approximately 2%. Over the next few years, an even higher percentage increase is anticipated in the case of special steel products. In particular, in the case of stainless steel products, forecasts point to a rise of approximately 5% per annum.

Within this rather optimistic framework, it must be stressed however that competition at an international level is becoming an increasingly significant factor. This is particularly true in view of the anticipated rise in output in the emerging Asian countries. Cheap labour and the application of innovative 'green field' technologies are together expected to see these countries succeed in appreciably lowering production costs and thus absorbing the majority of any future increases in market demand.

In addition to the on-going effort to reduce operational costs and to increase productivity, a further aspect which traditional manufacturers must consider carefully in order to be able to deal effectively with growing competition from 'newcomers', is how to ensure that the service to the customer, in its broadest sense, is continually improved and of how to establish a captive market share of around 50%. These particular issues are dealt with in greater detail at the third Working Group of the Conference.

To bring my speech to a close I would like to take this opportunity to express the hope that all the participants at this Convention will find their time here to be most profitable. I can see that representatives are taking part from all areas of the industry, from manufacturers, to plant machinery industries, end users at industry research centres, associations representing the industry, universities, representatives from the trade unions as well as national and supranational institutions.

I believe this conference to be one of those rare opportunities which must be exploited to the full to secure as many important and useful results as possible which can be added to the overall pool of knowledge and know-how which is available to the industry, as well as to promote progress within the bounds of the industrial sector within which we work.

IRONIES IN STEEL

P. F. MARCUS

Paine Webber Inc., 1285 Avenue of the Americas, New York, NY 10019, USA

INTRODUCTION

The steel industry abounds in incongruities. Outcomes are often unexpected. The reverse state of affairs is sometimes as common as the expected state of affairs.

Who would have thought just two years ago that the blast furnace/BOF steelmakers would be stealing scrap and steel scrap substitutes from the EAF producers? Or that the EAFs would be procuring sizable amounts of pig iron to charge in their furnaces? What an irony!

Indeed, there are plenty of ironies in steel. The invisible hand is in an iron glove.

The Youth Pill

The tradition-rich and tradition-bound steel industry took a youth pill about a decade ago. The pill was prescribed because the patient was weak after years of anemia. This condition had been brought on by attacks of complacency, dilapidated equipment, poor marketing efforts, bureaucratic organisations, inadequate responses to competitive threats and stable prices. The industry lacked iron in its blood – it needed a strong dose of Geritol (an iron-enriched elixir).

The first vital sign that the pill was working was the start-up of Nucor's thin-slab/flat-rolling plant in the United States in 1989. Subsequently, the steel industry scrapped its way to the reinvigorated competitor it is today.

What Caused the Industry to Grow Younger?

Besides the pill, it was the implementation of a multi-faceted rejuvenation program.

- *To lose weight*, it entered the cost cutting race. The race was not run of the mill. It was a run for survival.
- *To look better*, it colored its hair green. It invested heavily in measures to combat air and water pollution.
- *To clean up its house*, it swept its floor clean of old dust and scrap.
- *To gain strength*, it reduced its dependency on coke.
- *To stimulate its libido*, it courted new and attractive virgin iron units.
- *To learn how to dance with others*, it went to a hot strip mill.
- *To resurrect demand for its product*, it cut the price.
- *To stimulate its brain*, it submitted to electric arc shock therapy.

- *To widen its appeal*, it learned to play golf – the best business-to-business game in the world.
- *To lessen stress*, it gave its computers the responsibility for negotiating steel prices.

Main Themes in this Presentation:

1. Global steel demand outlook is outstanding.
2. Steel's technological revolution to deepen in the flat-rolling sector and broaden in the steel scrap substitute sector.
3. Global steel industry facing years of a tight metallics supply.
4. Most traditional BF/BOF steelmakers except those with extremely low costs, under siege. No matter what the scenario their perception of their prospects will darken.
5. Developments in China to exacerbate the shortfall of steelmakers' metallic and raw materials.
6. New winners and losers.
7. The truth and consequences.

1. – GLOBAL STEEL DEMAND OUTLOOK IS OUTSTANDING

Global steel demand prospects are unrestrained. The factors fettering steel demand during the past 20 years have been neutralised. We are now in an era in which demand forecasts are more likely to err on the low side (Fig. 1). Prospects are outstanding from many perspectives:

- Driving forces are looking more favorable. Reasons for the slow growth of steel demand from 1974 to the early 1990s have been reversed.
- Reduced inflation rates and interest rates could promote higher global growth for years to come.
- Steel intensity – consumption of steel per unit of industrial production – has been surprisingly stable since the early 1990s. Steel intensive sectors no longer seem to be losing out to other sectors of the economy at the same rate.
- Steel appears to be gaining market share from competitive materials in the construction sector. It is replacing wood in a number of instances. Also, the growth of metal commercial buildings based on steel has been substantial in the United States.
- As global economic activity expands at a high rate, the 'multiplier effect' which drives up steel demand proportionately faster kicks in.
  - There is little growth of steel demand when an economy grows 2% per year. Much of the expansion is in services.
  - Steel demand can be up 8% if an economy grows 5–6%. Most of the increment to growth occurs in consumer durables, capital spending and construction sectors.
- During the period from the late 1940s to the early 1970, global steel

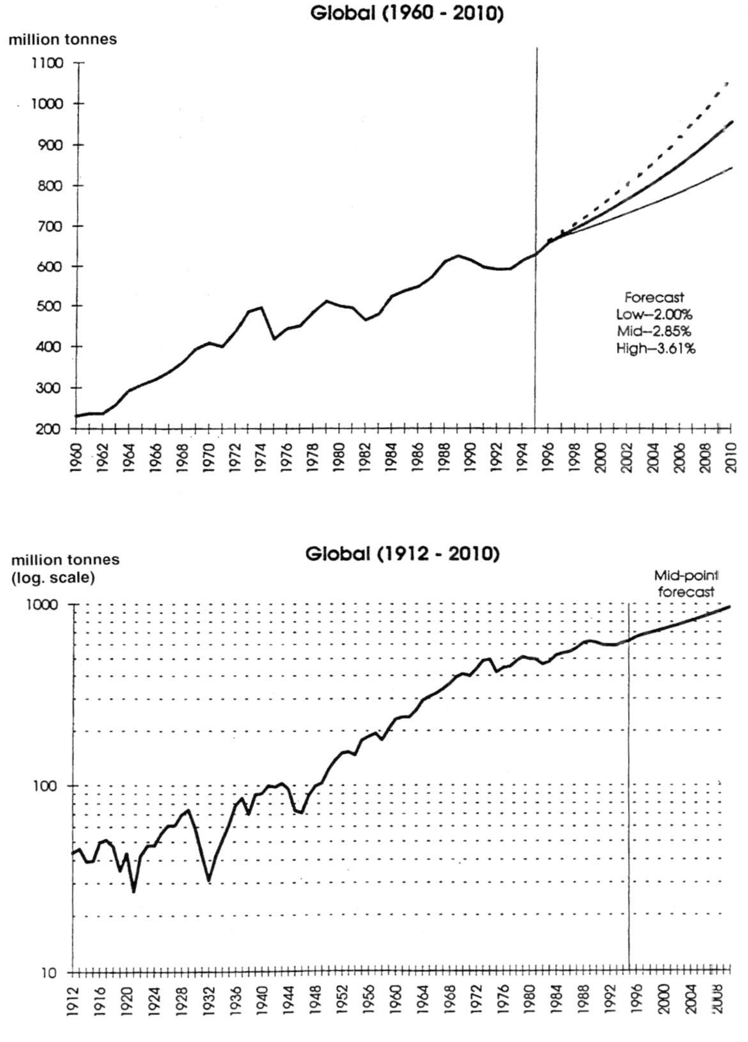

Fig. 1 Global apparent steel consumption (million tonnes).

products demand rose 6.8% per annum on a compound basis. This was a period with a number of characteristics similar to the present time, such as relative low interest and inflation rates. From 1973 to 1993, the growth of demand was just 1.1% per annum, compounded.

- Analyses of per capita trends tend to produce more optimistic demand forecasts. From 1995 to the years 2010, population growth is forecasted to be 1.8% per annum in the Atlantic Basin and 1.8% per annum in the Pacific Basin.
- An analysis of the steel demand outlook by region also tends to produce

high-end forecasts. This is because of the optimism for sizable growth in regions such as China, the Far East and South America.

- Pacific versus Atlantic Basin. In the next 15 years, World Steel Dynamics estimates that the compound growth rate of global steel demand will approach 3.0% per annum. Our single point figure is 2.86% per annum, including a sharp rebound in Chinese apparent steel consumption. Based on WSD's global metallics analysis to the year 2010 (Fig. 2 and Table 1):
  - Demand growth will be 4.6% per year in the Pacific Basin versus only 1.2% per year in the Atlantic Basin.
  - Demand growth will be 5.1% per year in the Developing World, 0.9% per year in the Developed World and 0.8% per year in the former Soviet Union and Eastern Europe.
- Developing World economies growing share of the global economy. They have grown so large that shifts in steel demand in the Developing World are just as important as those in the Developed World. The Developing World now accounts for about 44% of global GNP.

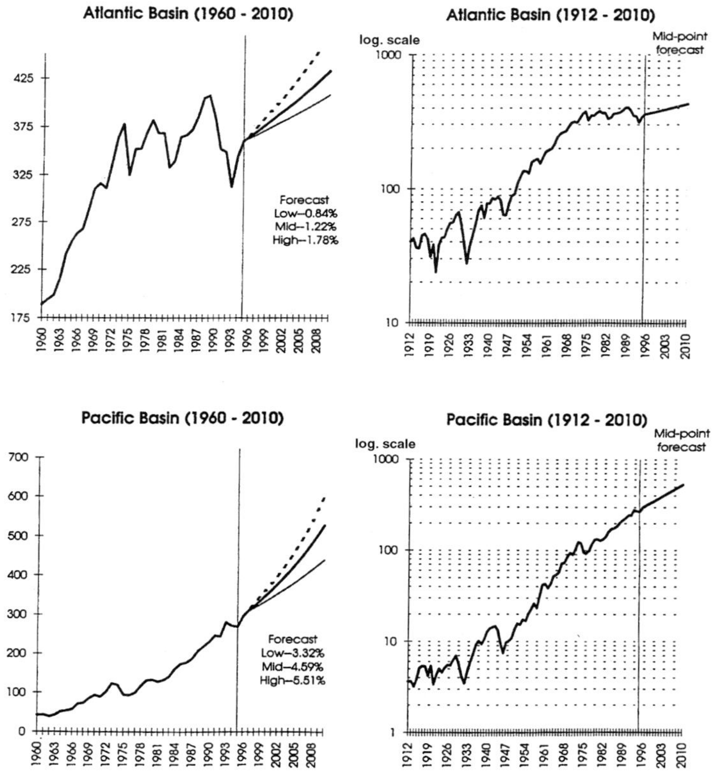

Fig. 2 Apparent steel consumption in the Atlantic and Pacific basins (million tonnes).

Table 1. Apparent steel consumption trends

	1995	2010					
		Low Case		Mid Case		High Case	
	Million tonnes per annum	Million tonnes per annum	CAGR	Million tonnes per annum	CAGR	Million tonnes per annum	CAGR
USA	105.0	111.5	0.4%	114.7	0.6%	117.5	0.8%
Japan	79.1	79.6	0.1%	82.6	0.3%	85.8	0.6%
European Union	108.5	117.8	0.6%	122.1	0.8%	124.1	0.9%
Other Western Europe	16.3	18.9	1.0%	19.7	1.3%	20.4	1.5%
Other Developing Nations	21.3	25.1	1.1%	27.7	1.8%	28.6	2.0%
South America	31.3	42.2	2.0%	47.3	2.8%	60.6	4.5%
Africa	11.6	20.9	4.0%	21.7	4.3%	22.4	4.5%
Middle East	20.6	29.8	2.5%	31.0	2.8%	34.4	3.5%
Asia	86.9	151.0	3.8%	163.0	4.3%	208.4	6.0%
CIS & Eastern Europe	66.0	66.0	0.0%	74.3	0.8%	88.8	2.0%
China and North Korea	81.6	183.2	5.5%	254.0	7.9%	278.6	8.5%
Developed Nations	330.1	352.9	0.4%	366.7	0.7%	376.4	0.9%
Developing Nations	232.0	427.0	4.2%	517.0	5.5%	604.5	6.6%
CIS and Eastern Europe	66.0	66.0	0.0%	74.3	0.8%	88.8	2.0%
Atlantic Basin	358.3	407.1	0.9%	431.7	1.3%	469.3	1.8%
Pacific Basin	269.8	438.8	3.3%	526.3	4.6%	600.3	5.5%
Global	628.1	845.9	2.0%	958.0	2.9%	1069.7	3.6%

CAGR – compound annual growth rate

- The increased transferability of technology is promoting capital spending. Developing countries are able to produce a high-quality, low-cost product using only moderately skilled workers.
- The extensive need for infrastructure around the world.

2. STEEL'S DOUBLE-BARRELED TECHNOLOGICAL REVOLUTION

The steel scrap substitutes (SSS) revolution raises to double-barreled from single-barreled the assault on traditional steelmakers. The rapid technological changes are attracting new capital to the steel industry. This new capital is often aligned with highly entrepreneurial business groups.

The first assault, the single-barreled one, was steel's thin-slab/flat-rolling revolution (Table 2). The latest versions of these units have the advantages of:

- Even lower capital costs per unit of capacity.
- Very low labor costs per unit of production. Man-hours per metric ton shipped are less than 0.5.
- Product characteristics which, in some respects, are superior to those of traditional units. For example, they are rolling to a thinner gauge than many of the traditional units.

Table 2. EAF-based TS/FR steel mills' share of the flat-rolled steel market*
(million tonnes, unless otherwise specified)

	1994				2000				2005			
	TS/FR Units	TS/FR Prod'n	Flat Prod Market	% Thin Slab	TS/FR Units	TS/FR Prod'n	Flat Prod Market	% Thin Slab	TS/FR Units	TS/FR Prod'n	Flat Prod Market	% Thin Slab
U.S.A.	2	2.0	55.8	4%	8	12.0	54.4	22%	8	14.0	57.0	25%
Japan	0	0.0	28.8	0%	0	0.0	29.8	0%	0	0.0	30.6	0%
EU	1	0.3	54.2	1%	3	4.0	55.3	7%	4	6.0	56.1	11%
Other W. Europe	0	0.0	7.8	0%	1	1.0	8.5	12%	2	3.0	8.9	34%
Other Dev. Countries	0	0.0	11.1	0%	2	4.0	10.5	38%	3	6.0	11.0	55%
Developed World	3	2.3	157.7	1%	14	21.0	158.4	13%	17	29.0	163.5	18%
Latin America	0	0.0	12.8	0%	3	4.0	14.2	28%	5	7.0	17.0	41%
Middle East	0	0.0	8.8	0%	2	3.0	9.2	33%	3	5.0	10.5	47%
Africa	0	0.0	4.1	0%	1	0.5	5.8	9%	3	2.0	7.5	27%
South East Asia	0	0.0	38.9	0%	8	9.0	57.5	16%	12	15.0	76.6	20%
Developing World	0	0.0	64.6	0%	14	16.5	86.6	19%	23	29.0	111.6	26%
CIS, Eastern Europe	0	0.0	32.9	0%	2	2.0	40.8	5%	6	8.0	44.2	18%
China, N. Korea, Vietnam	0	0.0	21.7	0%	2	2.0	38.3	5%	7	8.0	51.0	16%
Ex Communist World	0	0.0	54.6	0%	4	4.0	79.1	5%	13	16.0	95.2	17%
Atlantic Basin	3	2.3	163.6	1%	17	23.0	173.1	13%	25	38.0	183.2	21%
Pacific Basin	0	0.0	113.4	0%	15	18.5	150.9	12%	28	36.0	187.2	19%
World	3	2.3	277.0	1%	32	41.5	324.0	13%	53	74.0	370.3	20%

* TS/FR = Thin or thinnish slab (to 5" thick) flat-rolling units.
** Note: Flat products market is reduced 15% in all regions in order to exclude cut plate.

The second assault, making the threat double-barreled, is new production methods at the primary end – ironmaking and steelmaking side – of the business.

A sizable number of new processes are being developed in which the iron is extracted from the iron ore without the use of coke. EAF steelmakers in the years ahead will have the option to become 'integrated' by taking full or partial ownership positions in low-cost steel scrap substitute (SSS) facilities (Fig. 3).

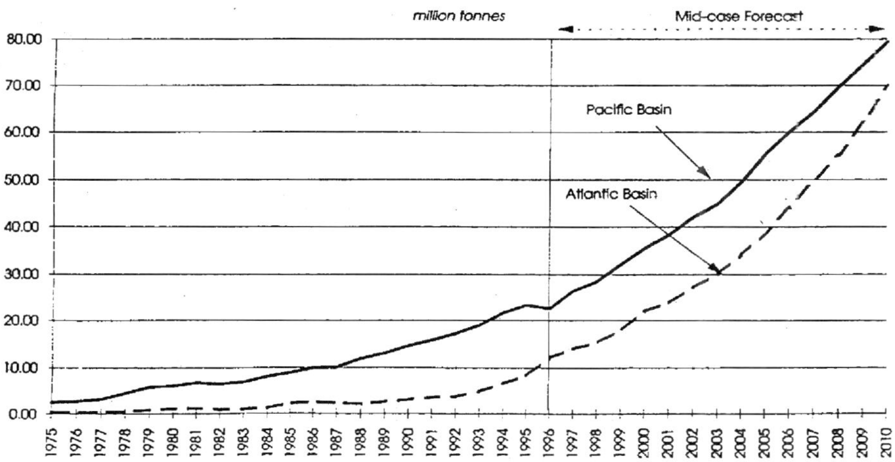

Fig. 3 Steel scrap substitutes and hot metal substitute output in the Atlantic and Pacific basin.

Say's Law, which is the principle that 'supply creates demand', is alive and well at the primary end of the steel industry. The greater the number of steel scrap substitute facilities that are built, the more rapid the construction of EAF steelmaking furnaces that will consume the steel scrap substitutes.

- Developments will include:
  (a) Coal-based SSS technologies that work with both lump ore and fines.
  (b) Natural gas based processes that work well with iron ore fines. This includes the iron carbide process and Circored.
  (c) Coal and natural gas based processes, which produce liquid pig iron either directly or indirectly. Already, the Corex process, which uses coal and lump ore or pellets, produces liquid pig iron. We will also see:
    - The conversion of solid SSS materials into a liquid hot metal (pig iron) in separate furnaces, such as submerged arc electric furnaces, before they are placed into steelmaking furnaces. This material will then be charged, often, into EAF steelmaking furnaces.
    - The direct production of liquid iron from iron fines, with the reductant being either coal or natural gas.

    - The HIsmelt process has for years been attempting to achieve this. One of the major problems, we understand, is the very abrasive nature of the materials that are produced, which causes high refractories wear.
    - A general problem with the coal-based process is accretion, which is coal sticking to and clogging up the processing equipment.
- Steel's technological revolution in both phases – the flat-rolled phase and the steel scrap substitute phase – will cause a significant shift in new capacity in the global steel industry towards the steel-scrap-using electric arc furnace (EAF) steelmaking route. It is highly threatening to the traditional blast furnace/basic oxygen furnace (BF/BOF) route.
  - Electric arc steelmaking furnaces can now convert steel scrap and steel scrap substitutes to liquid steel at a far lower cost than 20 years ago.
  - EAF steelmakers can now make a broad range of steel products on an effective basis. Hence, EAF steelmaking no longer has significant restrictions when it comes to the markets that it can serve.
  - In developing countries, the substantially lower capital costs and smaller economies of scale are promoting the installation of EAF units over the BF/BOF route.
  - As noted above (Say's Law), the more sizable the announcements of new steel scrap substitute units in the future, the greater the incentive to build EAF-based steelmaking units.

Incentives to Build SSS Units

The financial returns when planning a steel scrap substitute facility have improved substantially. In the early 1990s, when steel scrap prices were about $20 per ton lower than at present, the incentive was 'fair to good'. At current scrap prices, the incentive is 'very good'. If steel scrap prices move up another $20 per ton over the steel cycle, the incentive would become 'outstanding' (Figs 4 and 5).

The construction of SSS units has been hindered over the years for these reasons:

- BF/BOF producers were often not willing to participate in these ventures because they preferred the blast furnace route.
- EAF producers were reluctant to boost their fixed costs, especially when they were producers of long products. They had learned from their experiences, as well as from observing the major mills that fixed costs can become a devastating problem when the steel market is depressed. They also know that, at least historically, when steel scrap prices are high, so are the prices of steel finished products.
- The construction of a SSS unit takes two years or more. Negotiations can be lengthy when it comes to site selection.
- There are not many regions in the world at which natural gas is available at

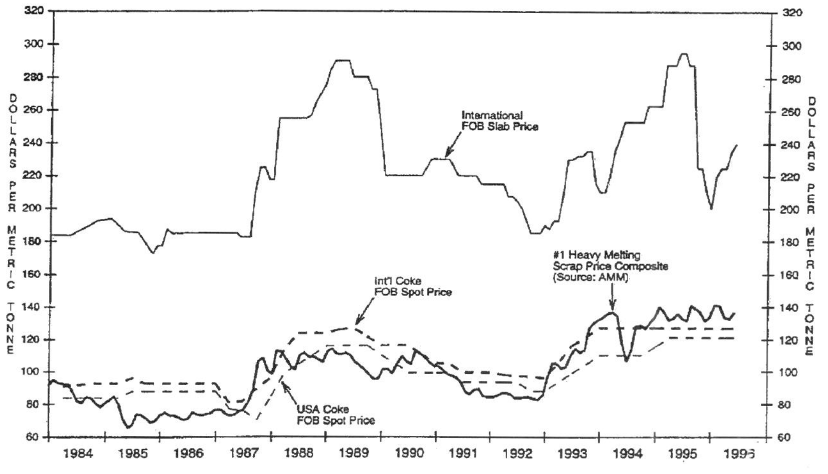

Fig. 4 Steelmaker's raw material prices.

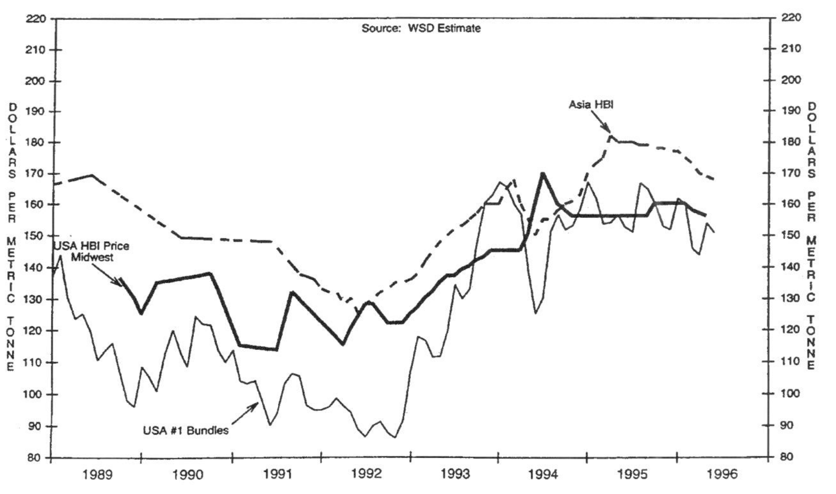

Fig. 5 WSD HBI delivered price estimates.

bargain prices. There are often extended discussions and changes in contract terms before prices are settled.

- Merchant DRI plants have been difficult to finance. This is because not many buyers were willing to sign up to purchase at a fixed minimum price, they perceived that the minimum price was too high. (Note: This is no longer the case, since scrap prices are higher and because of the rising demand for low residual scrap – tied into EAF-based sheet mills – has become abundantly evident).

- A shortage of iron ore pellets in 1995 and 1996 was a factor restraining the traditional DRI- and HBI-type plants.

Stage I – from 1989 to 1997
Key milestones included:

1. *Nucor's first thin-slab/flat-rolling plant in Crawfordsville, Indiana in 1989.* This plant opened the door for EAF-based steelmakers to get into the flat-rolled steel business on a low capital cost and low operating cost basis.
2. *The application of computer systems to increase the reliability of steelmaking and steel rolling processes.* Consequently, the generation of secondary steel was reduced and product quality improved.

Stage II – from 1998 to 2004
Important milestones will include:

3. *The commercialisation of a variety of new steel scrap substitute processes.* These will depend on iron ore fines rather than pellets as the feedstock and/or coal rather than natural gas as the reductant (which converts FeO to Fe).
4. *The advancement of the capability of thin-slab/flat-rolling facilities to the point that the product quality is excellent from a mechanical, chemistry and surface quality point of view.* In Phase I of the revolution, commodity-grade steels were produced at low cost. In Phase II, specialty steels will be produced at close to the cost of commodity steels.
5. *The production of 300 series stainless hot-rolled band at new facilities with a low operating cost and a low capital cost.* Japan's Nippon Steel Hikari plant, a $110 million capital cost new facility scheduled to produce about 400 000 tons per year of stainless hot-rolled band via a twin-drum continuous caster. This development lowers the barriers to entry to the stainless steel business and promotes demand for stainless steels.

Stage III – starting in 2005

6. *The use of the computer to control much of the steel business of an individual company except for the price setting.* The computer receives new orders from the company's customers (by computer, of course), determines the steel production and rolling schedules, controls the processing of the product and keeps the customer informed as to the status of each order.

Stage IV – starting in 2010

7. *Computers set the prices of steel.* The computers of all companies link up. They take over the price negotiating process with customers in both the spot and one-year contract markets. The result is an 'administered' pricing system never imagined in economic textbooks. Prices rise when volume declines because fixed costs are higher.

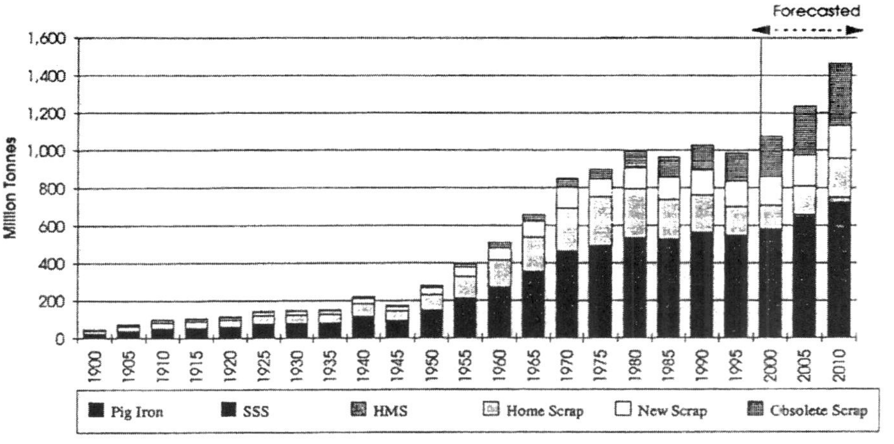

Fig. 6 Global metallics usage (million tonnes).

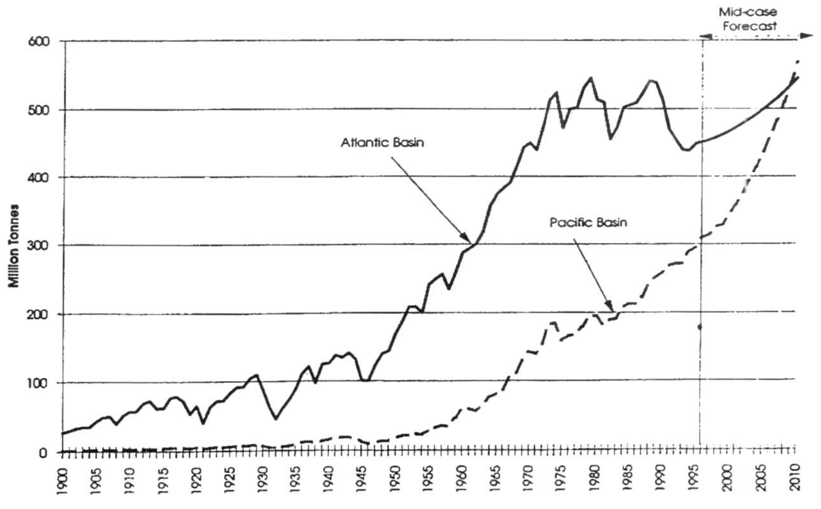

Fig. 7 Crude steel production for the Atlantic and Pacific basin.

3. GLOBAL STEEL FACING YEARS OF TIGHT METALLICS SUPPLY

World Steel Dynamic's ongoing analyses indicate that a global metallics crisis will be a factor dominating the steel cycle over the next decade, see Figs 6 and 7. Some general points:

- Similar to the copper, aluminum and nickel industries, it will be the supply/demand balance for metallics that will be driving the price of the finished products.
- On an after-the-fact basis, there is never a shortage on a physical basis in

the real world. The price of the product rises to 'clear the market'. This is what has happened with respect to steel scrap prices. High steel scrap prices are expected to produce a continuing frenzy to build steel scrap substitute plants.

Constrained Metallics

The World Steel Dynamics analysis indicates that there will be a shortage of steelmakers' metallics almost no matter which demand scenario is chosen. How is this possible?

1. Blast furnace output will not surge, no matter what the scenario. As will be discussed in the next section, most BF/BOF producers are in trouble on a long-term basis. The cost curve for the BF sector of the industry in steep; many facilities will be abandoned in the next 15 years. The location in the slow-growing Atlantic Basin is another negative for this group.
2. Swings over the cycle in BF/BOF output will be greater than those in EAF output.
3. The amount of steel scrap substitute capacity that will be attracted into the market will vary depending on the scenario. In 1995, SSS output was 32 million metric tons. By the year 2010, the output is forecasted to range from 120 to 180 million tons, depending on the ebullience of the industry environment in the ensuing years (see Fig. 6).
4. Open-hearth steel output will decline sharply in the years ahead, no matter what the scenario.
5. The higher the capacity to produce steel scrap substitutes, the greater the EAF steelmaking capacity that is attracted into the industry.
6. The reservoir from which obsolete steel scrap is derived will not vary significantly in size, no matter what the scenario.

Optimistic Case Shortfall

Granted steel scrap recovery at the high end of the possibilities – which we call the 'optimistic' forecast – by the year 2010 we see the steel scrap shortfall at about 62 million tons (Fig. 8 and Table 3). This calls for the recovery of obsolete steel scrap as a share of the reservoir which is, on average 8 to 30 years old, to rise from about 62% in 1995 to 82% in the year 2010.

Hopefully, recoveries at this level can be accomplished without steel scrap prices rising to extraordinary levels.

Conservative Case Shortfall

Granted our 'conservative' forecast of steel scrap recovery, the shortfall is on the order of 118 million tons.

Specifics Behind Our Mid-Case Global Metallics Balances Analysis

Assuming a global steel products demand growth rate of 2.86% per annum, compounded, from 1995 to the year 2010, the shortfall in the metallics supply

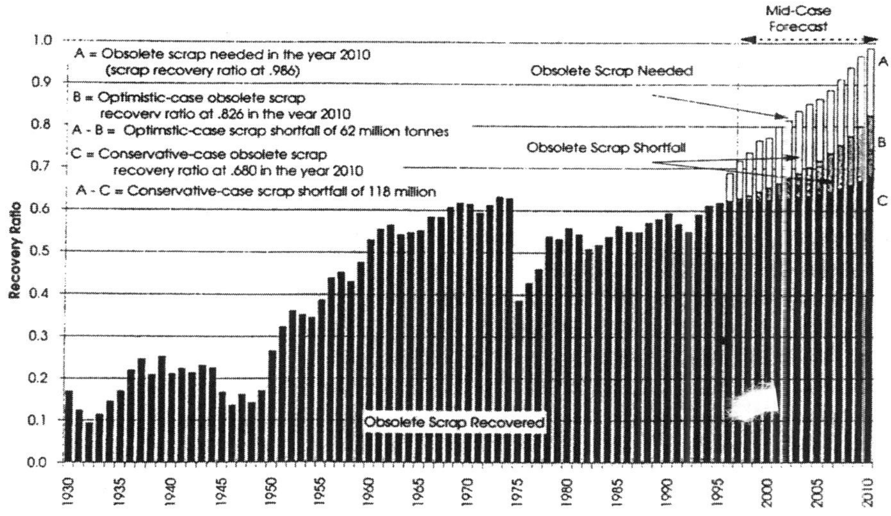

Fig. 8 Global obsolete scrap recovery versus reservoir 8–30 years old on average.

Table 3. Steel scrap and relative items compound growth rates

	1947 to 1973	1973 to 1994	1994 to 2010E
Steel products demand			
Atlantic Basin	6.08%	−0.26%	1.46%
Pacific Basin	10.30%	3.84%	4.22
Global Total	6.80%	1.13%	2.82%
Scrap reservoir avg. 8–30 yrs. old			
Atlantic Basin	3.44%	3.83%	−0.12%
Pacific Basin	5.62%	7.16%	4.86%
Global Total	3.72%	4.54%	1.64%
Obsolete scrap recovery requirements			
Atlantic Basin	6.64%	3.45%	1.83%
Pacific Basin	9.65	8.59	7.25%
Global Total	6.87%	4.23%	3.64%

ranges from 62 to 118 million metric tons depending on the rate of recovery of obsolete steel scrap from the steel scrap reservoir. Note: The obsolete steel scrap requirement is derived last, after the other sources of metallics have been estimated.

In 1995, the need for steelmakers' metallics, plus foundry production and unreported pig iron deliveries, was 979 million tons. In the year 2010, for our mid-case forecast, this will rise to about 1457 million tons. The ratio of steelmakers' metallics to total steel and foundry production in that year, plus unreported pig iron deliveries, will be about 1.36.

Assumptions and expected developments include the following:

- *A rise in the output of steel scrap and hot metal substitutes to 149 million tons by the year 2010 versus 31 million tons in 1995.* This is a fairly substantial figure when considering the time it takes to build these plants, the limited supplies of cheap natural gas, which are available at present and many other factors. Assuming that the average cost of this new SSS capacity is $300 per ton, this represents an investment over this period of $38 billion dollars. A rise of 128 million tons in output over a period of 15 years is 8.5 million tons per year.
- *A reduction in the percentage of steel scrap that is charged in EAF steelmaking furnaces.* The figure is 82% in the year 2010 versus 95% in 1995. More pig iron and SSS materials will be charged into EAFs; at the same time, more scrap and SSS materials will be charged into blast furnaces.
- *Slower growth of yields (the shipments to crude steel ratio) in steel plants.* From 1975 to 1995, the yield rose from about 67% to 84%. By the year 2010, the figure will be about 88%. Thus, steel production requirements will grow about 93% as fast as the demand for steel products. From 1975 to 1995, it only grew about 40% as fast. Also, home steel scrap generation grows only about 2% per year.
- *More efficient usage of steel by users.* The generation of 'new' steel scrap declines from about 14% of apparent steel consumption in 1995 to about 12% in the year 2010.
- *Restraints to BF/BOF.* These are considerable, no matter what the scenario. From 1995 to the year 2010, BOF output rises 1.3% per year compounded from 445 to 544 million tons (Fig. 9).
- *Stimulants to the EAF steel production route.* The incentives to boost EAF steel output will be substantial, no matter what the scenario. EAF steel output rises 5.7% per year from 243 million tons in 1995 to 556 million tons in the year 2010 (Fig. 10). This is an average gain of 21 million tons per year. The 313 million ton gain in EAF output compares to the 120 million-ton increase in SSS production. Incentives for this sector include:
  - Low capital costs to build new facilities.
  - The ever-more-efficient conversion of metallics to liquid steel in EAF steelmaking furnaces due to technological advances over the year.
  - Rising steel scrap substitute capacity
  - The ability to participate in new SSS units on an equity basis, thereby gaining access to a supply of this material at an attractive price.
  - Expected high returns when building EAF-based sheet mills.
  - Far lower fixed costs than those for BF/BOF producers.
- *Regional imbalances.* Growth of steel demand in the Pacific Basin will far outstrip that in the Atlantic Basin. Most of the shortfall of obsolete steel scrap occurs in the Pacific Basin (Fig. 11).
- *Key country developments*

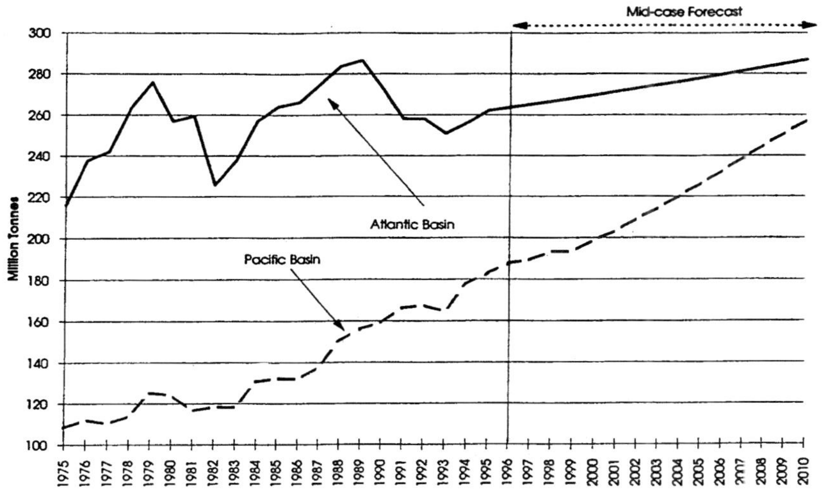

Fig. 9 BOF production in the Atlantic and Pacific basins.

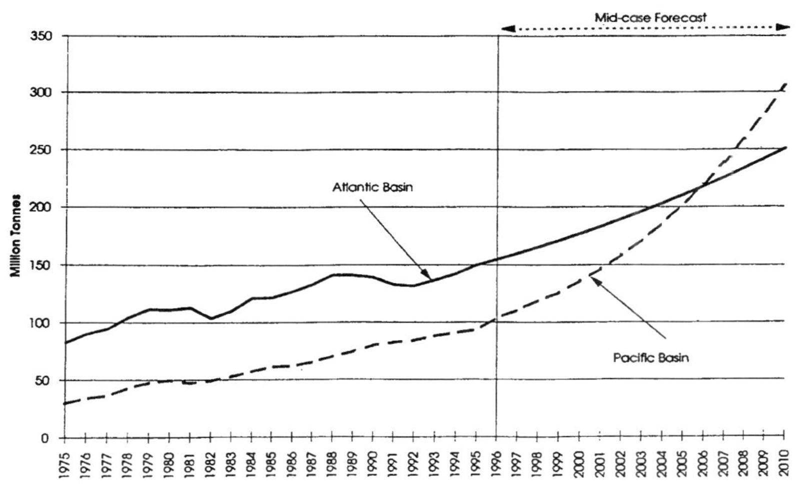

Fig. 10 EAF production in the Atlantic and Pacific basins.

- *United States*. From 1995 to 1999, new EAF steel plants, plus round-outs of existing EAF-based sheet mills, will require an additional 11 million metric tons of additional metallics. About 8–9 million tons of this material will need to be either low residual scrap or alternatives (such as pig iron and steel scrap substitutes). In 1994 the generation of low residual steel scrap in the United States was only about 24 million tons.
- *China*. Steel demand in China is expected to grow at a sizable rate in the

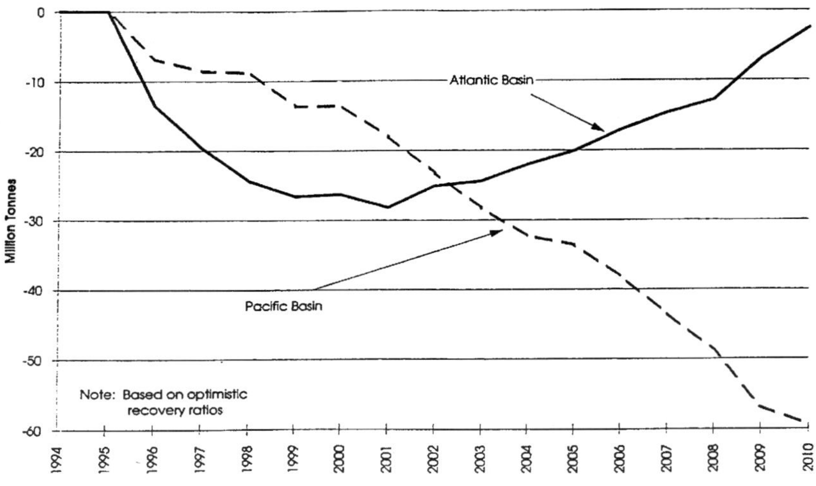

Fig. 11 Obsolete scrap balances for the Atlantic and Pacific basins.

decade ahead. However, the rate of addition to steelmaking capacity is limited at present. We see a massive shortfall.

- *India*. Steel products demand in India in 1995 was about 21 million tons. If steel demand in the next 15 years grows at a 7.0% per annum compounded rate, in the year 2010 it will rise to 58 million tons.

4. MOST BF/BOF STEELMAKERS UNDER SIEGE

The forces that are changing the structure of the steel industry are so powerful that they are independent of the steel cycle.

- Changes accelerate in good and boom times for the industry because of the surge in new investment.
- In bad times, changes also accelerate because most traditional steelmakers are performing poorly on a financial basis. Traditional steelmakers enter a 'dis-investment' mode. They pare back their primary operations to a level sufficient to support only a base load of shipments. They plan to purchase semi-finished steels during periods of stronger demand. We are witnesses to this phenomenon in 1995–96.

'Heads I lose; tails they win'. This is the new reality for many integrated steelmakers. The only exceptions are the ones with ultra low costs and unique product niches.

Only at high operating rates is there a profit boom for the typical integrated steelmaker. In the past, they had a good possibility to generate good profits at medium-high operating rates.

Many developments will cause BF/BOF steelmakers to grow at only a moderate rate in the next decade:

- 'Death spirals' and 'tsunamis' mark flat-rolled steel pricing patterns when oversupply is only moderate.
- More steel is being sold in many countries on a spot/dealer-pricing basis. Even in Japan, the Big Buyer pricing system has started to break down.
- Major mills are under attack in the one-year contract market, which is the basis on which they serve most of their largest OEM customers. Customer loyalty is often a thing of the past.
- Technological threats to BF/BOF producers are extreme, as pointed out earlier.
- Growth of SSS facilities demoralises traditional BF/BOF producers. Each expansion drives a wooden stake – or, perhaps a steel stud – into the heart of BF/BOF steelmakers. It provides evidence that their longer-term prospects are poor.
- Sharply rising EAF-based TS/FR (and thinnish slab) capacity is especially threatening to established steelmakers.
- The threat from EAF-based sheet mills grows even more extreme when they are tied into low cost steel scrap substitute facilities.
- A whole series of new SSS processes are in the process of being commercialised, such as those which use steam coal as the reductant and/or depend on iron ore fines rather than pellets.
- Coke shortages may be developing.
  - Many integrated steelmakers will face the need to build new coke ovens or make costly investments in older ones to meet environmental regulations. The life of many coke ovens in the United States is not clear.
  - Chinese cokemakers in 1995 exported more than 8 million tons of medium quality coke (high ash content) to buyers in the West. With about 25 million tons of beehive coke oven capacity in China mandated by the central government to be closed in the next few years, this supply of coke is in question.
- Most of the world's blast furnace capacity is located in the Atlantic Basin.
- A number of BF/BOF facilities are tied to inefficient and/or threatened downstream facilities. This is particularly the case in Russia/Ukraine and is also true in North America, Western Europe and Japan.
- The increased availability of semi-finished steels permit integrated producers to turn to this source to meet their peaking requirements.
- BF/BOF steelmakers have massive replacement and modernisation costs. We doubt that steelmakers will maintain marginal blast furnace capacity just to sell pig iron to third parties.
- The cost to build new BF/BOF facilities is huge. A typical SSS unit will

have a cost of about $250 per ton, while a blast furnace, including coke ovens, a sinter plant, etc., will have a cost of $500 per ton or more. In India, several promoters are considering the construction of BF/BOF facilities connected to a hot strip mill, with a production capacity of about 2.0 million tons per year. The capital cost is $1.5 billion. If $1.0 billion of this amount is borrowed at an interest cost of 18%, the interest expense per ton shipped amounts to $90. In addition, the return on equity investment must be more than 30% to justify such a project.

- It is high cost to maintain existing BF/BOF units. The expense to reline and upgrade a blast furnace can easily be $50 to $90 million.
- Breakdowns at existing facilities are common – and expensive. In the attempt to boost output, temperatures and pressures are sometimes increased to levels at which there is greater risk of a catastrophic failure. In the United States, a blast furnace explosion occurred because of undetected excessive wear of refractories near a tap hole. The cost of this problem was $100 million.
- The 'reconstitutions' of old steel mill – 10 years ago often via the bankruptcy route and now via the privatisation route – is threatening to established producers.
- The World Cost Curve is steep for both blast furnaces and BOF steelmakers. Based on WSD's analysis about 18% of surveyed global capacity has an operating cost greater than $150 per ton. This is further persuasive evidence that further sizable BF close downs are in store.
- When it comes to considering a new blast furnace or a new SSS unit, the SSS route has more to offer.
  - A SSS unit with an operating cost before interest expense of $80–100 per ton, and a capital cost of perhaps less than $250 per ton.
  - A blast furnace with a cash cost of $125–140 per ton, and a cost of capital of a least $500 per ton. (Note: Liquid hot metal is probably $15 to $20 per ton more valuable than a steel scrap substitute).
- Foreign exchange movements can be particularly threatening to integrated steelmakers. This is because a higher proportion of their costs is in the home currency.
- The steel demand outlook is always uncertain, steel is a derived-demand industry – it depends on activity levels in steel-using industries.
- Poor cash flows lead to reduced capital spending. Capital spending lags cash flow by about 1 and a half years. Flat-rolled steelmakers can only make good profits when operating rates are especially high.
- Poor price/earnings rations for steel stocks limits steelmakers' access to capital. Many of the steel company stocks have been poor performers.
- Many established mills have sizable legacy costs – including pollution control and retired worker liabilities.
- Management attitude plays a role. Many steel company managements feel defeated.

5. CHINA IS THE KEY TO THE MANY DEVELOPMENTS IN THE GLOBAL STEEL INDUSTRY

We find it hard to believe that steel production in China will keep up with the sizable growth of steel demand in the next decade.

- According to our projections, the medium-case scenario assumes apparent consumption of finished steel products grows at about a 6% per annum rate from about 98 million tons in 1995 to about 175 million tons by 2005. (Note: Apparent consumption of steel in China in 1995 was only about 81 million tons because of massive inventory liquidation by steel users). If crude steel production in the year 2005 is 125 million tons, output of finished steel products in that year would amount to only about 109 million tons (assuming a shipment-to-production yield of 87.2%). Under this scenario, the gap between demand and production reaches 66 million tons.

  This rate of demand growth is not out of line with historical levels – apparent steel consumption grew at a 7.8% annual rate between 1980 and 1995. With economic growth looking solid, it seems unlikely that steel demand will also not grow at a high rate.
- In comparison, the Chinese government projects just 2.2% per annum growth in steel production in the latest five-year plan. This would be just 105 million tons of crude steel production in 2000. The plan also calls for the closure of all open-hearth furnaces by 2000. This steelmaking route accounted for perhaps 15 million tons or 16% of total steel production in 1995 of 94 million tons.

In its global metallics analysis for China/North Korea/Vietnam, WSD estimates that steel consumption in China will expand at a 7.8% per annum rate. Accordingly, steel production in the year 2010 rises to 253 million tons, versus 98 million tons in 1995. This is a growth rate of 6.5% per annum (which may not be achievable). BOF steel production rises at a 5% per annum rate, while EAF output rises at about a 12% per annum rate. The growth rates for steel production are quite massive, which raises the strong possibility that there will not be sufficient capital to meet these requirements.

6. WINNERS AND LOSERS

Characteristics of Winners Will Include:

- Low costs
- Ongoing cost-cutting efforts
- No employee loyalty
- Domestic markets have features that permit a better price
- Possession of niche markets
- Good financial positions
- Ongoing product mix improvements

- Downstream outlets for the steel
- Possession of a 'natural hedge', whereby revenues are in dollars and costs in the home currency, that is prone to devaluation

Characteristic of Losers Will Include:

- Slow decision-making process and bureaucratisation of the management group
- Lack of low-cost raw materials
- Commodity-type product mix
- Mediocre to low customer standing
- Leveraged financial positions
- Sizable legacy costs
- Inability to lay off workers when demand weakens
- Medium to low quality product
- High costs

Winners:

- Vendors of new flat-rolling and SSS technologies
- Low-cost steelmakers in the developing world whose revenues are in dollars and much of their costs in the home currency
- EAF steelmakers that have equity position in low-cost SSS facilities
- Steel processors that can take advantage of the rising supply of hot-rolled band (especially in the United States)
- Iron ore suppliers that have iron ore that is suited for the production of steel scrap substitutes
- Sellers of coke
- Sellers of steel scrap
- Sellers of SSS materials
- Sellers of steel slabs
- Steel buyers, particularly larger OEM buyers of sophisticated steel. They will continue to receive an ever better product at a reasonable price
- Steel traders
- Steelmakers in the Pacific Basin
- Companies that sell their steel on a spot basis. While they will suffer when demand is weak, they will be able to quickly maximise price when demand is strong.

Losers:

- Exporters of steel to the United States
- Traditional BF/BOF steelmakers with medium and high costs
- Sellers of steel to major OEM manufactures on a one-year contract basis
- Traditional Chinese steelmakers. Their costs are too high and their sales are tied into the spot market in China
- High cost steelmakers. There is no room for high costs given steel's never-ending cost reduction race.

- Many Russian steelmakers. Similar to the Chinese, their costs will not be favourable as they purchase more of their steel on a dollar-denominated basis
- Sellers of steel on a one-year contract basis to major OEM steel buyers

Not Yet in the Winners' or Losers' Circle

The decision is still out as to whether the following groups will be in the winners' or losers' category

- Japanese steelmakers. On the positive side, they are benefiting from huge cost-cutting efforts and a weaker yen versus the dollar. On the negative side, they will probably suffer from a further deterioration of Big Buyer steel prices, which have protected them to a sizable extent in the past two decades.
- Those mills which will benefit from, or be hurt by, swings in the value of their currency. This is an important factor especially for integrated steelmakers.
- If steel scrap prices rise too sharply, all steel mills will be losers. This will push up product prices to levels that dampen demand. Steel would be more vulnerable to the loss of markets to competing materials.

7. TRUTH AND CONSEQUENCES

For each truth in the steel industry there can be many consequences. A steel price increase, for example, is good for steel companies' profits; however it also attracts new capacity to the industry and increases the odds that the markets will be lost to competing materials.

At times, it even becomes difficult to separate cause from effect and to comprehend the indirect consequences of an event.

In this section, for a number of possible events, we present some of the contradictory consequences.

Event 1: Flat-Rolled Steel Prices Fall Sharply Even Though the Steel Industry's Operating Rate is High

Consequences:

- Steel gains markets from other materials
- BF/BOF steelmakers abandon capital improvement plans at an accelerating rate
- Steel stocks are poor performers on many bourses/exchanges.

Event 2: Steel Scrap Prices Surge $50 to $200 per ton

Consequences:

- High profits for EAF steelmakers as their customers panic. Customers anticipate large price increases.

- Steel demand is dampened, especially for long products.
- The frenzy to build steel scrap substitute facilities becomes even more intense.
- Scrap exports are banned from the United States.
- Massive cash flows for steelmakers lead to huge expenditures to add to steelmaking capacity.

Event 3: Steelmakers' Coke Prices Take Off

Consequences:

- The supply from China does not rise as a number of beehive coke ovens are closed due to severe air pollution problems.
- Only a few new coke ovens are built because of the high capital intensity.
- The Chinese begin to build new cokemaking capacity at a high rate.

Event 4: Oil Prices Take Off

Consequences:

- Electricity prices rise in countries where the dependence on oil to generate electricity is substantial, such as Japan.
- The rise of steel demand is substantial for oil country tubular goods and related items.
- Rising global inflation rates drive down economic activity.

Event 5: Hot-Rolled Band Capacity is so Substantial in the United States that Prices don't go up Significantly Even During Periods of Steel Shortage
There has always been excess hot-rolled band capacity in steel shortages. It is tight supply at the primary end that drives up the price.

Consequences:

- BF/BOF producers close down about 20% of their capacity
- Losses are massive. Bankruptcies occur.

Event 6: An Imbalance Between Steelmaking Capacity and SSS Capacity Addition

Consequences:

- Steelmaking capacity falls far short of demand; yet, additions to SSS capacity are substantial.
- There is an oversupply of SSS. The price is low for steelmakers' metallics, yet high for finished steel products.
- Steelmakers will find ways to boost production in such an environment, in part by paying more for incremental addition to the metallics supply.

Event 7: China Closes Down its Beehive Coke Ovens in the Next Three Years

Consequence:

- A shortfall of cokemaking capacity becomes pronounced.

Event 8: Blast Furnace Output Rises 100 Million Tons More Than Expected by the Year 2005. It Amounts to 674 Million Tons Versus our Estimate of 574 Million Tons and 1995 Output of 256 Million Tons.

Consequences:

- The average output per blast furnace rises 10%.
- Steel scrap prices rise. This is because at least two-thirds of the increase is due to the charging of sizable quantities of steel scrap and steel scrap substitutes into these furnaces. In other words, of the 100 million-ton rise in global BF output, steel scrap and SSS materials are used to provide 67 million tons of the increase.
- Flat-rolled steel prices go up.
- This scenario has a low probability because of limitation to the supply of steelmakers' coke.

Event 9: Fifty New EAF-Based TS/FR Plants Come into Production Globally in the Next Decade

Consequences:

- Hot-rolled band prices rise because of the surge in steel scrap prices.
- Steel scrap prices go up.
- BF/BOF steelmakers cut back their capacity at marginal facilities.

Event 10: The Supply of Obsolete Steel Scrap Rises Sharply at Current Price Levels

Consequences:

- There is not a metallics shortfall. World Steel Dynamics is wrong.
- The return on investment in new SSS facilities is poor.

Event 11: Steel Demand Rises Far More Than Expected in the Next Decade – About 3.5% Per Annum

Consequences:

- The price of steel scrap rises $50 per ton
- BF/BOF steemakers still become more negative about their future because so many new SSS plants are announced. A rising number of EAF-based TS/FR producers become 'integrated'.
- Announcements of new EAF-based steelmaking capacity are fast and furious. The reason is that, although steel scrap prices are high hot rolled band prices rise even more.
- The shortfall of metallics is extraordinary

Event 12: Steel Demand Grows Only 1% Per Year Over the Cycle in the Next Decade.

The 'disinflationary' forces, which dampened global economic growth from the mid-1970s to the early 1990s, remain in effect.

Consequences

- Steelmakers head into tough times.
- BF/BOF steelmakers cut back capacity because of low prices.
- The cost of money is so substantial that Blast Furnaces, along with supporting coke ovens, are not built.

Event 13: Demand for Steel in China Grows 9% Per Annum Over the Next Decade

Consequences:

- China, almost all by itself, creates tight supplies for steel steelmakers' metallics and nonferrous metals. Steel demand grows from about 97 million tons in 1995 to 225 million tons in the year 2005. The need for crude steel would be about 250 million tons, which compares with the Chinese official forecast for steel productions in the year 2000 of about 110 million tons.
- Investments in China in the steel sector become extraordinary once again.

Event 14: Steel Continues to Gain Share of the Construction Market From Competing Materials

Consequences:

- Steel demand is boosted an extra 0.3% per annum, all other things being equal.
- More commercial buildings are built from steel.
- More homes are built with steel studs and wood loses out to a greater extent.

Event 15: The Japanese Yen Weakens Sharply

Consequences:

- The Japanese, once again, are counted among the lowest cost steelmakers in the world. Plus, their product quality will be the highest.
- The Japanese companies start to expand their steelmaking capacity once again.
- The Japanese round out their steelmaking capacity in order to ship more steel to China.

Event 16: Steel Scrap Exports are Banned by the US Government Because the Material is Considered a Natural Resource
(Note: This is now the case for several CIS countries)

Consequences:

- Scrap prices skyrocket. Bootlegging of scrap, from the U.S. to foreign markets, is reported.
- The frenzy to construct SSS plants intensifies, especially in the Pacific Rim.
- The 'integration' of the EAF-based TS/FR producers is stepped up.

Event 17: Global Trade Wars Break Out

Consequences:

- Poor demand growth and bad times for steelmakers.
- High tariffs become the norm in most regions/countries. Economic growth is reduced

Event 18: Steel Scrap and Hot-Rolled Band are Listed as Commodities of the London Metal Exchanges (LME)

Consequences:

- Trading of these commodities exceeds global production by 10 to 20 times, which is the current case for the major nonferrous metals.
- A difficult proposition, as there are too many grades of steel scrap and hot-rolled band. Hence, the material held in LME inventories cannot be properly certified and classified.

Event 19: The Commercialisation of Several Now Coal-Based SSS Technologies Occurs in the Next Five Years
The product is liquid pig iron.

Consequences:

- BF/BOF steelmakers become more negative.
- EAF-based sheet makers become more optimistic.

Event 20: In 1999, a New Steelmaking Process is Commercialised by Nucor and the USS Group of USX Which Depends Principally on Iron Carbide and Oxygen
Savings are about $20 per ton in operating costs.

Consequences:

- BF/BOF steelmakers turn even more bearish their prospects.
- The price of Nucor common rises.
- Other producers rush to adopt the technology.

Section I
Raw Materials

TRENDS IN RAW MATERIAL SUPPLY

J. AYLEN

Department of Economics, The University of Salford, Salford, M5 4WT

Trends in raw material supply have been favourable to the steel industry over the past two decades. The *real* price of ferrous raw materials has fallen since the mid-1970s. Efficiency of energy use has improved markedly too. The cost of capital has dropped and the range of technologies has broadened. A picture emerges of gradual and continuous savings in the cost of materials and wider and wider process choice.

An underlying theme of all these papers is the growth of electric arc steelmaking. As Lüngen documents (Fig. 3) the share of electric arc steelmaking has risen from well below ten per cent of world steel output in 1950 to 33 per cent by 1995. Biamonte and Hinder focus on the technological opportunities brought by large scale electric arc steelmaking, including development of 'integrated' electric arc based plants with iron based feedstocks such as DR and, potentially, directly smelted material. After a period of technical stability, a wide range of alternative process routes are opening up for new steel plants. It is evident that scrap will be of growing importance as a feedstock for the expanding arc sector. At the moment, commercially successful DR and smelting processes rely on lump iron ore, or expensive pellets as a feedstock. The challenge for the next decade is to develop a commercially successful smelting process which use iron ore fines as a raw material.

Yet as Lüngen emphasises the established blast furnace-BOS route has also improved dramatically. The amount of reducing agents such as coke, oil and directly injected coal needed to make a tonne of iron has been cut by more than a half in Germany between 1950 and 1995. Figures are even lower for some countries: notice the combined coke and oil rate for Finland in 1995 is below 450 kg/tonne in a league table of comparisons. Given the steady progress in fuel consumption worldwide, it seems unduly cautious of Lüngen to project a coke shortage on the basis of no further decrease in worldwide average coke consumption in blast furnaces.

As a result of these improvements, the blast furnace remains a highly cost competitive steelmaking route. Indeed, novel Corex plants cannot compete with established ironmaking technology, with Corex requiring more reductants than a blast furnace. Again Corex plants also require more expensive lump ore or pellets. Given continuing improvements in the established blast furnace-BOS route, the prediction by Biamonte and Hinder that conventional plants will only account for a small share of new steelmaking capacity between the years 2000 and 2010 looks a bold forecast. They foresee almost half of all new capacity being supplied by new smelting reduction processes which might feed any steelmaking route. Evidently, technical progress is still required if these new processes are to prove superior to the conventional blast furnace in terms of energy efficiency. As Biamonte and Hinder remind us, choice of process route will ultimately depend upon local circumstances. Capacity, product range and availability of local inputs will dictate process choice rather than the efficiency of one particular process stage or another.

The rise in global electric arc production has not brought a rise in real scrap prices according to Aylen and Albertson. They show there is one global market for ferrous scrap. British and European scrap markets closely follow price movements in the USA. Scrap prices have fallen in real terms throughout the century, but there are marked cyclical swings in prices worldwide which are driven by US steel production. The US market also brings seasonal changes in prices throughout the year.

The global scrap price is certainly volatile, but in the long run it has trended downwards as supply of scrap has more than kept pace with the rise in demand. According to Genevaz, the rate of growth of scrap supply has exceeded the rise in scrap use in the European Community, the USA and Japan since 1966. The uncertain feature in the global scrap balance is the level of exports from Eastern Europe and the CIS. Transport problems and disorganisation hinder development of a scrap industry in these locations. Although as Genevaz points out, the cost of entry to scrap trading is low and the technology of scrap preparation is fairly simple.

There is a single world market for all ferrous material since iron ore prices are determined by recent movements in the scrap price. The price of iron ore mirrors the price of scrap, with a lag of between one and two years. So the process choice between the conventional blast furnace/BOS route and scrap based electric arc route will never be settled on the basis of ferrous material. For as Aylen and Albertson conclude, there is no cheap source of ferrous material. In the long run, scrap prices and iron ore prices always move together, although iron ore prices follow the lead set by scrap.

The world market for merchant iron is highly volatile, with major suppliers entering and leaving the market. DR material has overtaken pig iron as an alternative to scrap, but very little DR iron is traded. Virtually all directly reduced material is produced close to the point of use. Biamonte and Hinder see this trend continuing as arc based plants are increasingly 'integrated'.

The 'cost of capital' has dropped since 1980 too, partly because the cost of finance has fallen and partly because capital equipment has got cheaper. New designs such as Schloemann-Siemag's Compact Strip technology and Arvedi's ISP technology deliver 'more bang per buck'. Biamonte and Hinder pursue improvements in process routes to their logical conclusion. They argue plant suppliers will become involved in 'financial engineering' to arrange the flow of capital funding to new projects, as well as procuring the equipment itself.

The outlook for raw materials supply beyond the millennium is essentially optimistic from the steelmakers point of view. Continuing improvements in existing processes will curtail pressures on coke supplies. There seems little prospect of a scrap shortage. In turn, lower real scrap prices will damp down iron ore prices. Steelmakers will have a wider range of processes to call upon and a broader range of energy sources. But, it is by no means a foregone conclusion that DR or direct smelting will predominate since established processes such as conventional arc steelmaking from scrap and the blast furnace/BOS route continue to improve.

TECHNOLOGICAL INNOVATIONS IN IRON AND STEELMAKING AND THEIR EFFECTS ON COAL, COKE AND IRON ORES

H. B. LÜNGEN

Verein Deutscher Eisenhüttenleute (VDEh), Postfach 1051 45, D-40042 Düsseldorf, Germany

INTRODUCTION

The effects of technological innovations in iron and steelmaking on the demand for coal, coke and iron ores may be concluded from

- the expected world crude steel production and the share of Oxygen to Electric steelmaking influenced by the amount of continuous casting and by the development of near net shape casting technology,
- the further development of direct reduction processes and
- the further development of smelting reduction processes as well as
- innovations in blast furnace technology and cokemaking technology.

WORLD CRUDE STEEL PRODUCTION AND SHARE OF OXYGEN TO ELECTRIC STEEL

The development of the percentages held by the different steelmaking processes used in the total world production is shown in Fig. 1. In addition to 60% Oxygen steel and 32% electric furnace steel in 1994, there still existed a share of 8% held by the open hearth steelmaking process because of the slow pace of change in Eastern Europe.

For the near future experts predicted an increased recycling and reuse of scrap and an increasing share of electric steelmaking in the world. On a world average, the hot metal to crude steel ratio has indeed decreased marginally from approximately 710 kg per tonne in 1970 to approximately 685 kg in 1992 but came back to a ratio of 708 kg in 1995, Fig. 2. Hot metal charges are on high levels between 690 and 720 kg/t crude steel for the average of the European Union, Great Britain, Germany and Japan. USA and Italy are on a hot metal charge level of 557 and 450 kg/t crude steel respectively. Only in Germany and USA an increase occurred in 1994 compared to 1993. In countries such as Brazil, Korea, Taiwan and the People's Republic of China, new steelmaking capacities were and are built mainly in accordance with the blast furnace/oxygen converter technology and require a corresponding production of hot metal. In

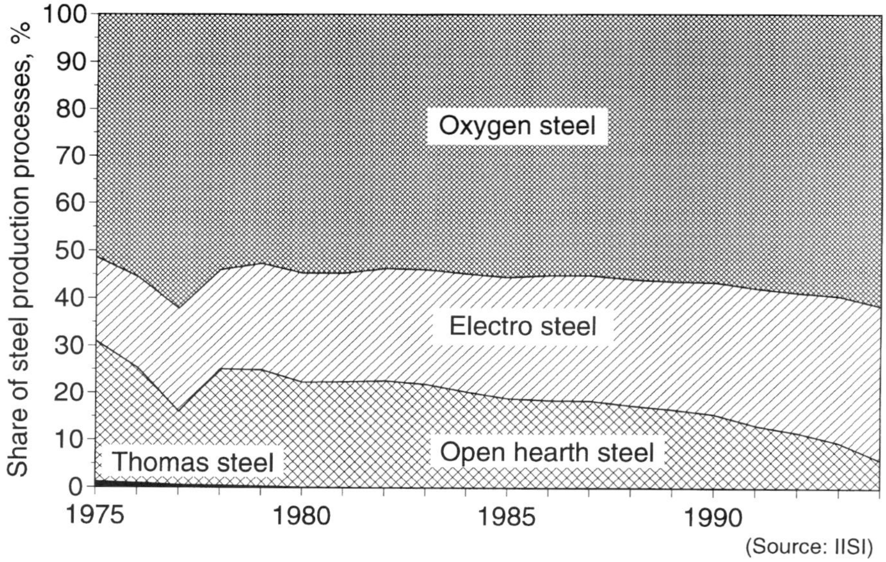

Fig. 1 Development of the worldwide share of steel production processes.

developing countries but also in some industrial countries a switch over to electric arc furnaces and minimills was observed.

On the other hand a further worldwide increase of the share of continuous casting and the implementation of near net shape casting technology influences

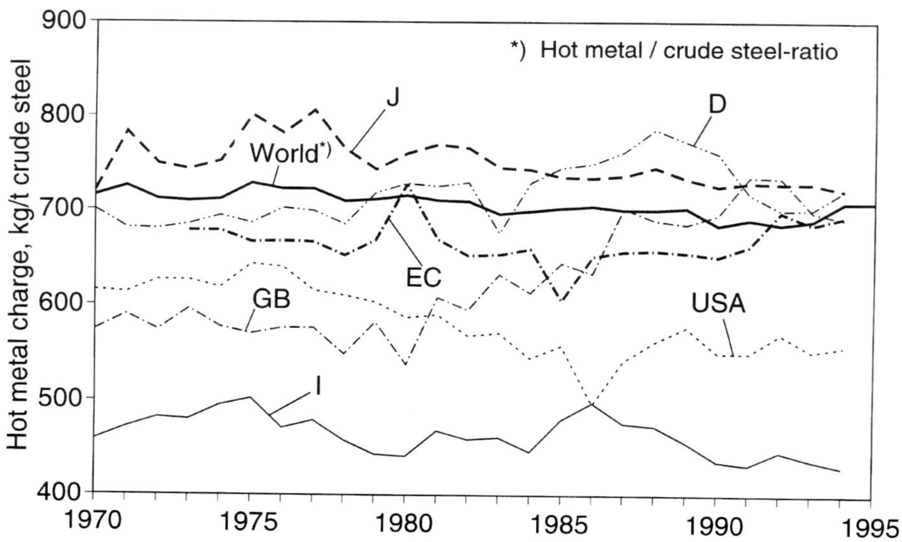

Fig. 2 Hot metal charge for crude steel production.

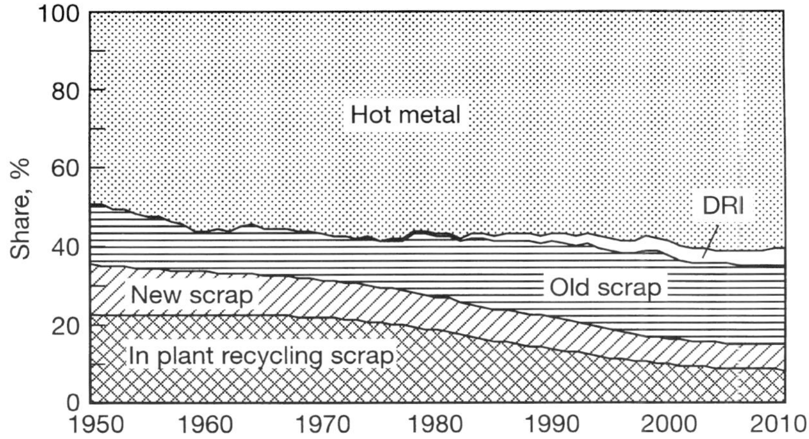

Fig. 3 Share of Fe carriers on the total charge for steelmaking.[1]

the availability of scrap on the market. Assuming that there will be a growth of 1.5% per year in worldwide crude steel production, as well as an increase of the continuous casting share from today 70% to 95% worldwide in 2010 and that the near net shape casting technology will be implemented at certain industrial plants, a further decrease of availability of new in-plant recycling scrap is forecasted,[1] Fig. 3. In addition to this, the decreasing share of new scrap due to better utilisation of steel will slightly continue. This will result in an increased demand for hot metal and directly reduced iron (DRI) to supply the required amount of Fe for steelmaking. These results, which are based on model calculation[1] lead to the prognosis of Fig. 4 which shows that the importance of the electric steel process will slightly increase due to the increase of old scrap available and due to the decrease of the share of the open hearth furnace technology. The share of the oxygen steel process will also increase. This forecast for the year 2010 gives a 70% share of oxygen steel and a 30% share of electric steel. A shortage in scrap availability, especially high quality scrap, influences the hot metal and DRI to scrap price ratio.

FURTHER DEVELOPMENT OF DIRECT REDUCTION

In general today direct reduction processes are not seen as an alternative to the blast furnace, because they produce solid sponge iron, but as a competition or supplement to the use of scrap in electric steel production especially in the case of the production of high quality steel made via the electric steelmaking route. The development of the worldwide capacity and production of sponge iron is shown in Fig. 5. It can be seen that a steady growth in capacity and production occurred year by year. The utilisation of existing capacity increased remarkably over the past three years. In 1992 the plants were only utilised up to 60% with a production of 20.7 Mt of sponge iron and a capacity of around 34 Mt, whilst in

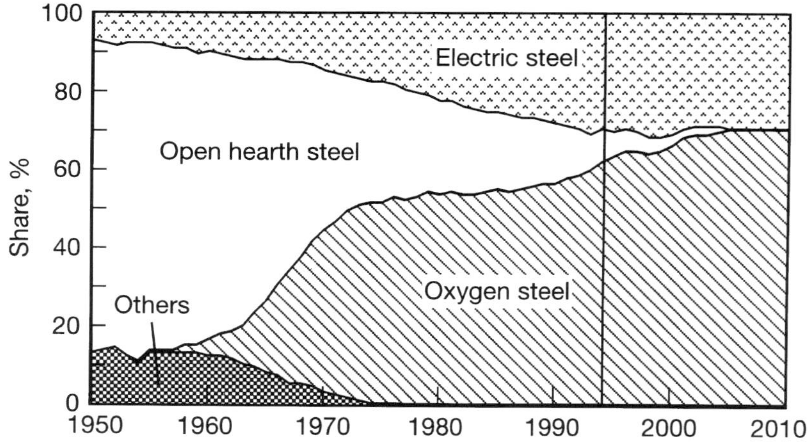

Fig. 4 Share of steelmaking processes on worldwide crude steel production.[1]

1995 utilisation increased to 83% with a sponge iron production of 30.7 Mt and a capacity of 36.8 Mt. In 1995 7.7 Mt of sponge iron corresponding to 25% of the total production was not processed at the plants directly[2] but traded on the market.

Nevertheless a further growth of the sponge iron capacity worldwide of at least 12 Mt is predicted for the year 2000. The extent to which the available capacity of around 46 Mt is then utilised depends on how the prices for the reductants and scrap qualities develop in the individual regions, how much time it takes in some developing countries and countries on the verge of economic

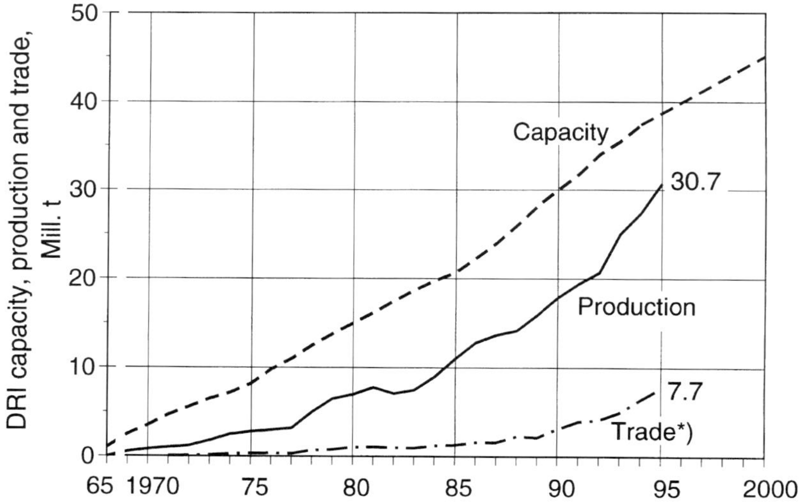

Fig. 5 DRI capacity, production and trade.

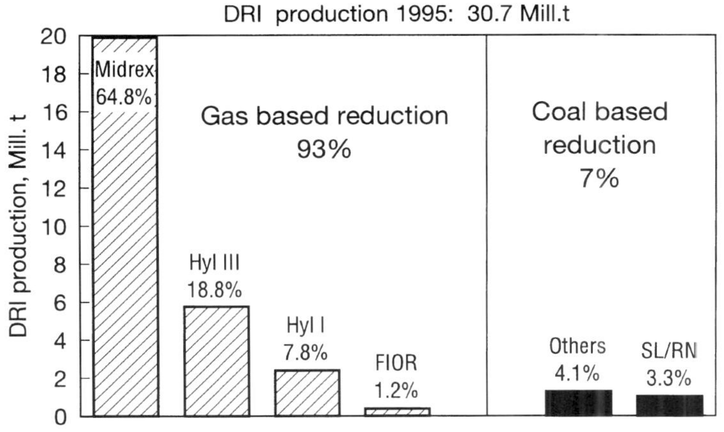

Fig. 6 Share of direct reduction processes 1995.[2]

growth to reach the designed output, how the economic situation of the companies develop and to what extent these capacities are affected by political unrest in unstable regions. The increase of sponge iron production will generally require higher amounts of high grade iron ores low in gangue contents from the world market.

Figure 6 shows the share of the different direct reduction processes in world sponge iron production for 1995. Gas based reduction accounts for 93%. Only the Fior fluidised bed process produces 1.2% using fine ores. So 91.8% of production comes from shaft furnaces which require coarse charge materials like lump ores and pellets. Coal based rotary kilns are more flexible compared to shaft furnaces and fluidised beds with regard to the physical properties of the charged iron ores. Pellet and lump ore charges are common, but operation with fine ores is also possible. New processes based on fine ores are commercialised and under optimisation like iron carbide, or under construction, like Finmet, or planned for industrial scale, like Circored.[3]

SMELTING REDUCTION PROCESSES

The only alternative to the blast furnace is smelting reduction. Smelting reduction processes also produce a liquid hot metal but without any dependency on coke. Of all smelting reduction developments, the only process progressed to full scale industrial feasibility is the Corex process depending on lump iron ores. As Fig. 7 shows the Corex process principally divides the blast furnace into two parts. This avoids the cohesive zone of the blast furnace in which coke guarantees the permeability of the ferrous burden materials during softening and melting and also acts as a sprung roof for the dry burden column.

As a comparison, Fig. 8 shows the size of the 2 000 t/d Corex Pohang

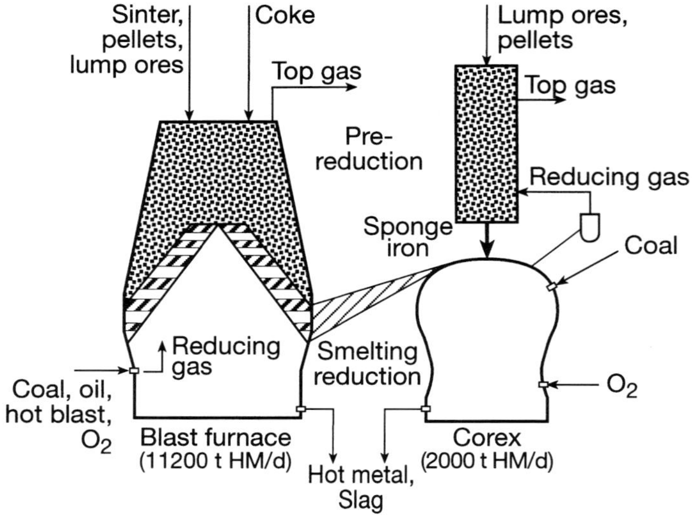

Fig. 7 Avoidance of the blast furnace cohesive zone using the corex process.

configuration and the 11 200 t/d Schwelgern 2 blast furnace of Thyssen Stahl AG. The Corex plant projects above the blast furnace, without taking the charging devices and the top gas pipe into consideration, by more than 20 m. From this comparison, it is worth mentioning that it requires five such Corex plants to produce the same amount of hot metal as the Schwelgern 2 blast

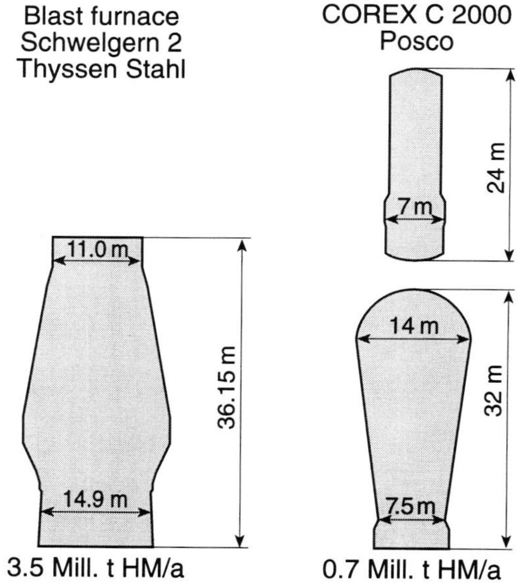

Fig. 8 Size comparison between blast furnace Schwelgern 2 and COREX C 2000.

furnace. The investment costs for a Corex plant according to a European feasibility study are in the range of 235 ECU/t·a excluding infrastructure and oxygen plant and for a blast furnace in the range of 134 ECU/t·a.[4] The future field for smelting reduction processes is seen in mini-mills, or as replacement for small blast furnaces in integrated iron and steel works.

Table 1 summarises Corex plants in operation, in construction and planned.[3,5,6] The yearly capacity of a small blast furnace is reached with the C 2000 configuration. All these plants are based on coarse ores like lump ores and pellets. The fine coal injected into the melter gasifier need not be metallurgical coal.

It has to be mentioned that all smelting reduction processes based on fine ores are still at the development, pilot or demonstration stage. The critical problems encountered in these investigations relate to the mutual coordination of the fine ore and smelting reduction stages. Concerning iron bath reactors the control of the required post-combustion rate in the reactor, and the development of slag control which reduces the wear of the refractory lining are critical factors.

Compared to the blast furnace, including its upstream coking and sintering plant, all smelting reduction systems today are characterised by a higher demand of reductants and energy carriers.[6] The total use of the export gas and the achievment of high gas credits is a prerequisite for Corex plants to operate in competition to the blast furnace.

INNOVATION IN BLAST FURNACE TECHNOLOGY AND COKEMAKING TECHNOLOGY

Great improvements in economy and efficiency of the blast furnace process have always been aims of blast furnace operators. This was and is directly linked to decreasing reductants consumption, especially coke consumption, for hot metal production. Figure 9 shows this development for example for German blast furnaces and the most important plant and process measures which have contributed towards this. From the early fifties onwards, the quality of the iron bearing burden materials was most important to reduce the total reductant consumption. This was achieved by an increased charge of agglomerated burden materials, sinter and pellets. Other developments in blast furnace technology like hot blast temperatures of more than 1200 °C, oxygen enrichment of the blast, use of elevated top gas pressure, burden distribution, monitoring of blast furnace operation and injection of auxiliary reductants contributed to the decrease of the total reductant rate to below 500 kg/t hot metal. Table 2 lists for some countries yearly average fuel consumption for all blast furnaces to demonstrate, that in most of these countries a rate of below 500 kg/t HM was achieved. Through the injection of coal and oil the coke rate was reduced to nearly 350 kg/t HM as an average for some countries. The coal injected need not be metallurgical coal.

Table 1. Corex plants in operation, in construction and planned (May 1996).[3,5,6]

Corex plants	Amount	Year of commissioning	Ore charge	Capacity, Mill t/a	Gas utilisation	Remark
Iscor, Pretoria Works; South Africa	1	1989	Lump ores	0.34	In plant	Hot metal charge to electric arc furnace
Posco, Pohang Works; South Korea	1	1995	Lump ores Pellets	0.60 to 0.80	Power station	Hot metal charge to LD converter
Hanbo Steel, Asan Bay; South Korea	2	1997	Pellets (50%) Lump ores (50%)	1.50	Direct reduction (0.80 Mill t DRI/a)	Hot metal charge to electric arc furnace
Jindal Vijayanagar Steel, Toranagallyu; India	1	1997 1998	Lump ores Pellets	0.64 to 0.80	Power station; Oxygen production	Hot metal charge to LD converter
Saldanha Steel, Saldanha Bay; South Africa	1	1998	Lump ores	0.65	Direct reduction (0.80 Mill t DRI/a)	Hot metal charge to electric arc furnace

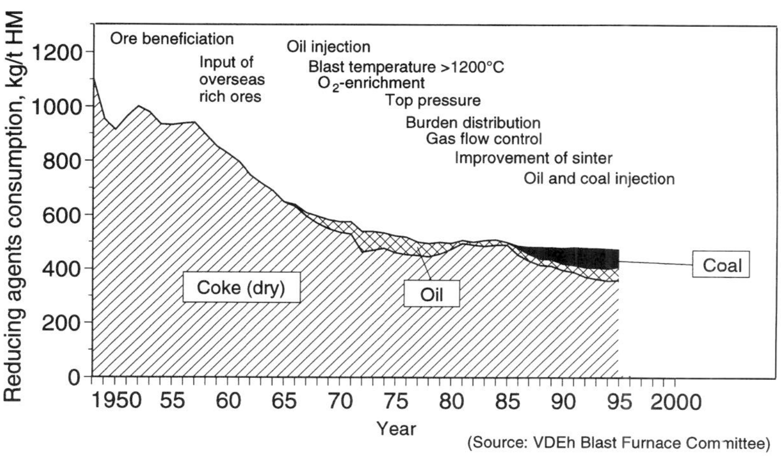

Fig. 9 Average consumption of reducing agents in German blast furnaces.

The average coke consumption of blast furnaces operated in the European Union decreased from 430 kg/t HM in 1988 to 376 kg/t HM in 1995, a 12.5% reduction which corresponds to a coke production of approximately 5 Mt/a. At individual blast furnaces coke rates below 300 kg/t HM have been achieved through the injection of very high amounts of oil and coal with oxygen enrichment of the blast or oxygen injected simultanously via lances.

Coke consumption of 297 to 272 kg/t HM have been achieved at coal injection rates of 194 to 212 kg/t HM,[7,8] Fig. 10. The coke/coal replacement ratio achieved was in the range 0.9 to 1.08.

Through increasing the oil injection rate to 143 kg/t HM the coke rate was reduced to approximately 294 kg/t HM,[9] Fig. 11. The coke to oil replacement ratio was determined to be 1.2. At this furnace about 30 kg/t HM small coke (8 to 30 mm) was mixed with the ferrous burden, the coarse coke consumption was only 260 kg/t HM.

Such excellent blast furnace results are only achievable with excellent burden quality and especially excellent coke quality. The coke should have a good stability, good CSR values (coke strength after reaction with CO_2) and low sulphur and alkali values.

These examples show that there are still potentials to further decrease the average coke demand worldwide, as today the average coke consumption for all blast furnaces in the world is estimated to be 500 kg/t HM.[10] To what extent the coke rate will be reduced depends on how far the coke rate in the blast furnace can be decreased, on the price ratio of coke/auxiliary reductants, on the available coking plant capacities and the advancing age of the coking plants with all the required capital-intensive measures for environmental protection.

The criticism raised against the coking industry by the advocates of alternative ironmaking processes are usually related to the increasing age of today's coking facilities, Fig. 12,[10] environmental problems, which have to be solved at

Table 2. Consumption of total reduction agents in blast furnaces as countries average.

Country	Average consumption						No of BF taken under consideration
	Year	Coke kg/t HM	Coal kg/t HM	Oil kg/t HM	Others kg/t HM	Total kg/t HM	
Japan[1]	1995	414	98.8	–	1.2	514	30
USA[2]	1995	413	36.0	13.0	39.5	501.5	47
Belgium[3]	1995	382	123	2	–	507	8
Finland[3]	1995	375	–	74	–	449	3
France[3]	1995	351	125.6	3.5	–	480.1	11
Germany[3]	1995	359	51	63	–	473	20
Italy[3]	1995	353	129.5	15	–	497.5	5
Netherlands[3]	1995	357	141	–	–	498	2
United Kingdom[3]	1995	394	43	55	–	492	8

[1] Average for Jan.–Sept. 1995 (Source: T. Fujiwara: IISI International, 1996, **36**(4), 370).
[2] Source: AISI Ironmaking Committee.
[3] European Blast Furnace Committee.

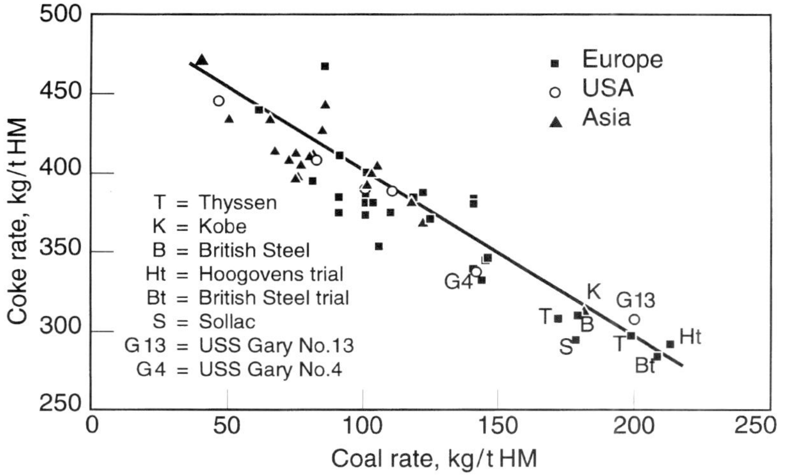

Fig. 10 Comparison of world-wide coal and coke rates in blast furnaces.[7,3]

older plants, and the high investment costs for new cokemaking facilities. Major steel producing countries such as Italy, France, Japan and the USA all have an average age of the coking plants in excess of 20 years.

Examples of the most modern and fullfilling all regulations for environmental protection for conventional cokemaking plants using the multi chamber system are given by the coking plants of Hüttenwerke Krupp Mannesmann and Prosper in Germany, Kwangyang in South Korea and Kaiserstuhl in Germany, Fig. 13. The multi-chamber cokemaking concept reached the peak of its development with the construction of the new Kaiserstuhl coking plant

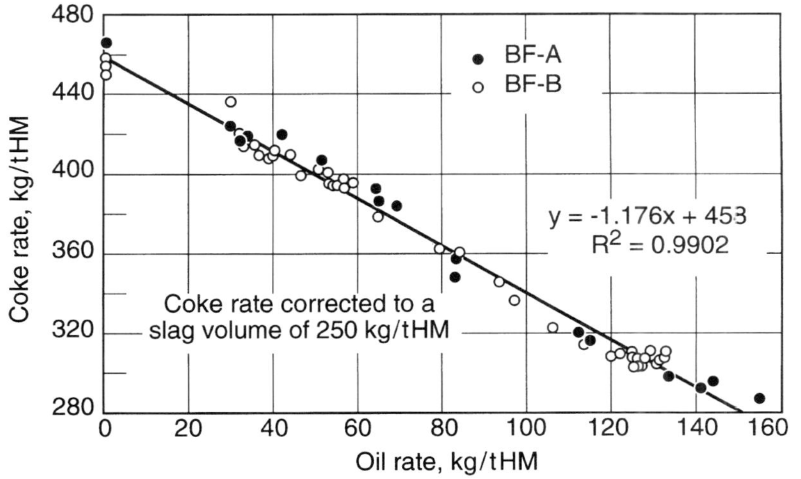

Fig. 11 Coke and oil rate at the blast furnaces A and B of HKM.[9]

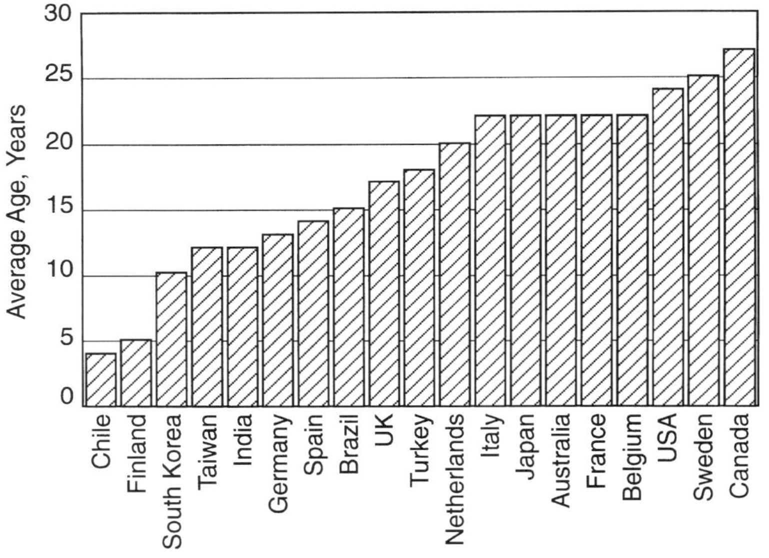

Fig. 12 Average age of coke facilities in 1994, selected countries.

Fig. 13 Coking plant Kaiserstuhl.

commissioned in 1993. This plant comprises 120 chambers each with a volume of 78 m^3 and is designed for an annual coke production of about 2 Mt.

On the site of the Prosper Coking plant in Germany, the new cokemaking system 'single coking reactor', Fig. 14, is being tested on a demonstration scale.[11] The process is based on independent single chambers, and one of the remarkable features is a greater independance on raw materials, this allows the use of non-coking, low-cost coals as a major portion. The industrial application can be expected at new locations in competition to the conventional cokemaking system.

INFLUENCE ON THE COKE MARKET

A cokeless operation of the blast furnace is not possible for physical reasons. The future coke demand will be influenced more by the fact of how the coke consumption of the blast furnace can be decreased rather than by the commissioning of smelting reduction processes as a possible replacement for small blast furnaces.

In 1993 world coke production was around 320 Mt, produced in approximately 250 coke plants having a total capacity of 370 Mt, Table 3.[12,13] The total utilisation was 86% but varies substantially in different regions. In 1993 a surplus of cokemaking capacity is to be seen in all regions considered.

Dealing with the question of how the coke demand and supply will change in the future, one has to keep in mind that not all coke produced today is exclusively used for blast furnace ironmaking. 50 Mt were consumed in 1993

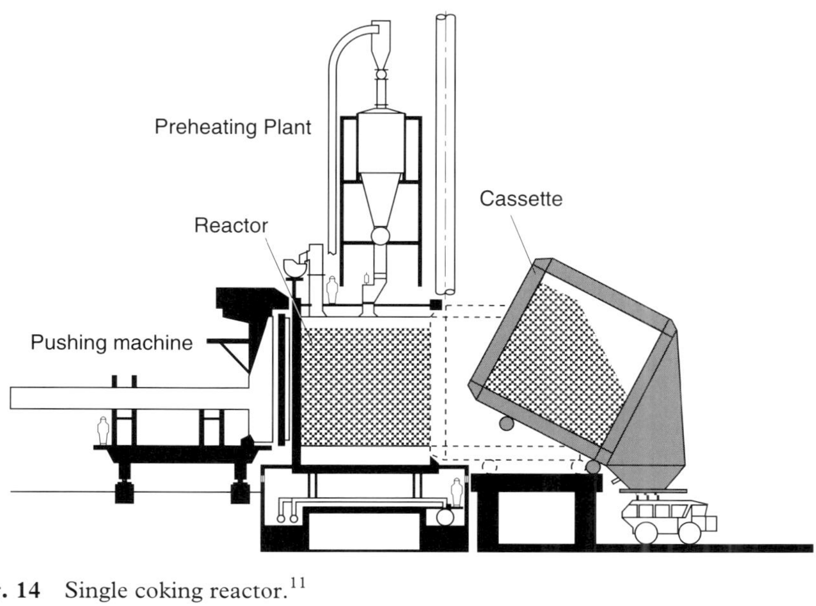

Fig. 14 Single coking reactor.[11]

Table 3. Worldwide cokemaking capacity/Worldwide coke production 1993.[12,13]

Region	Capacities Mt	Production Mt	Utilisation %
Western Europe[1]	54	46	85
Japan	47	43	91
USA	24	22	92
CIS	60*	54*	90*
Eastern Europe	25	20	80
China	90*	80*	88*
South Korea	11	10	91
Latin America	12	11	91
Others	47*	34*	72*
Total	370	320	86

[1]EC plus Turkey and Ex-Yugoslavia; * Estimation.

worldwide for other purposes than steelmaking so that the remaining amount used for ironmaking was 270 Mt, corresponding to a coke/hot metal ratio of approximately 540 kg/t including coke breeze for sinter production.

Table 4 shows the results of an estimation done by CRU International[10] based on a worldwide survey in 1994 of effective cokemaking capacity by region and forecasts for the year 2005. Effective capacity means that the age of individual coking plants was taken into consideration. 52% of the capacity was located in Eastern Europe, CIS and China in 1994. The forecasts only see cokemaking capacities grow in China and other Asian countries. By now China has undertaken extensive deliberations, but a concrete project has yet to start.[11] On the basis of the CRU study Table 5 shows an estimate of the coke balance by the year 2000. The need for coke imports is seen for Western Europe, Latin America and North America, with the highest coke deficit ocurring in Western

Table 4. Effective cokemaking capacity 1994–2005 (CRU survey in 1994).[10]

Region	Estimated	Forecast		
	Mt 1994	1995	Mt 2000	2005
North America	25.20	25.20	23.60	21.32
Latin America	12.16	12.16	12.03	11.73
Western Europe*	42.44	41.19	35.16	32.03
Eastern Europe & CIS	85.44	81.24	57.84	57.21
China	100.00	104.00	124.00	140.00
Japan	45.72	45.72	39.27	34.79
Other Asia**	37.62	37.62	41.75	46.23
Africa and Oceania	11.44	11.44	10.71	10.16
Total	360.03	358.58	344.36	353.46

* incl. Ex-Yugoslavia; ** incl. Turkey.

Table 5. Coke balance: Effective capacity less total consumption 1994–2005 (CRU survey in 1994).[10]

Region	Estimated	Forecast		
	Mt 1994	1995	Mt 2000	2005
North America	−0.80	−0.39	−0.61	−0.43
Latin America	−1.08	−1.24	−0.87	−1.54
Western Europe*	−4.94	−5.17	−6.09	−7.55
Eastern Europe & CIS	20.44	21.31	8.98	8.26
China	5.69	6.85	8.24	8.86
Japan	6.90	6.19	3.32	2.12
Other Asia**	3.41	1.70	0.22	0.00
Africa and Oceania	2.52	2.43	1.65	0.32
Total	32.15	31.69	14.83	10.04

* incl. Ex-Yugoslavia; ** incl. Turkey.

Europe.[11] The corresponding prognosis for world hot metal production is given in Table 6. It has to be mentioned that hot metal production was 517 Mt in 1995, nearly 20 Mt higher than expected by this study.

This CRU forecast will become pure speculation if economical reasons change production levels, or if the coke rate in the blast furnace is reduced more drastically than expected. Based on the assumption of a yearly 1.5% growth in steelmaking and that the hot metal to crude steel ratio remains the same in 2000 as for 1995 (a certain amount of hot metal may be produced by smelting reduction), a shortage of available coke for ironmaking will occur in 2000 if there is no further decrease in the worldwide average coke consumption in blast furnaces. 15% of the worldwide cokemaking capacities will then have an age of over 38 years and 40% of 24 to 38 years. The average lifetime of a coking plant

Table 6. World hot metal production by region (CRU survey in 1994).[10]

Region	Production Mt		Forecast Mt			Increase %
	1993	1994	1995	2000	2005	
North America	56.8	56.2	55.9	56.5	53.7	−0.4
Latin America	22.8	23.9	24.7	25.4	27.3	1.6
Western Europe*	92.7	97.5	96.0	87.7	85.9	−0.6
Eastern Europe & CIS	90.8	80.8	75.1	62.7	66.2	−2.3
China	87.3	96.5	100.9	126.5	148.8	5.9
Japan	73.7	73.3	75.2	72.1	67.5	−0.7
Other Asia**	50.5	51.8	54.6	65.0	74.9	4.0
Africa and Oceania	15.9	16.0	16.2	17.0	18.9	1.6
Total	490.6	496.0	498.5	512.9	543.3	0.9

* incl. Ex-Yugoslavia; ** incl. Turkey.
Source: CRU 1994; IISI 1995.

is normally 25 years.[14] Depending on future economic and political developments in China, there is a distinct possibility that the world will face a severe coke shortage at the turn of the century,[10] especially if the forecast of capacity growth in China does not become a reality.

It is a decision for the individual steel producers whether they believe in the availability of coke on the world market or whether they modernise an existing or build a new coking plant. Building a coking plant has the disadvantage of high investment costs but the advantages that on the one hand the coke is consumed directly where it is produced avoiding long transportation distances and that on the other hand a coking plant is generally an excellent fit into an energy network of an integrated iron and steelworks. Apart from the fact that the coke rate used in the blast furnaces will be further decreased it will be necessary to make every effort to diminish anticipated bottlenecks in supplies of coke now. This must be realised by preservation of existing capacities, by growth in capacity on existing coke plants and by the construction of new coking plants.[11]

INFLUENCE ON THE COKING COAL MARKET

In the past two years, the demand for coking coal has risen on international markets. The consumption at present obviously stabilises at a high level, Fig. 15. In 1995 the quantity of coking coal offered on world markets was approximately 225 million tonnes. There is a demand for approximately 185 Mt, resulting in a surplus of 40 Mt. These quantities are mainly sold on the market as pulverised injection coal or power station coal.[11] On a worldwide

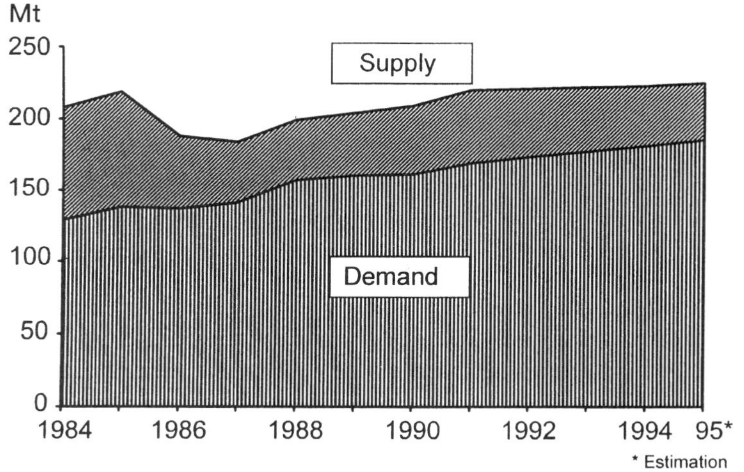

Fig. 15 Supply and demand for coking coal on world markets.

level, however, there are some acute supply bottlenecks in some quality fields.[15]

Growth in coking coal consumption in the future will be expected exclusively in the Asian countries.

The injection of coal into the blast furnaces, which need not be coking coals, and an increasing charge of semisoft coking coals in cokemaking technology further decreases the demand on coking coals.

INFLUENCE ON THE IRON ORE MARKET

Experiences gained in 1994 and 1995 have shown that existing iron ore capacities were fully utilised when worldwide steel production was high.[16] Any additional requirements for iron ore were only partially or not available at all.

Figure 16 shows that the development of lump ore production in the Western World decreased from 163 Mt in 1960 to 123 Mt in 1994, whilst fine ore production increased from 163 Mt to 345.5 Mt and pellet production from 13.6 to 148 Mt during the same period. Due to the nature of the existing iron ore mines, in the western world, an increase of lump ore production is not expected so that the trend to a higher production of sinter and pellet fines will continue.[16,17]

Table 7 summarises the most important iron ore producing countries as well as their amount of seaborne export.[16] From this it becomes clear, that only 45% of iron ore production is traded on the seaborne market and that also the number of suppliers is limited, the main suppliers being Brazil and Australia.

Countries like Brazil, Australia, India, South Africa, Sweden, Canada, USA, Russia and the Ukraine are self-sufficient concerning iron ores. If the forecasted growth of Chinese steel production of up to 110 Mt/a in 2000 occurs, China is

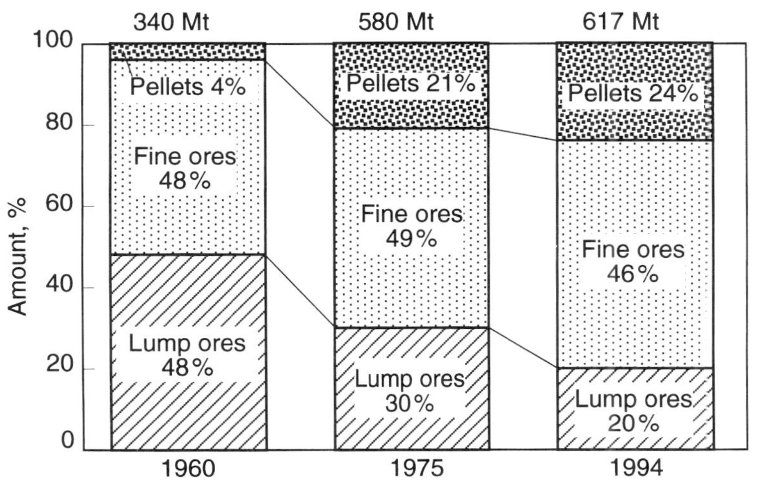

Fig. 16 Development of iron ore production in the Western World.

Table 7. The most important iron ore producers in 1994.[16]

Country	Iron ore production Mt/a	Export Mt/a	Share on total seaborne export %
Brazil	153	130	33
Australia	131	121	31
China	102	–	–
Russia	78	–	–
Ukraine	69	–	–
USA	58	–	–
India	58	29	7.5
Canada	36	21	5.4
South Africa	27	20	5.3
Venezuela	23	16	4.1
Sweden	19	17	4.1
Mauritania	10	10	2.5
Others	110	26	6.6
Total	874	389	100

expected to import up to 50 Mt from the world market.[18] In 1991 the proven iron ore reserves in China were 50 billion t out of which only 53.5% is mineable. Furthermore, of the mineable reserve, the low grade iron ore (Fe content $<$ 35%) accounts for 97.7% and the richer grade (Fe content $>$ 50%) for only 2.3%.[18]

Studies concerning the future iron ore demand come to the conclusion, that expansion of existing infrastructure is necessary to supply required amounts.[16,19] Table 8 lists two examples of forecasts for iron ore demand and supply on the overseas world market. The optimistic forecast is given by an iron ore supplier, assuming that expansion of existing mines is carried out. The pessimistic study considers the developments in steel production in the different regions of the world and the increased iron ore demand of Asia. The forecasted demand differs by only 5% whilst the forecasted supply differs by 14%.

The production of 30 Mill t of sponge iron in 1995 required some 44 Mt of iron ore, approximately 92% being lump ore or pellets respectively, low in

Table 8. Forecasts on supply and demand of iron ores on the world market in the year 2000.

	Case 1 (16)				Case 2 (19)
	Total Mt	Lump ores Mt	Pellets Mt	Fine ores Mt	Total Mt
Supply	404	69	69	266	462
Demand	423	74	73	276	446
Balance	−19	−5	−4	−10	+16

gangue contents for direct reduction shaft furnaces. However, 40% of the DRI capacities are dependent on a supply by bulk carrier transported ores, but concerning the market for pellets these DR plants have a great importance as their burden composition is on average 80% pellet. In 1994 30% of the pellets traded on the world market, which was in total 54 Mt, have been DRI pellets.[20] On the basis of a DRI production of 40 Mt in 2000 an overseas pellet demand for DRI in the range of 20 Mt[20] is expected.

As long as fine ore smelting reduction is not developed to an industrial scale and smelting reduction processes constructed to the Corex process depends on a lumpy ore charge, this will further affect the balance of pellet and lump ore supply.

A worldwide decrease in sinter production could also occur if sinter facilities are shut down because they do not fullfill the increasing environmental requirements. In this case, as good blast furnace results are in correlation with high amounts of good quality agglomerated burden, pellets will have to replace sinter production. As a consequence additional pellet facilities will have to be installed which require a planning and construction time of three to five years.

To fullfill future demand for iron ore apart from the extension of existing mines in the Western World the iron ore reserves in Russia and the Ukraine could also be considered for use. In terms of known reserves, Russia remains one of the leading sources of iron ore.[21] But Russian iron ore deposits are on the whole characterised by low grade quality, the average iron content being 35.8%. Currently the only co-operation with European countries is for the low grade ore deposit at Kostomuksha because of its location near the border to Finland.[16] Another ore reserve of some interest for the European market is Krivoi Rog in the Ukraine, but to export this ore Odessa harbour would need enlargement. These ores also need quality improvements through beneficiation.

CONCLUSIONS

As a conclusion the following can be summarised.

- The forecasted share of Oxygen to Electric steelmaking by the year 2010 will be 70/30 as a consequence of the decreasing availability of in-plant recycling scrap assuming a yearly growth rate in crude steel production of 1.5%.
- The decreasing availability of high quality scrap will require a higher production of hot metal or a higher production of scrap substitutes, like DRI. As an alternative further activities could lead to development of new scrap beneficiation processes.
- Smelting reduction processes are currently only developed to industrial scale on the basis of lumpy ores. The future field of smelting reduction processes is seen in mini-mills, or as a replacement for small blast furnaces in integrated iron and steel works.
- Worldwide hot metal production from blast furnaces will not decrease.

The demand of coke availability depends much more on the future coke consumption of the blast furnaces than further construction of smelting reduction plants. If the forecasted increase in China's coke capacity does not occur there is a distinct possibility that the world will face a severe coke shortage at the turn of the century. Therefore it will be necessary to make every effort now to diminish the anticipated bottlenecks in supplies of coke.

- Coking coals in sufficient quantity are available on the market to satisfy the needs of the iron and steelmaking industry at present. Additional efforts will have to be taken to assure the supply of the required coking coal quality in the long-term.
- Iron ore capacities are fully utilised today if iron and steel production is at a high level. A shortage for iron ores in the western world iron ore industry is not expected if the efforts in expansion of existing mines are realised.
- Mini mills will have their future predominantly in developing countries on the basis of direct reduction or smelting reduction and near net shape casting technologies but also in industrialised countries if new plants are built on a green field basis.
- As a lot of capital is tied up in modern plants of existing integrated iron and steel works of industrialised countries these will remain dominant.

The most discussed innovation in the production of steel at the moment is the near net shape casting technology, i.e. thin slab casting and strip casting. These processes in principle can be supplied by steel produced via the hot metal or scrap route. Local conditions will influence the decision on the general production route which will have no influence on the conclusions given above.

REFERENCES

1. H. Wienert: 'Zur gegenwärtigen und künftigen Schrottverfügbarkeit in der Welt', *Fachausschußbericht No. 1.023/2.033 of the VDEh*, February 1996 or *Stahl u. Eisen*, 1996, **116**(5), 110–114.
2. *World direct reduction statistics*, Midrex, 1996.
3. K.-H. Schubert, H. B. Lüngen and R. Steffen: *Stahl u. Eisen*, 1996, **116**(8), 71–79 .
4. 'European Direct Steel Feasibility Study', *ECSC contract 7210-AA/401*, August 1995.
5. N.N.: *Steel Times International*, January 1996, 41.
6. H. B. Lüngen: *Stahl u. Eisen*, 1995, **115**(3), 45–55.
7. T. W. Oshnock: *Iron & Steelmaker*, December 1995, 15.
8. K. H. Peters, H. B. Lüngen and H. Mohnkern: *Proceedings of the First International Congress of Science and Technology of Ironmaking*, Sendai, Japan, 1994, 493–99.
9. R. Altland, M. K. Beseoglu, T. Broch, W. Lanzer and K. P. Stricker: *Stahl und Eisen*, 1995, **115**(8), 83–91.

10. P. Cleary: *Cokemaking International*, 1996, **8**(1), 28–32.
11. H. Bertling and U. Strobel: *Cokemaking International*, 1996, **8**(1), 20–24.
12. *World Cokemaking Capacity*, IISI, Brussels, 1993.
13. J. F. King: *World Capacity and Production Report Coke-Oven-Coke 4/93*.
14. G. Nashan: 'Assessment of Future Market Changes and the Demand for Coking Coal', *Symposium on Use of Coal in the Steel Industry*, Pittsburgh, USA, 14–15 September 1994.
15. H. Gruß: 'Entwicklung und Perspektiven für Angebot und Nachfrage auf dem Steinkohlenmarkt', *Zeitschrift für Energiewirtschaft*, Ausgabe 1/92, 1991.
16. K. Rösener and C. N. A. Siebel: *Stahl und Eisen*, 1996, **116**(1), 29–34.
17. O. Weil: 'Gegenwärtige Kapazität und künftige Entwicklungen im Eisenerzbergbau der westlichen Welt', *Fachausschußbericht No. 1.020 of the VDEh*, October 1994.
18. Z. Dong: 'Current situation and future outlook of iron ore market in China', *Fachausschußbericht No. 1.021 of the VDEh*, January 1995.
19. S. Walsh: 'Iron ore demand and supply', *Skilling Mining Review*, December 9, 1995, 4–8.
20. H. G. Wens: 'Angebot und Bedarf an Pellets im Überseemarkt', *Fachausschußbericht No. 1.022 of the VDEh*, September 1995.
21. L. Antonenko: 'Uncertain future for Russian ore', *Metal Bulletin Monthly*, February 1996, 31–33.

THE WORLD MARKET FOR SCRAP AND IRON

J. AYLEN and K. ALBERTSON
Department of Economics, University of Salford, Salford M5 4WT, UK

INTRODUCTION

In round terms, the world uses one quarter of a billion tonnes of scrap and just over half a billion tonnes of pig iron to make three-quarters of a billion tonnes of steel. We can only give approximate figures since there is little information on use of ferrous scrap in foundries and scrap use is under-recorded. At least thirty per cent of ferrous units used for world steelmaking came from recycled scrap in 1995, after taking due account of the role of directly reduced iron.

WORLD SCRAP USE

Ferrous scrap is a global industry. Some 310 million tonnes of scrap was used to make new steel and castings in 1995. Scrap trading has a worldwide turnover of around forty billion US dollars a year. About 14 per cent of all scrap is traded across international borders. In round terms, the four major trading areas of the world used four-fifths of the world's scrap in 1995 (Table 1), with North America the largest scrap consumer.

The USA is now the world's biggest single source of scrap and the biggest exporter (Table 2). Some 70 million tonnes of scrap a year are processed in America, ranging from junk cars shredded in California to demolition scrap from old rust-belt factories in Ohio. Much of the scrap goes to feed the

Table 1. World Ferrous Scrap Use

(million metric tonnes)	1995
USA & Canada	67.7
European Union	80.8
Japan, South Korea and Taiwan	69.7
Eastern Europe and the former USSR	52.2
China	30.0
Other Western Europe	11.5
Latin America	11.9
Approximate World Total	309.6

Source: International Iron and Steel Institute (1997, page 16)
Note: including iron foundries

Table 2. World Scrap Exports

(million metric tonnes)	1995
USA	10.4
Germany	8.0
France	3.7
Netherlands	3.7
UK	3.4
Canada	2.1
Belgium-Luxembourg	1.9
Japan	0.9
Approximate World Total	39.6

Source: International Iron and Steel Institute, (1997, page 16)
Note: Figures for Germany, Netherlands and Belgium include entrepôt trade

Table 3. World Scrap Imports

(million metric tonnes)	1995
Turkey	6.9
Italy	6.3
South Korea	5.0
Spain	4.9
Approximate World Total	43.1

Source: International Iron and Steel Institute, (1997, page 16)
Note: The higher total for world scrap imports than total exports reflects discrepancies in customs statistics

emerging US steel mini-mills which melt down local scrap in electric arc furnaces to make new steel products, particularly for the construction sector.

There is a growing international trade in scrap. The mature industrialised economies are the main exporters. Over 43 million tonnes were traded internationally in 1995. The main trade flows are from Northern Europe into Spain, Italy and Turkey and from the USA outwards to Europe and towards the Newly Industrialising Economies of the Pacific Rim. New trade flows are emerging: Japan has switched from being a scrap importer to a scrap exporter as the economy has matured. The Ukraine is becoming a leading supplier to nearby Turkey. The UK has begun to export to the new mini-mills around the south-western USA via the Port of New Orleans. But for the most part, scrap supplies are predominantly local with transport costs typically representing 10–20 per cent of a steel mill's scrap purchasing budget.[1]

DEMAND FOR FERROUS SCRAP

Scrap demand is surprisingly stable and barely fluctuates from year to year. Demand for scrap is fairly unresponsive to price. In a good year, high scrap prices are covered by high final product prices. Conversely, in recession years low steel prices feed through to depress prices in the scrap market. According to Ref. 2, price elasticity of demand for US scrap is only minus 0.04. So a 10 per cent increase in price only curtails scrap demand by 0.4 per cent. This implies steelmakers are reluctant to alter scrap ratios in BOS vessels or use substitute pig iron in electric arc furnaces just because scrap prices have risen. Increasing availability of traded pig iron and directly reduced iron for electric arc furnaces, combined with more flexible BOS operation may have increased the responsiveness of scrap demand to price changes over time, but not by much.

Evidence on scrap use is seldom available, mainly because of limited data on foundry scrap consumption. Electric arc steelmaking is the main user of ferrous scrap, accounting for maybe 60 per cent of scrap use worldwide. The share of scrap going to electric arc steelmaking varies markedly from place to place, ranging in 1994 from 56 per cent of consumption in the USA and France, to 61 per cent of scrap use in the UK and 71 per cent in Japan.

Oxygen steelmaking converters are the other user of scrap for steelmaking. Systematic evidence on scrap/hot metal ratios is not available. In principle, ratios ought to vary according to the relative price of scrap compared with the marginal cost of hot metal. Some scrap is required as a coolant for the powerful exothermic reaction between tonnage oxygen and impurities in the molten iron. However, cheap iron ore often functions just as well as a coolant. The move to low silicon iron making outside North America also permits higher hot metal ratios where it is advantageous on cost grounds and where crane capacity can handle heavier iron ladles. Oxygen steelmaking seems to account for just over 20 per cent of all scrap used in the USA, UK and France in 1994 – but the figure for Japan was only around 14 per cent. Iron foundries account for the remaining scrap consumption. Some 3 per cent of all scrap is scavenged in blast furnaces in the USA, a route that has disappeared elsewhere.

GLOBAL SCRAP SUPPLIES

Evidence on scrap supply is limited. Judging by French evidence for 1992, one third of scrap supplies are circulating or process scrap. This scrap is either generated during steel production or discarded from manufacturing facilities such as car industry press lines. Perhaps one quarter of total supply is collected by dealers from the highways and byways, or from municipal refuse and similar sources. Perhaps a fifth comes from demolition of buildings, especially factories of an earlier industrial era. The remaining key source of scrap is car shredders which accounted for a further 14 per cent of French supplies in 1992. This leaves perhaps 6 per cent deriving from other diverse sources and some 4 per cent which is specifically stainless and other high value alloy steels, for example from discarded military or industrial equipment.

The scrap collection industry has a tiered structure. Typically, a multitude of small collectors supply a much smaller number of medium sized yards. In turn, they feed just a few large dealers.[3] Although there are a few major global dealers, no single supplier is able to influence the world price. There are signs of global consolidation as companies merge or develop multinational operations.[4] Nevertheless, the scrap market remains a perfectly competitive commodity market.

Scrap supplies are fairly responsive to price. Barnett and Crandall (1986) estimate an overall elasticity of supply for purchased scrap of around 0.35 and suggest that the responsiveness of scrap supplies to price is growing over time. For obsolete scrap alone, they suggest a supply elasticity of around 0.8. That is to say a 10 per cent price increase results in a rise in supply of 8 per cent. Anecdotal evidence suggests factory demolition and sale of ships for scrap respond to changes in scrap price.

SCRAP PRICES

Ultimately, global scrap prices are set by supply and demand worldwide. But world prices for scrap closely follow the huge US market. In turn, American domestic scrap prices closely follow domestic steel production. Prices of a marker grade such as No. 1 Heavy Melting Scrap reflect supply and demand on a daily basis. In this way, the upswings and downswings of the American economy echo through the global recycling market. Similar cyclical patterns are evident in the market for scrap aluminium or waste paper.

Scrap trading is a volatile market in which prices regularly halve and double (Fig. 1). Prices move in line with demand and supply and currency shifts. Unlike many conventional commodity markets there is no futures market, arguably due to the difficulties of defining quality.[5] In the USA, scrap processors speculate on future prices by building up and running down stocks.

With few exceptions, scrap prices worldwide always move in step. Like accomplished ballroom dancers they move together although the outline pattern of the dance is not always readily discernible. In the jargon of time-series statistics, scrap prices worldwide are 'co-integrated'. Take, say the US and UK scrap prices (Fig. 2). They are both volatile series. They swing up and down in an erratic and jumpy way. Yet the two series seldom drift too far apart. If they did part company, a profit minded scrap trader would seize the opportunity to supply scrap from a low price location to a high price location, thereby damping demand in the place where prices had risen and increasing demand where prices had slumped. One reason they do appear to drift apart is measurement error which caused *Metal Bulletin* to suspend publication of the UK OA series for a while. So, in the long run, there is a persistent relationship between these two price series in a global market going back for at least 30 years.

Scrap trading is a highly competitive and sophisticated market. Inter-dealer

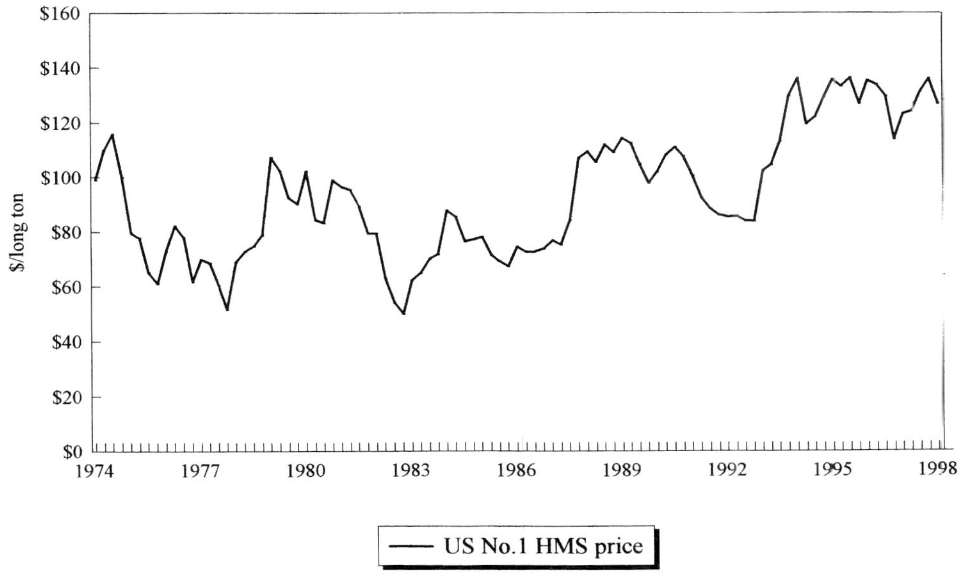

Fig. 1 US Domestic Composite Scrap Price – US No. 1 HMS quarterly average per ton.

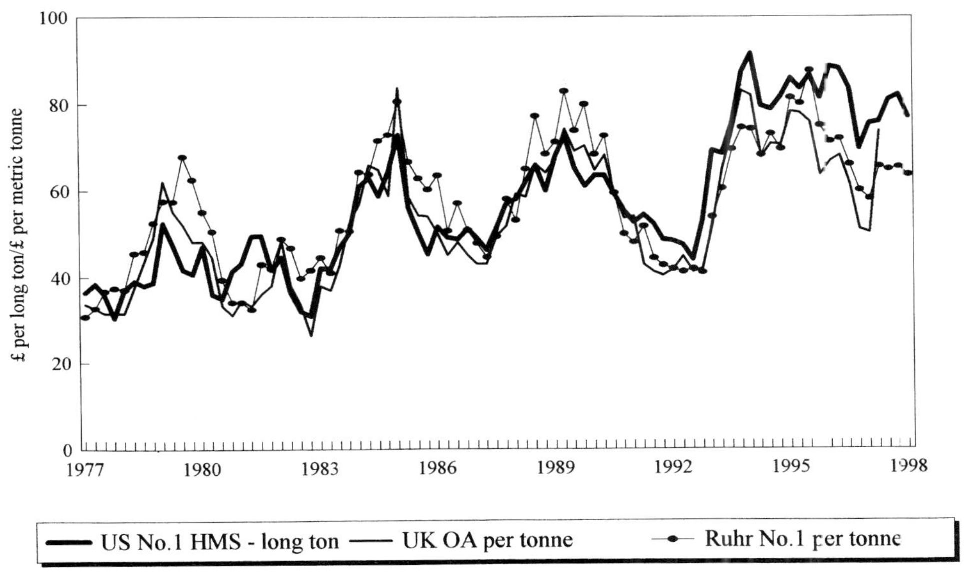

Fig. 2 US Domestic, UK and German Scrap Price – US No. 1 HMS, UK OA and Ruhr No. 1.

trading in export markets help set prices. Dealers trade between themselves to complete cargoes for export. There are major export flows to Spain, Italy and Turkey and eastward to Asia out of the long industrialised countries such as the USA, Germany and the UK. While it is hard to visualise, traders in key ports such as Tilbury, Birkenhead, Amsterdam and the US East Coast ports set global prices for ferrous scrap in just the same way that dealers on the London Metal Exchange determine aluminium or copper prices.

Domestic transactions in the large US market and in the UK tend to follow recent export prices. Monthly supply contracts in the home market follow the lead set by the dockside price. The US market is becoming more fragmented due to the growth of mini-mills away from the traditional steel producing centres. The huge size of North America militates against cross-hauling of scrap, giving rise to occasional and localised price differences. The German market is more controlled. Published prices appear to be works of fiction, but transactions in the crucial 'grey' market take place at global prices.

LONG RUN TRENDS IN THE SCRAP MARKET

Over time, scrap price movements display three key features. There is an underlying *downward* trend in real scrap prices. There are marked year to year cyclical fluctuations. Finally, there are seasonal patterns within each year. We examine these issues in turn.

SCRAP IS GETTING CHEAPER

It is possible to show almost any trend in a commodity market if the time period is chosen with sufficient care. To resolve this dilemma, we take the longest available scrap price series for a single grade, US No.1 Heavy Melting Scrap, where data exists back to 1907 in nominal terms and 1909 in real terms.[6] This long run series provides evidence there has been a gradual decline in real scrap prices (Fig. 3). Real scrap prices are not a stationary series. The real annual price of US No. 1 Heavy Melting Scrap has dropped by 0.65 per cent a year compound since 1909. The real price of scrap is now half its 1909 level. The nominal price trends upwards with inflation by 2.5 per cent a year between 1907 and 1993.

It is evident from Fig. 3 that prices have fluctuated widely in response to supply and demand shifts. Wartime peaks are apparent. Notice too the recession from 1929 onwards. Finally, the rise of the US flat product mini-mill in the 1990s has not perturbed the real US scrap price. Growing scrap supply from a mature economy has damped down US prices since 1981.

Prices are volatile over the trade cycle. Prices regularly double and halve. Typically peak prices are five years apart. Prices closely follow the cyclical pattern of US steel production. Since the US is now the world's largest scrap supplier and consumer, global prices tend to follow the American market. It follows that high scrap prices during a boom in the USA may coincide with recession in other steel

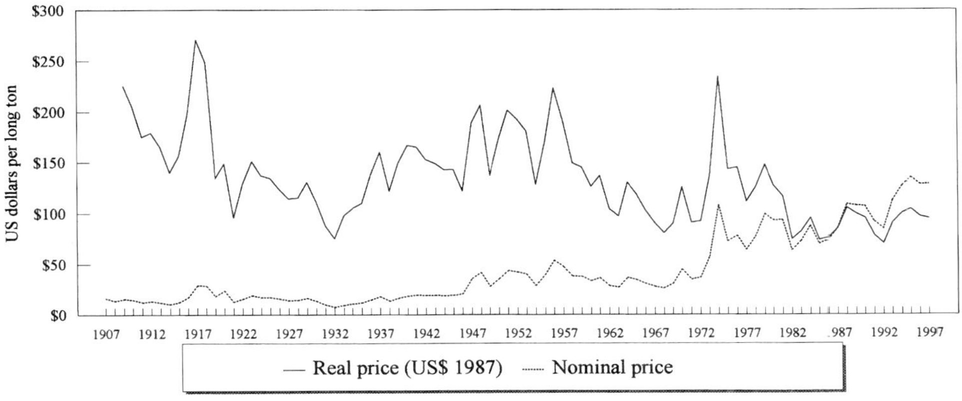

Fig. 3 Real and Nominal Scrap Prices – US domestic composite price No. 1 HMS.
Source: US Bureau of Mines and Metal Bulletin

markets. In these circumstances electric arc producers suffer a double squeeze: high global scrap prices coinciding with low local product prices.

THE GREAT LAKES FREEZE

The scrap market is influenced by seasonal effects for two reasons; the weather in North America and industrial production. Global scrap prices rises when the Great Lakes freeze in North America and fall when the locks at Sault Ste. Marie re-open in the Spring. Albertson and Aylen[8] show that seasonal effects cause global scrap prices to rise by 2 per cent in the first quarter of the year, only to fall back by 1.3 per cent in the Spring. In truth, closure of the Great Lakes to navigation is just one seasonal effect at work in the huge US market. Severe winters affect rail transport and inhibit collection of obsolete scrap. Conversely the spring and autumn are more conducive to scrap collection from the highways and byways of North America. The rhythm of industrial production and the accounting year have their effect too. Steelmakers run down inventories ahead of the financial year end in December only to rebuild stocks during the first quarter. The whole global economy takes a holiday break in July and August.[9] All these factors echo through as seasonal influences on the global scrap market.

FORECASTING SCRAP PRICES

There are two approaches to forecasting future trends in the scrap market: extrapolating the past into the future using statistical techniques of time-series analysis, or explaining the behaviour of markets with a causal econometric model.

Time series techniques, such as ARIMA models, provide an elegant summary of past market movements. Time series models project past trends, cycles and seasons into the future assuming no change in the underlying behaviour of the market.[7] Implicitly, the past replicates itself into the future. Albertson and Aylen[8] suggest an ARIMA model with long lag lengths is required to capture the wide cyclical swings of the US scrap price. They use quarterly data on the US No. 1 Heavy Melting Price for forecasting. Taking account of seasonal effects improves forecasting accuracy when using quarterly data. Just recently shredded scrap has become the most widely traded grade in the US. Shredded scrap might prove a better candidate for future forecasting. One advantage of naive time series techniques is their parsimony in the use of data.

The behaviour of the scrap market can also be explained with a simultaneous equation regression model of the sort estimated by Barnett and Crandall.[2] The central problem here is that scrap prices are explained by variables including US steel consumption and US steel output. Consequently, these independent variables themselves have to be forecast before arriving at a forecast for scrap prices. In effect, the forecasting problem is pushed one stage further back. Forecasting problems are compounded by shortcomings in data availability, notably scrap consumption. Use of annual statistics on scrap use, which include estimates of foundry scrap consumption, helps overcome some data problems when forecasting. The USA is blessed with extensive data on the scrap market from the US Bureau of Mines, now the US Department of Geology. It is doubtful whether similar econometric models could be estimated for other countries.

IRON ORE MARKETS AND SCRAP PRICES

Iron ore prices are determined on a different basis to those for scrap. Dockside scrap prices move daily. Iron ore prices are negotiated on an annual basis for the year ahead through a process known as the 'mating season'. Two countries and two sets of consumers dominate world seaborne trade in iron ore. The two leading suppliers are Brazil and Australia which jointly command 61 per cent of world exports in 1996 (Table 4). The leading consumers are Western Europe

Table 4. World Iron Ore Exports

(million metric tonnes)	1996
Brazil	129.7
Australia	128.6
India	31.7
Canada	28.0
Ukraine	20.6
South Africa	19.3
Sweden	16.1
Russia	11.3
Approximate World Total Exports	439.5

Source: IISI (1998, table F2)[16]
Note: trade by weight of ore, not iron content

Table 5. World Seaborne Iron Ore Imports

(million metric tonnes)	1996
Pacific Basin	
Japan	119.2
P.R. China	43.9
S. Korea	34.5
Taiwan	10.1
Atlantic Basin	
Germany	39.3
UK	20.3
France	17.7
Italy	16.0
Belgium-Luxembourg	12.9
Netherlands	8.3
Spain	6.1
Approximate World Total of Seaborne Trade	380

Source: IISI (1998, table F3)[16]
Notes: (excluding water borne transport across Great Lakes)

and Japan which together account for almost 60 per cent of world imports (Table 5). Essentially, the Australians and the Japanese lead on prices in the Pacific Basin while the Brazilians and Europeans follow in the Atlantic Basin. Russia and the USA are also major ore users. They play no part in world price negotiations since they rely on local ore. South Korea and Taiwan are emerging as significant buyers too, purchasing some 10 per cent of world imports. Essentially the mating season is a square-off between Europe and Japan in one corner and Brazil and Australia in the opposite corner.[10]

The resulting annual price movements are given in Fig. 4 in terms of 1992

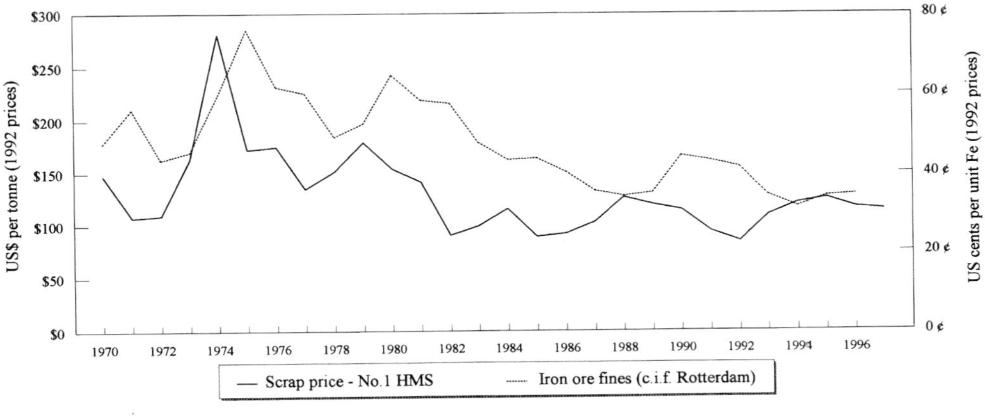

Fig. 4 Scrap Prices versus Iron Ore Prices.
Sources: TEX Reports, US Bureau of Mines

prices. These are for a typical marker grade, Hamersley fines delivered, cash-insurance-freight paid at Rotterdam in US cents per unit of Fe. It is clear that real iron ore prices have been falling since 1975. Each cyclical peak is lower than the last.

Seaborne trade in iron has grown, approaching 400 million tonnes in 1995. Yet, falling iron ore prices have meant that investment in new greenfield mines has not been justified.[11] Rather supply has increased through 'brownfield roundouts' adjacent to existing mines and around existing major rail lines in Brazil and Australia.[12]

The iron ore and scrap markets move in step. Changes in the global scrap price feed through to the subsequent mating season in the iron ore market. Statistically we can be confident that there is a long run relationship between scrap prices and iron ore prices using the standard cointegration tests for a bivariate model, the Cointegrating Regression Augmented Dickey Fuller and the Cointegrating Regression Durbin Watson tests suggested by Engle and Yoo.[13] There is clear evidence of cointegration between the two real price series. That is to say, annual average US domestic scrap prices and subsequent world iron ore prices move in step (Fig. 4).

The dynamic structure of the relationship is by no means constant – the lag between changes in scrap prices and the subsequent change in iron ore prices varies over time. If anything the lag length increased over time from one year towards two years in the early 1980s. This complicates tests for causality. The relationships fail weak exogeneity tests: such as an LM exogeneity test (see Ref. 14). In effect, there is a common source of underlying errors influencing both the scrap and ore price series, maybe exchange rate movements or the rhythm of global steel production. Yet, in terms of *Granger causality* there is very strong evidence that real scrap prices do 'cause' subsequent iron ore prices. The relationship does not work in reverse: there is no evidence that the outcome of iron ore negotiations influence scrap prices.

It follows there is no cheap source of ferrous material. In the long run, scrap prices and iron ore prices always move together. There may be temporary advantages for one ferrous feedstock or the other. But, ultimately, costs of iron inputs for the electric arc route and the integrated blast furnace/BOS route will always come into line in a single global market for ferrous material when ore prices finally adjust.

THE PIG IRON AND DR MARKETS

Cold pig iron is a substitute for scrap. It is hard to be precise, but perhaps 25 million tonnes of cold pig iron is used worldwide, quite apart from the 500 million tonnes of iron in molten form used in BOS converters. We can estimate that approaching 5 per cent of iron made worldwide is used in cold form. Some 15 million tonnes of merchant iron is exported or imported each year (Table 6), although customs statistics are not reliable for this product. In addition, there is

Table 6. World Pig Iron Exports

(million metric tonnes)	1996 *(1995)*
P.R. China	3.6
Brazil	2.5
Russia	2.1
Japan	1.0
Ukraine	0.6
India	*0.4*
South Africa	*0.3*
Approximate World Total	*14.8*

Source: IISI (1998, table A9)
Note: *figures in italics relate to 1995*, all grades of merchant iron, including foundry iron

Table 7. World DR Exports

(million metric tonnes)	1993
Venezuela	0.4
Trinidad & Tobago	0.2
Approximate World Total	0.8

Source: UNCTAD (1995, table 26)

a revival of domestic merchant iron production in the USA for electric arc use following the virtual disappearance of pig manufacture by the end of 1990. Pig iron is easier to transport than some grades of DR which are potentially pyrophoric. Consequently, the merchant iron trade dwarfs any seaborne trade in DR material (Table 7).

Between two-thirds and three-quarters of world cold pig iron is used in foundries. Electric arcs account for between a third and a quarter of world pig use.[15] An unknown amount finds its way into ageing open hearth furnaces where they survive in Eastern Europe and India. There are few statistics on iron usage in foundries. Even the output of castings is not known with accuracy. So it is difficult to assess how much foundry iron goes into the iron castings sector.

The pattern of pig iron consumption is complicated by use of foundry grades of merchant iron for steelmaking. Foundry iron is a valuable feedstock for an electric arc when used in small amounts since the high silicon and carbon contents generate exothermic reactions in the bath, thereby reducing electric power consumption. Merchant pig iron in all forms is free from residuals found in scrap, such as copper. Typically, pig iron low in phosphorous is required for steelmaking since an arc furnace is not good at dephosphorising. Certain specialised grades of pig, such as those used for making SG iron end up almost exclusively in the foundry sector.

Perhaps 5–6 million tonnes of merchant iron are currently used for steelmaking. Against this, total world production of DR iron in 1996 was 32.8 million tonnes.[16] Arguably then, DR iron was six times more significant than merchant iron as an electric arc feedstock. Virtually all DR iron is used close to where it is made.

Pig iron and scrap are very close substitutes. Global pig iron prices closely follow world scrap prices. For example, the UK price for 2.5/3.0% silicon pig tracked the 9A cast iron scrap price – with a lag – from 1977 until domestic production came to an end in 1992. High scrap prices encourage supply of merchant iron, but cheap scrap depresses merchant iron production. Consequently, world pig iron trade has been a volatile market, arguably peaking in the late 1980s. During the 1980s merchant iron was among the fastest growing products in world steel trade. A subsequent drop in world cold iron output during the early 1990s followed supply upheavals in the former Soviet Union and Brazilian government restrictions on virgin forest destruction for charcoal ironmaking.

Rapid growth in merchant iron exports came to an end in the early 1990s. Nevertheless, pig iron trade eventually recovered to reach a new peak in 1995, the last year for which we have fairly complete figures. Broadly speaking, China has supplanted Russia and Brazil as the leading commodity supplier of merchant pig iron. Chinese exports are volatile, having risen from a third of a million tonnes in 1993, peaked at 5.4 million tonnes in 1995 and since dropped back to 3.6 million tonnes in 1996. In general, merchant pig iron tends to be made in economies with low wages, cheap carbon sources and excess furnace capacity.

CONCLUSION

The picture that emerges is of a scrap market where small scale supply decisions of local 'rag-and-bone-men' form part of a huge, sophisticated global market for ferrous material. The outcome is a volatile world market with strong cyclical swings and mild seasonal fluctuations dominated by the US. During this century, the long run trend in real scrap prices has been downwards, suggesting growing supply has readily matched growing demand. In turn, scrap price fluctuations feed through to influence the global iron ore market and world trade in pig iron.

ACKNOWLEDGEMENTS

The financial support of British Steel-Engineering Steels is warmly acknowledged. We are grateful to Mr Don Turner, Managing Director of BSES for permission to publish this paper. Earlier research on the global pig iron market was supported by the International Pig Iron Secretariat. The technical advice of the David J. Joseph Company in Cincinnati on the workings of the US scrap market is much appreciated.

REFERENCES

1. S. Wulff: 'Ferrous Scrap: an involuntary by-product that has become an essential raw material', paper to *AISE Conference on the future of flat rolled steel production*, mimeo, 1995.
   S. Wulff: 'The coming metallics surplus: heresy or inevitability'. *Steel Times International*, November 1997, **21**(6), 19–23.
2. D. F. Barnett and R. W. Crandall: *Up From the Ashes: the Rise of the Steel Minimill in the United States*, Washington D.C., Brookings Institution, 1986.
3. International Iron and Steel Institute: *Scrap and the Steel Industry: Trends and Prospects for Solid Metallics*, Brussels: IISI Committee on Raw Materials, 1987.
4. T. Gill: 'Big is beautiful in N. American ferrous scrap', *Metal Bulletin Monthly*, May 1997, 71–75.
5. J. K. Stone: 'The iron and steel industry and the commodity market', *Iron and Steel Maker*, February 1977, 36–38
6. U.S. Bureau of Mines: *Metal Prices in the United States through 1991*, Washington D.C.: Bureau of Mines, U.S. Department of the Interior, 1993.
7. R. S. Pindyck and D. L. Rubinfeld: *Econometric Models and Economic Forecasts*, Boston: Irwin McGraw-Hill, 4th ed., 1998.
8. K. V. Albertson and J. Aylen: 'Modelling the Great Lakes Freeze: forecasting and seasonality in the market for ferrous scrap', *International Journal of Forecasting*, September 1966, **12**(3), 345–359.
9. J. J. Beaulieu and J. A. Miron: 'A cross-country comparison of seasonal cycles and business cycles', *Economic Journal*, 1992, **102**, 503–534.
10. J. Aylen: 'Mating time – Iron ore and exchange rates', *Steel Times International*, November 1991, **15**(6), 40 and 48.
11. P. Doolan: 'Raw materials strategy', paper to *Institute of Materials, Annual Conference of Iron and Steel Division*, Wishaw, Birmingham, 1996
12. U.S. Department of the Interior, U.S. Geological Survey: *Iron Ore, Annual Review – 1996*, Reston, Virginia: November 1997.
13. R. F. Engle and B. S. Yoo: 'Forecasting and Testing in Cointegrated Systems', *Journal of Econometrics*, 1987, **35**, 143–159.
14. J. Johnston and J. DiNardo: *Econometric Methods*, New York: McGrawHill, 1997.
15. J. Aylen: 'Pig puzzle', *Steel Times International*, November 1993, **17**(6), 31 and 44.
16. International Iron and Steel Institute: *Supplement to the Steel Statistical Yearbook, 1996–1997 update*, Brussels: IISI, Lotus 1–2–3 file, 1998.
17. International Iron and Steel Institute: *World Steel in Figures*, Brussels: IISI, 1997 edition.

THE SCRAP DEBATE: POSSIBLE GLOBAL SUPPLIES

P. GENEVAZ

Euroconsult, 71–79, Bd Richard Lenoir, 75011 Paris, France

THE SCRAP DEBATE

Drastic changes in scrap resource and consumption are shaping the international scrap trade. This will have major consequences on the world trade balance at the beginning of the next century. Europe has become self-sufficient, the OECD countries have become a zone of increasing surplus, and Asia has become a zone of increasing deficit. The potential for exports from the ex-Soviet zone is anyone's guess.

As statistics for scrap consumption and resource are either unreliable or inexistent in many countries, the scenario method is very helpful as a forecasting tool. Basically, we shall test two different scenarios. In the first scenario, the price of scrap will continue to fluctuate within the traditional cycle (see Graph 1). In the second scenario, we suggest that the world price of scrap may remain high (say around $130/t for the composite price) for a long period.

This paper examines how the world scrap market could adapt to the growing

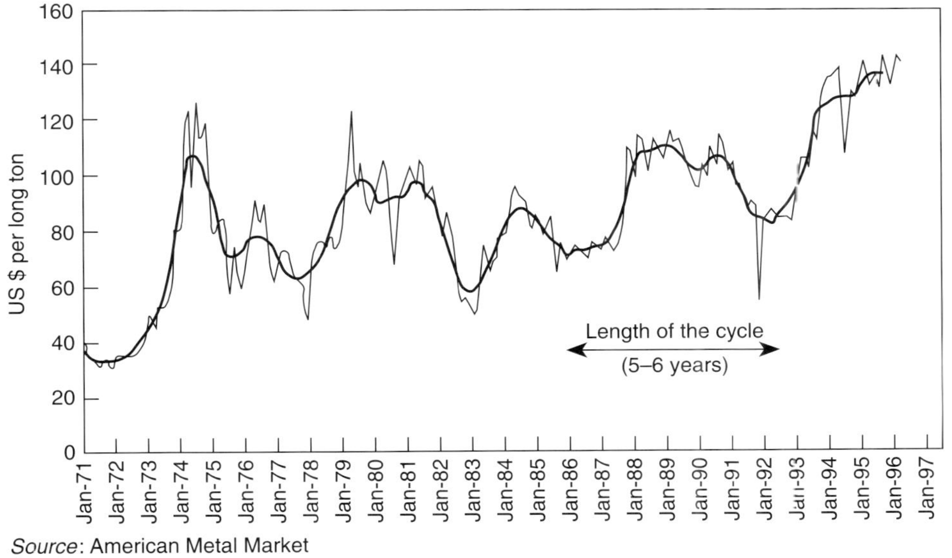

Source: American Metal Market

Graph 1. US: The Scrap Cycle. Monthly composite price, with 12-month moving average, 1971–96

needs of the Asian zone. It is based on two multi-client studies published by Euroconsult: 'Iron Ore/Scrap 1995–2005: the Alternative' and 'Steel Industries and Markets in Asia: Prospects to 2005'. It is also based on other studies carried out for specific clients, including the European Commission.

THE OECD SURPLUS

The scrap surplus within the OECD zone has been increasing since the mid seventies. Growth has been fairly regular since 1983, reaching almost 3 million tonnes in 1985, over 7 million tonnes in 1990 and almost 8.5 million tonnes in 1993.

A declining deficit from the European Union (which changed to a surplus since the beginning of the 1990s) and Japan, coupled with a high export level from the United States explains the rising surplus from the OECD zone. Germany, which quadrupled its export surplus since 1980, is the leading exporter of scrap from the European Union. On the other hand rising imports from Turkey have increased the deficit of the rest of Europe six fold since the beginning of the 1980s.

Forecasting foreign trade is a tricky exercise for two reasons. First, strong variations from one year to the next have been the rule in the past. Second, the long term evolution of scrap trade does not show any statistical links with resources or consumption. We expect the OECD surplus to stabilise over the next 2–3 years, and remain at around 10 million tonnes per year. This will be due to a decrease in the US surplus which will be compensated by a surplus increase in Europe (see Graph 2). In 2000, the EU surplus could level out at

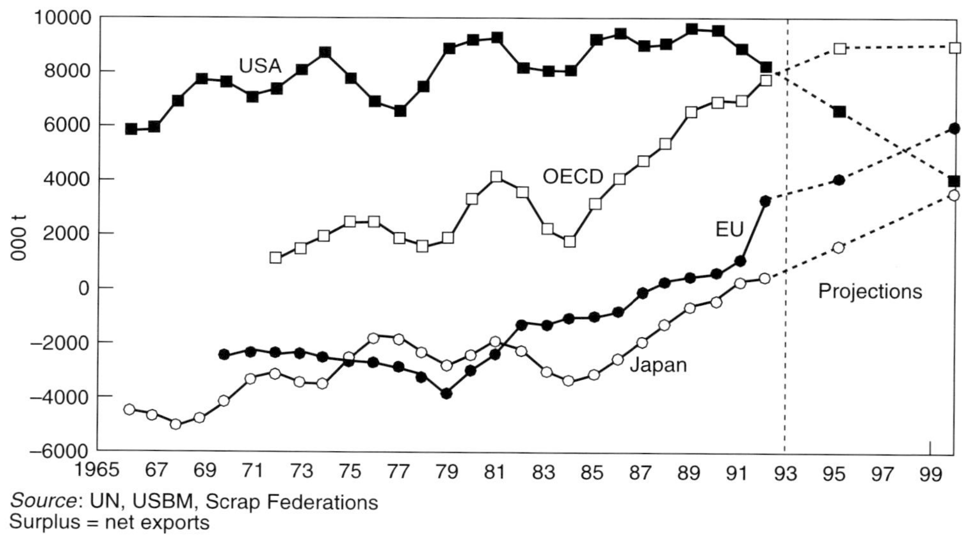

Source: UN, USBM, Scrap Federations
Surplus = net exports

Graph 2. OECD: Scrap Surplus Forecasts (3-year moving average)

Table 1. USA: Trends in Steel Scrap Trade

000 T	1987	1988	1989	1990	1991	1992	1993	1994	1995	2000 p	2005 p
Exports	9405	9160	11140	11580	9345	9260	9870	8810	10400	8000?	7000?
Imports	765	940	1015	1310	1075	1315	1545	1730	2090	3000?	3500?
Balance	8640	8220	10125	10270	8270	7945	8325	7080	8310	5000?	3500?
Exports to:											
Canada	300	570	937	939	806	1163	1435	1680	1650		
China	226	28	44	65	37	215	515	445	200		
Korea & Taiwan	2272	2746	3079	3412	3534	2540	3910	2710	4320		
Japan	894	589	434	481	340	460	535	535	710		
Italy & Spain	537	448	744	406	152	194	50	135	215		
Turkey	2045	1801	2926	2748	1776	2019	1310	1180	1390		
Imports from:											
Canada			863	1090	893	910	1150	1340	1620		
Mexico			52	64	64	65	95	115	170		
Venezuela							220	90	145		
Japan			60	67	52	85	40	60	65		
Pig iron							830	2490	2100		
DRI							1015	1120	1250		

Source: USBM.

about 6 million tonnes, the US surplus at 4 million tonnes and the Japanese at 3–3.5 million tonnes. At the same time, the deficit from other European countries will also slightly decline, as growth in domestic supplies within Turkey stabilise imports, as was the case in Italy and Spain.

All of these projections are conservative; many experts predict a Japanese surplus of 5 million tonnes in 2000, and the decline of the US surplus is by no means a certainty. Actually, the latest data available suggest a lasting scrap surplus of 7–8 Mt in the US trade (Table 1). Despite its booming EAF capacity, the US has not significantly increased its scrap consumption, except in 1993. Rather, it is increasing its consumption of scrap substitutes (DRI, pig iron), and this trend is set to continue, as DRI capacity is following suit: no less than 6–7 million tonnes of new capacity (DRI and IC) are projected either in the US or in South America to supply the US market.

THE FORMER SOVIET BLOC

Statistical data from the countries belonging to the ex-Soviet bloc are scarce, and often unreliable. Until the end of the 1980s, most Eastern European countries were either self-sufficient in scrap, or small importers from the USSR. At the same time, the USSR declared net exports of about 3 million tonnes per year, mainly to these countries. Since 1990, Slovakia, the Czech Republic and Hungary, have started exporting scrap on a significant scale.

All countries belonging to the ex-Soviet bloc were heavy consumers of scrap, as their steel industries were largely based on open hearth furnaces. Particularly

Table 2. The Rate of Growth of Scrap Consumption has been Consistently Inferior to that of Resource in the Past

		1966–73	1973–93	1983–93
	EC (7)	3.5%	−0.2%	2.8%
Resource	US	3.1%	−2.0%	1.6%
	Japan	9.9%	−0.2%	2.9%
	EC (8)	3.4%	−2.3%	0.2%
Consumption	US	1.7%	−3.7%	0.9%
	Japan	9.1%	−0.9%	0.6%

NOTE: Average annual rate of growth, based on 3-year centred average.

in the CIS, many of these countries could develop a considerable surplus. The CIS is bigger than the US; it has an important steel base, and a history of enormous steel consumption. The current reduction in steel production, particularly from OH furnaces, is reminiscent of the situation in the US after the first oil shock. Disorganisation and transportation problems are certainly major hindrances to the development of the scrap industry in these countries, which is not yet as developed as the West. But the cost of entry is low, and the technology is fairly simple.

There is certainly a large potential for scrap exports from this zone, but this is almost impossible to forecast. To overcome the difficulty one can suggest three possible surplus scenarios from the zone from 2000 onward: low (surplus of 0–2 million tonnes), medium (surplus of 6 million tonnes) and high (surplus of 10 million tonnes or more). It should be also underlined that the scenarios for each zone (i.e. Eastern Europe and the CIS) are independent: one can consider an

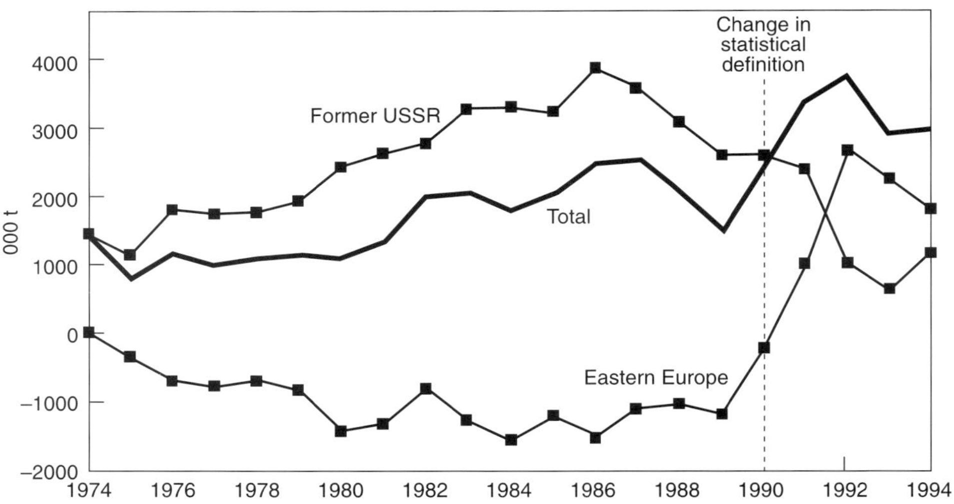

Graph 3. Eastern Europe: Trends in Scrap Trade (1974–94)

optimistic scenario for Eastern Europe combined with a pessimistic one for the CIS.

ASIA

When analysing world trade, Japan and Australia should be included in the framework of Asia, despite their industrial maturity. Graph 4 shows that the Asian deficit in scrap, including these two countries, increased considerably in the first half of the 1980s, after which it stabilised until 1993. The greater deficit which occured in 1993 has been considered by many observers as a characteristic of this zone, with threat of a major imbalance in world trade, and a high price for scrap. Two contradictory factors were at work in 1993: a low level of resource in Japan, which was directly connected with the recession, and China's entry on to the world market.

It is hard, for a western observer, to believe the level of scrap prices in Asia. When the US composite price is about US$ 130/t, the price of imported scrap at the Asian plant is at least US$ 180/t. The price differential can be even higher in certain countries and major importers, like China or India, can withdraw from the market if a price threshold is crossed. Furthermore, India has a very particular type of regulation for its scrap market. In this case, part of the scrap from ship-breaking is sold directly to re-rollers. There are at least a thousand small re-rollers in India, producing about 4 million tonnes of steel products per year. When the price of scrap gets too high, scrap re-rolling is less expensive than re-melting.

Asia will remain a major scrap importing zone for at least the coming decade, but the level of imports will mainly be price-regulated. Were the world price of scrap to remain at a high level (e.g. US$ 130/t) for a long time, the incentive for

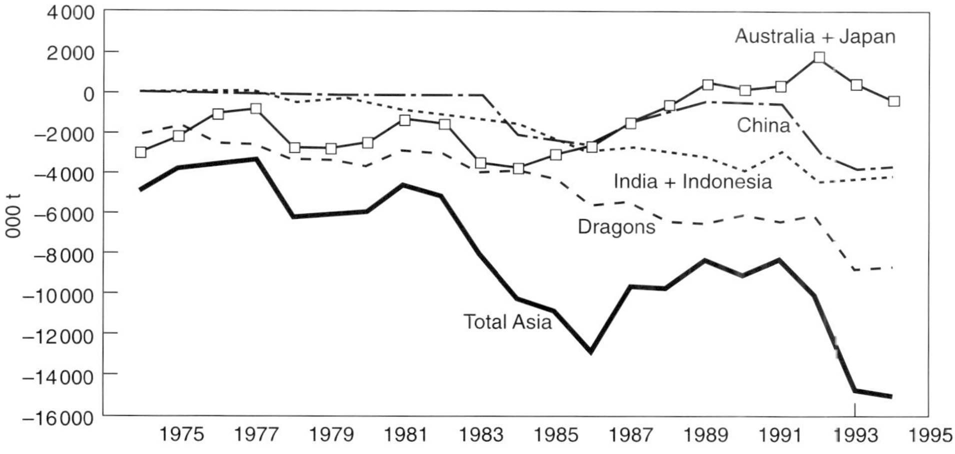

Graph 4. Scrap Deficit in Asia (including Japan and Australia, 1974–94)

both local scrap collection and the use of substitutes (pig iron, DRI) would be very high, thus strongly reducing the import level. In fact this healthy reaction already happened in 1994 and 1995. The production of DRI in Asia more than doubled between 1992 and 1995 (from 3.4 million tonnes to 7.2 million tonnes), due for the most part to new capacity construction in India. In China, the production of pig iron increased by over 20 million tonnes between 1992 and 1994 (+28%), with the construction of 3,400 new blast furnaces. Scrap substitutes will be in increasing supply: important DRI capacity projected in Australia, mini BFs in India specialised for merchant pig iron, and significant Corex capacity being commissioned or under construction in Korea.

Finally, the scrap resource itself is increasing rapidly, particularly in Korea, China and Taiwan (Graph 5).

OTHER COUNTRIES

We do not forecast a major increase in scrap imports from other countries. In South America, the BF/BOF route benefits from a distinctive cost advantage with cheap, local high grade ore. Moreover, gas producing countries can produce low cost DRI, which is an export product, but at the same time a direct competitor to expensive imported scrap. Imports from South America will probably remain below 1 million tonnes in 2000. In the Middle-East, the ready availability of gas has favoured the development of a DRI-based steel industry, particularly in Iran, Saudi-Arabia and Qatar. Scrap imports from Egypt and North Africa are small. In South Africa, available local scrap suffices to supply a strong electric-arc based industry.

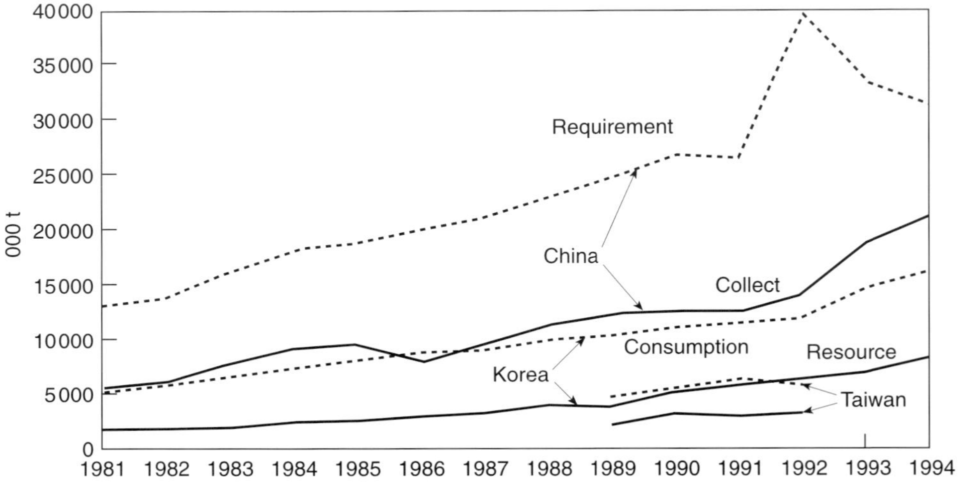

Graph 5. China, Korea and Taiwan: Scrap Consumption and Resource (1981–1994)

Table 3. The Rate of Growth of Scrap Resource has been Very High in Rapidly Industrialising Countries

		1973–93	1983–93
Resource	Italy	5.5%	3.3%
	Japan	2.7%	2.9%
	Korea	19.0%	9.9%
Consumption	Italy	1.1%	1.4%
	Japan	1.7%	0.6%
	Korea	14.3%	8.0%

NOTE: average annual rate of growth, based on 3-year centred average.

WORLD BALANCE

Table 4 summarises in simplified terms the different possibilities suggested above for the end of the century. As far as exports are concerned, total world exports of about 12 to 16 million tonnes, including 6–10 million tonnes from the OECD (except Japan and Australia) and 6 million tonnes from Eastern Europe and the CIS, represent the most likely scenario. In the worst case, the total world surplus could be limited to 8 million tonnes and in the best case, the world surplus could reach 20 million tonnes.

On the import side, if the price of scrap remains high, imports could decline to 8 million tonnes by 2000, leading to a large world surplus. In the case of a normal cycle of scrap, total imports could reach 16 million tonnes or more. The most likely scenario corresponds to one where there are total imports of 14 to 16 million tonnes, roughly in balance with total exports.

It is thus very unlikely that the price of scrap will remain at a high level for a long period. Such an event would inevitably result in a major fall in prices at the end of the century. On the other hand, Asia will undoubtedly remain a source of tension on the world market – a problem which could become permanent if the surplus from Eastern Europe and the CIS remains limited. It should be underlined that a surplus from Eastern Europe and the CIS could alter any

Table 4. Scrap World Trade Balance in 2000

	Scenario 1 (Traditional scrap cycle)		Scenario 2 (Structural high scrap price)	
World exports	12	20	8	20
OECD	10	15	10	15
OECD excl Japan & Australia	6	10	6	10
East Europe/CIS	6	10	2	10
World imports	16	16	8	8
South America	1	1	1	1
Asia (incl Japan/Australia)	15	15	7	7
Tension	High	Low	Low	Price fall

existing balance: in the case of a high level of surplus, the world market would be over-supplied. In the case of a low level, tensions would be inevitable.

As in the past, the 'world' price of scrap will be tied to the economic cycle, and mainly to that of the US. For if Turkey is included in the European zone, as it should be, the net export potential of Europe remains small. In the past, the world price was the result of transactions between US exporters and European importers; it was an Atlantic price. In the 1990s and at the beginning of the next century, the world price will be the result of US exporters and Asian importers, as these transactions are becoming predominant.

THE ENGINEERING BUSINESS IN THE EARLY DECADES OF THE NEXT CENTURY

E. BIAMONTE and A. B. HINDER

Innovation and Technology Management SA, 7, route de Drize, CM 1227 Geneva, Switzerland

1. THE CONTEXT

During the last few years there has been continued growth of the world steel market, although at a rather sluggish pace, below 1% per year on average. But, a number of structural changes have considerably modified the world's iron and steel industry over the past decade. These recent developments follow a long period of relative stability in terms of technologies and markets.

There has been a shift in the centre of gravity of the industry away from traditional steel producers (North America, Europe, CIS) towards Asia and other newly industrialised countries. In fact, major growth areas have been markets in China, South East Asia and Latin America. Figures 1 and 2 highlight recent trends in steel production, including forecasts up to the year 2010, for respectively the world and Asia.

The steel market has been characterised by an increasing demand for higher value added products (such as thinner gauge strip, galvanised and organically coated steels). On the supply side there has been strong growth in electric arc steelmaking, with larger capacity mini mills competing directly with conventional plants for certain flat products. Regional scrap shortages and higher quality requirements are contributing to the development of new processes

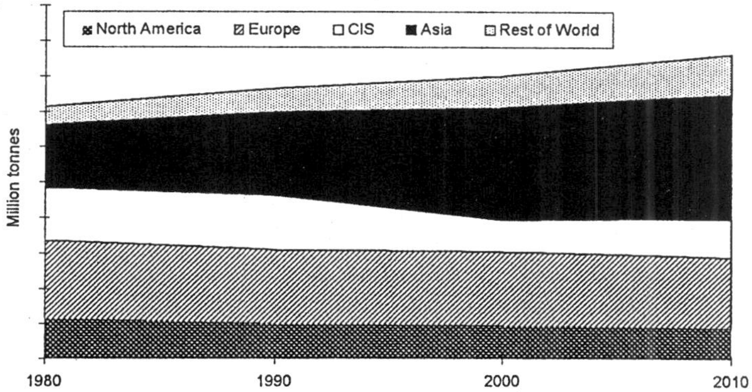

Fig. 1 World liquid steel production.

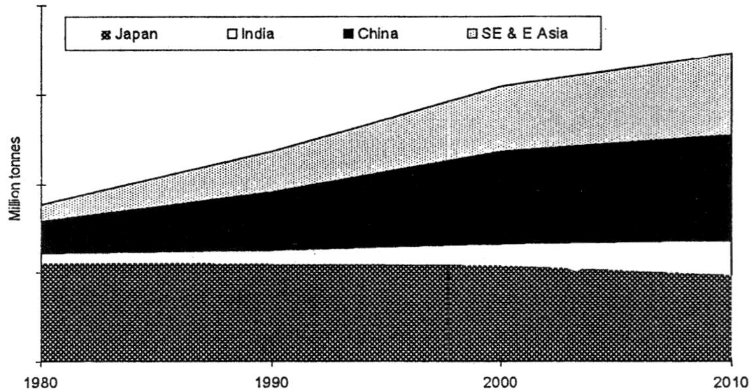

Fig. 2 Steel production in Asia.

supplying alternative EAF charges (Direct Reduced Iron, Hot Briquetted Iron, smelting reduction and iron carbide). In addition environmental considerations are having a growing impact on steelmaking technologies (Fig. 3).

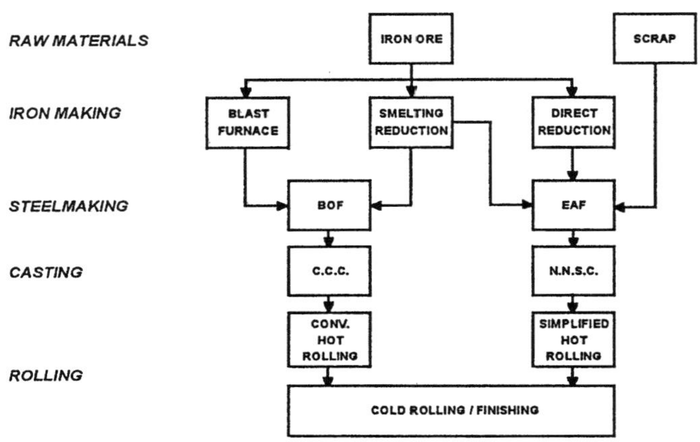

Fig. 3 Alternative iron and steelmaking routes.

Direct reduction processes have been developed using either natural gas or coal to reduce iron ore in the form of lump, pellets or fines to an iron rich charge material for electric arc furnaces. Two processes predominate: Midrex and HyL, both of which use lump ore or pellets and gas as a reductant. Together these account for over 90% of world DRI output (Fig. 4).

Iron carbide is a new charge material made from iron ore fines, using natural gas. Commercial scale production has only recently started. There are a number of other processes currently under development to use iron ore fines. Gas based

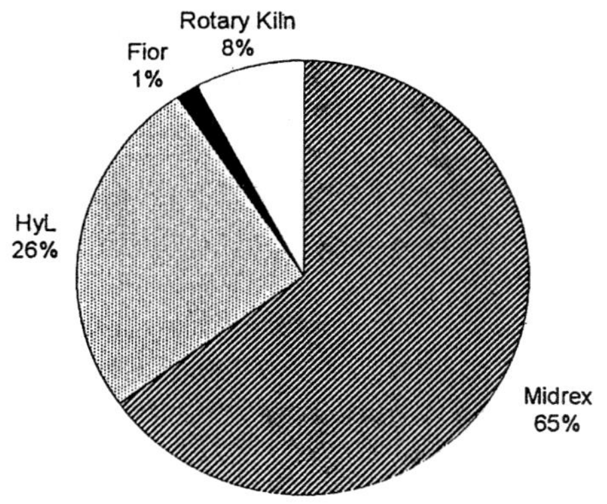

Fig. 4 World DRI production.

processes include Circored and Finmet and coal based processes include Circofer and Fastmet.

Smelting reduction is an alternative approach to iron making which uses coal to reduce iron ore in the forms of lump ore, pellets or fines to molten iron. There is only one commercial process currently operating which is Corex, based on either lump ore or pellets developed by Voest Alpine. There are a number of fines based smelting reduction processes currently under development. These include:

- AISI
- Ausmelt
- CCF
- DIOS
- Finex
- Hismelt
- Romelt
- Tecnored

Figure 5 highlights recent and projected use of ferrous raw material for steelmaking, anticipating the development of smelting reduction iron as a feedstock from the year 2000 onwards. The share of blast furnace iron, on the other hand, is projected to decrease beyond this date. Figure 6, on the other hand, shows world steel production trends by process route. The Electric arc furnace is expected to continue to increase its share of world steel output. It is evident from Fig. 7 that the competitiveness of the conventional oxygen steelmaking route varies from location to location. Environmental costs are likely to be particularly severe in Japan and Western Europe.

Current conventional casting plants typically have a high capacity of 3–4 M/t year and are associated with oxygen steelmaking units at integrated plants. Near Net Shape Casting processes include CSP, Conroll, Danieli, INNSE, ISP,

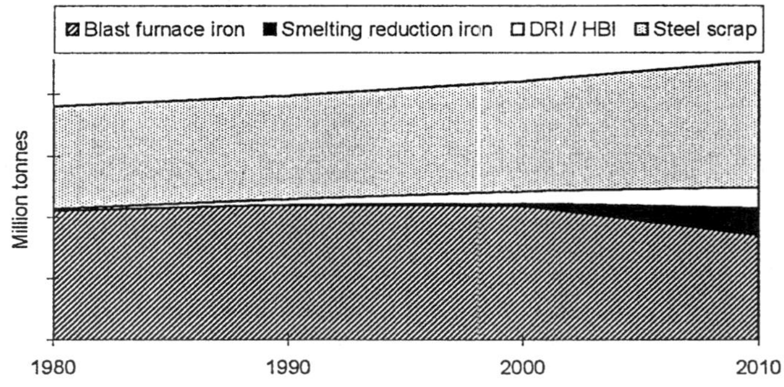

Fig. 5 World primary metal consumption.

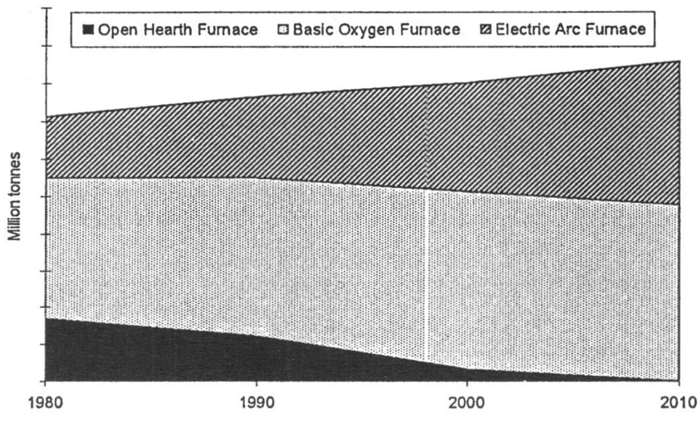

Fig. 6 World steel production by process route.

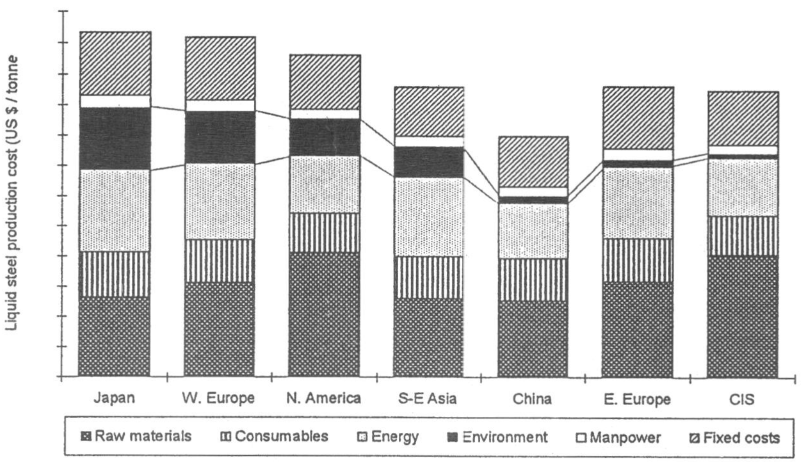

Fig. 7 Oxygen steelmaking: cost structure.

Sumitomo and TSP. Capacity of these plants is typically around 1 M/t year which is suitable for larger EAF plants (e.g. Nucor, Crawfordsville). A further development has been that of 'integrated' electric steelmaking plants using smelting reduction and/or DRI, plus EAF plus either conventional or thin slab casting to produce quality flat steels (Hylsa, Hanbo, Nippon Denro Ispat).

2. THE OPPORTUNITIES

This is only the beginning of a new era. The changing face of the industry offers numerous opportunities for adapting and improving existing processes, as well as developing new technologies. The number of available process routes and alternative plant configurations will continue to grow.

At the primary metals stage there are opportunities such as improved scrap quality control, use of mini blast furnaces for ironmaking, commercial development of new smelting reduction processes, and adoption of improved direct reduction technologies. Recent developments in electric arc steelmaking include DC arc furnaces, continuous charging, charge preheating and post combustion. After the year 2000 smelting reduction processes are likely to account for a markedly increased share of new ironmaking capacity. Very few conventional blast furnace BOF plants are likely to be built (Fig. 8).

The optimum process route for a given project depends on the planned capacity, target product range, and proposed location for the new plant. Process selection will also depend on whether it is a completely new plant, or a revamped plant at an existing site. There are no hard and fast rules: each project must be evaluated individually to determine the best combination of primary metal, steelmaking, casting, hot and cold rolling and finishing processes. With a growing number of technologies becoming available, customers will need

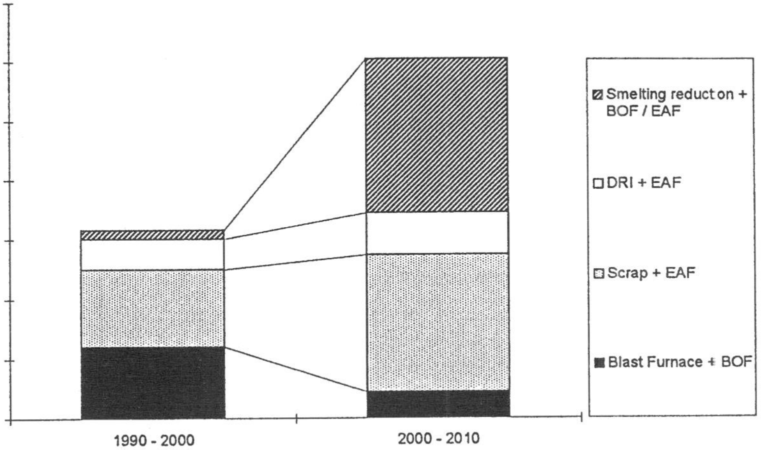

Fig. 8 New steelmaking capacities by process route.

Table 1. Process routes: tentative regional guidelines

Region	New Plants	Revamping
North America	Scrap/purchased DRI/Smelting reduction + EAF + NNSC	Smelting reduction + BOF +CCC
Western Europe	Scrap/Smelting reduction + EAF + NNSC	Smelting reduction + BOF + CCC
Eastern Europe	Scrap/Smelting reduction + EAF + NNSC	Smelting reduction or existing BF + BOF +CCC
CIS	Scrap/DRI + EAF + NNSC	Smeting reduction or exisiting BF + BOF + CCC
Asia	Local or imported DRI/Smelting reduction + EAF +NNSC	Smelting reduction, existing or new mini BF + BOF + CCC or NNSC

objective expert advice in order to make their choice. A tentative forecast of the technologies likely to be used in the main producing regions is provided by Table 1.

3. THE CHALLENGE

Faced with an increasing demand for their services, engineering companies will also be under pressure to enhance the range of skills and services offered. This will require new modes of operation, as well as major cultural and organisational changes within process plant suppliers. Firms will work closely in a consultancy capacity with customers, raw materials and utilities suppliers to develop optimum technological solutions. Given the volume of investments required early next century (see Fig. 9) and the inevitable competition between projects for available funds, the provision of global 'financial engineering' services will

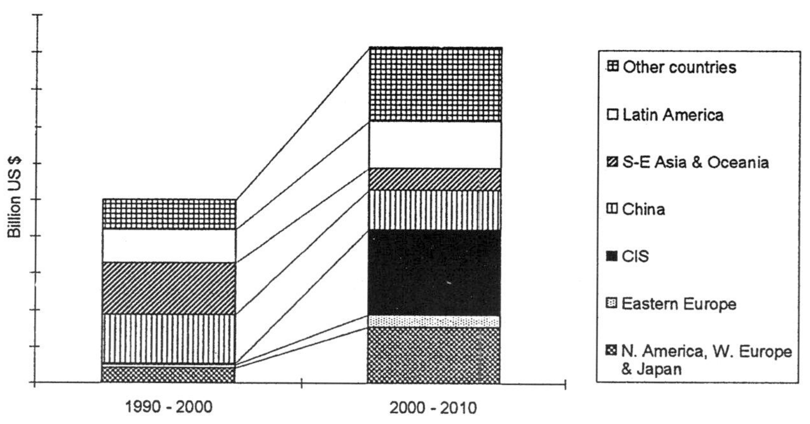

Fig. 9 Projected investment requirements by region.

also be a key function. This means engineering companies will increasingly be called upon to provide support to their customers during business planning and negotiations with their key business partners, for instance to secure long-term contracts with steel customers, or power suppliers.

Engineering companies will have to develop closer relationships with investors and customers and provide them with support in negotiating long term contracts. At the business level, engineering companies will need to become more involved in joint ventures and partnerships in the countries where they operate. Such joint ventures and partnerships with other companies will be required both to comply with local market conditions and to help customers secure project finance. Finally in their enhanced role of 'general consultant' to the industry, engineering companies will need to carry out systematic technology watching and acquire skills to assist customers in selecting the optimum process for their projects.

Section II
Changing Technologies
An Overview

PARADIGM SHIFT IN THE STEEL INDUSTRY

A. TOMIURA

Nippon Steel Corporation, 6–3 Otemachi, 2 Chome, Chiyoda-Ku, Tokyo 100–71, Japan

1. INTRODUCTION

The world steel industry now finds itself facing various difficulties. Although we have experienced difficulties from time to time in the past and have overcome them by developing and applying new technologies, the changes in the corporate environment facing us now are apparently different from those in the past.

The first characteristic of these changes is the realisation of a borderless world. As the movement of capital, people, goods, etc., has become freer, political, economic and social issues, which directly influence the steel industry, have arisen.

The second characteristic is the concern about global issues such as overpopulation, food and energy shortages, and environmental pollution. As the steel industry is a material and energy intensive industry, we should make a commitment to solve these problems, but, at the same time, we should also try to break down these complicated interrelated issues into the steel-related problems, and then identify our tasks.

The third characteristic is the diversification of people's desires. Once the need for daily necessities is satisfied, people want something different. As a result, many more different types and smaller units of products are manufactured. This change in people's desires also relates to the request for shorter working hours, more comfortable working conditions, and so on.

The circumstances which have led the steel industries of developed countries to such distressing conditions can be summarised as follows,

(1) The growth rate has been saturated, and some countries are suffering from excessive capacity.
(2) Mega competition in the world market has already started, and thus international competitiveness is one of the governing factors for management.
(3) The mass production system is collapsing due to the diversified market requirements.
(4) Renovation or closing down of existing facilities has already begun to take place because of stringent environmental regulations.
(5) Due to the rapid growth of mini-mills, the oligopolistic structure of the integrated mills has started to collapse.

In many developing countries, on the other hand, steel products are still in demand. Difficulties, however, may exist as regard the following points,

(1) Shortage of capital for construction of new facilities.
(2) Underdeveloped market.
(3) Unavailability of appropriate technologies.

Although all of these facts seem to be unrelated phenomena, the background of the facts are deeply rooted in the above mentioned three characteristics.

The IISI Technical Committee conducted a questionnaire survey in 1992 concerning what technical issues would require development efforts towards the next century, and reported its results at the following annual meeting of the IISI.[1] Since this survey, some obvious changes about such issues have begun to emerge. The author will try to identify the common issues, and will discuss general ideas to solve them.

2. IDENTIFICATION OF THE ISSUES

2.1 Steel Consumption and Global Issues

As I pointed out, the key global issues will be over-population, food and energy shortages, and environmental pollution. Each of these issues is interrelated to each other as described in Fig. 1.[2] Admitting that one of the features of this century is the prosperity of advanced countries, it may be said that the next century will be characterised by the levelling of living standards all over the world through the growth of developing countries. As the world population standing now at 5.8 billion is expected to reach 8.5 billion in 20 years, and this enormous increase of population will be largely attributable to developing countries, one of the most urgent tasks will be the development of the infrastructure in developing countries, which will require large quantities of steel products.

As shown in Fig. 2, steel consumption per capita in advanced countries has been declining in the past 20 years, but the discrepancy in steel consumption between developed and developing countries is still large.[3] Such a gap cannot be compensated for by international trade because the production capacity in developed countries will be restructured due to such reasons as environmental problems, superannuated facilities, labour problems and so on. Furthermore, from the viewpoint of developing countries, there is a desire to produce steel by themselves with the aim of achieving a quick take-off into the ranks of advanced countries. However, the market size in these developing countries is not large enough to utilise the capacity of the blast furnace based integrated steelworks, and, at the same time, developing countries usually suffer from a shortage of capital for the installation of large facilities. Even in developed countries, once they close down existing facilities for various reasons, they usually need some supplemental steel capacity. Therefore, development of appropriate technologies to make steel in both developed and developing countries will become one of the significant tasks.

2.2 Environmental Protection and Energy Conservation

Since the United Nations Conference on Environment and Development in Rio de Janeiro in 1992, the environmental issue, especially as it focuses on the

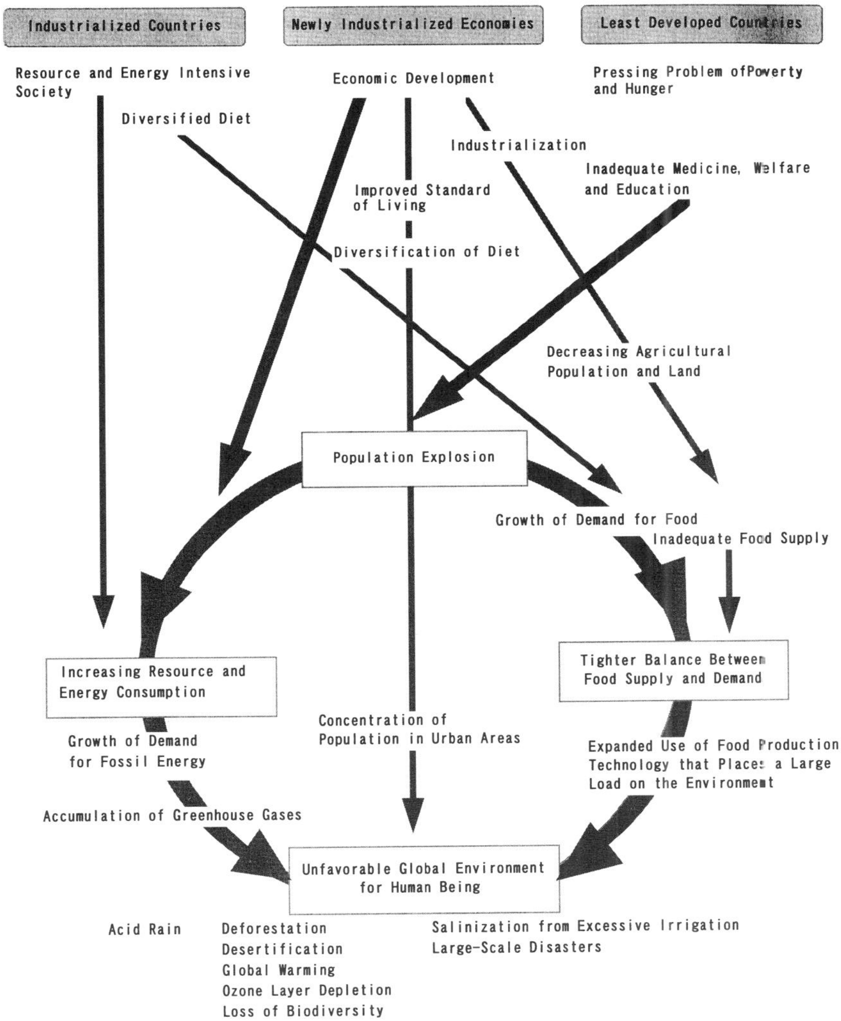

Fig. 1 Comprehensive diagram of global-scale issues (by courtesy of Science and Technology Agency of Japan).

emission of carbon dioxide, has been assuming increasing importance. Unlike localised pollution problems, the global environmental issue must be solved by considering the entire process from the production of materials to recycling of the products in ecologically harmonious systems. As global warming due to carbon dioxide has a serious, widespread effect, the reduction of carbon dioxide emissions is an important problem of common interest to mankind regardless of country. According to an estimation by a simple calculation, the carbon dioxide emissions by the steel industry accounts for 6 per cent of the total carbon dioxide emissions from all economic activities. This amount seems small, however, it is to be noted that some countries have already legislated or are considering legislation for a carbon tax which was or will be several tens of dollars per ton of steel.

Life Cycle Analysis has become popular as one of the measures for making ecological assessments. The Environmental Committee of IISI is now carrying out this study involving many member companies. Because the purpose of Life Cycle Analysis is to find areas to be improved, we should not discuss which material is superior to others based on this analysis.

As long as effective countermeasures to fix gaseous carbon dioxide remain undeveloped, technologies for energy saving, recycling and design of long life products will be important tasks to solve the environmental issue. Above all,

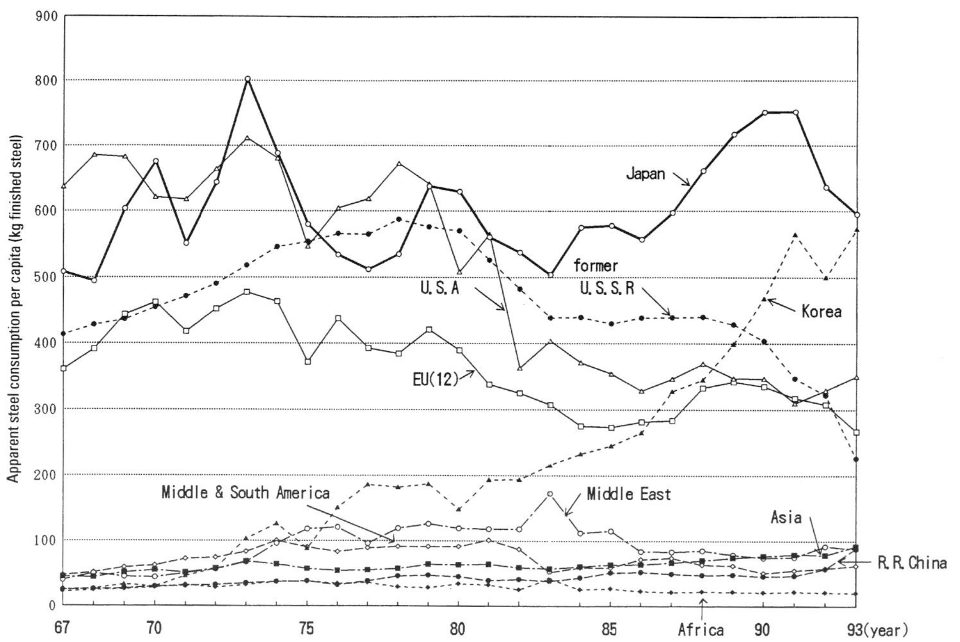

Fig. 2 Steel consumption in the world (source: IISI Yearbook).

recycling is an important issue in view of scrap supply and intermaterial competition.

2.3 Competitive Process and Product

While the expenditure on process research and development can be recovered by our own use of its results, that of product research and development cannot always be recovered. As already discussed, since the mass production system is collapsing due to diversified market requirements, production and investment are being directed toward a downstream process for the manufacture of high value-added products which have not brought about cost reduction nor improvement of productivity. It might be suspected that invested capital could be recovered by selling these high value-added products. Although there is no denying that product development is an important task, developed products should have the level of quality that customers want; otherwise mutual benefit for both producers and customers cannot be guaranteed.

The mass production system has undoubtedly contributed to the manufacture of products of uniform quality at low cost, but has lost flexibility. Therefore, one of the most urgent tasks today is to develop a flexible manufacturing system capable of making various products in small production units. The secondary refining facilities, thin slab casting and strip casting are good examples, but are not yet sufficient. What factors hamper a reduction in production lead time, schedule free production, straight flow of products, long service time of facilities and so on? We notice the presence of many problems such as surface defects, cycle of periodic maintenance, life of rolls, wear of refractories and so forth, which are known to everyone but remain unsolved because of their difficulty.[4] We should pay more attention to these inconspicuous yet basic problems.

2.4 Intellectual Productivity

Cost competitiveness is always the most significant concern in any company. Recently, mini-mills have made rapid strides in the flat rolled product area by means of achieving strong cost competitiveness over integrated steelworks. When we carefully examine the cost structure of a mini-mill, we find that the slab or billet cost of the mini-mill is higher than that of the integrated mill, while the final product cost of the mini-mill is apparently lower than that of the integrated mill, as illustrated in Fig. 3.[5] One of the reasons for this is the difference of the product mix between mini-mills and integrated steelworks, but the main reason is that the costs for sales and administration are lower for mini-mills than for integrated mills. Here lies a lesson that we should learn.

While the direct labour productivity in the integrated mills has been significantly improved by the introduction of large scale and high speed facilities, mechanical automation systems and computer control systems as shown in Fig. 4, the productivity of white collar people has not risen but has rather decreased.[6] There have been a lot of discussions about how to reduce the white collar force

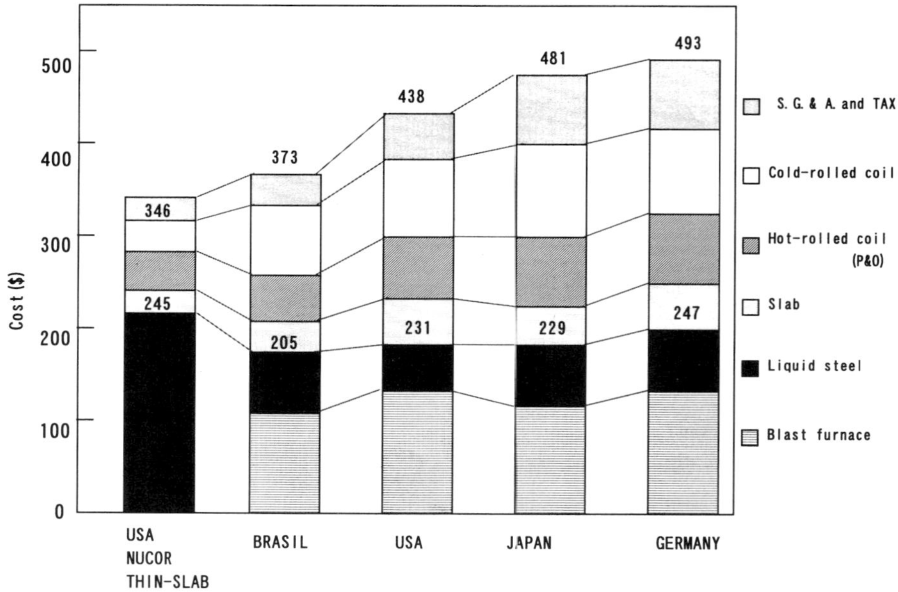

Fig. 3 Cost structure of cold rolled product (Paine Webber/World Steel Dynamic).

by such as organisational restructuring, outsourcing, full utilisation of computer networks, joint work and so on, but improvement of intellectual productivity should be emphasised.

In order to cope with this paradigm shift, there have been a lot of discussions, proposals and projects which have been implemented. The author does not

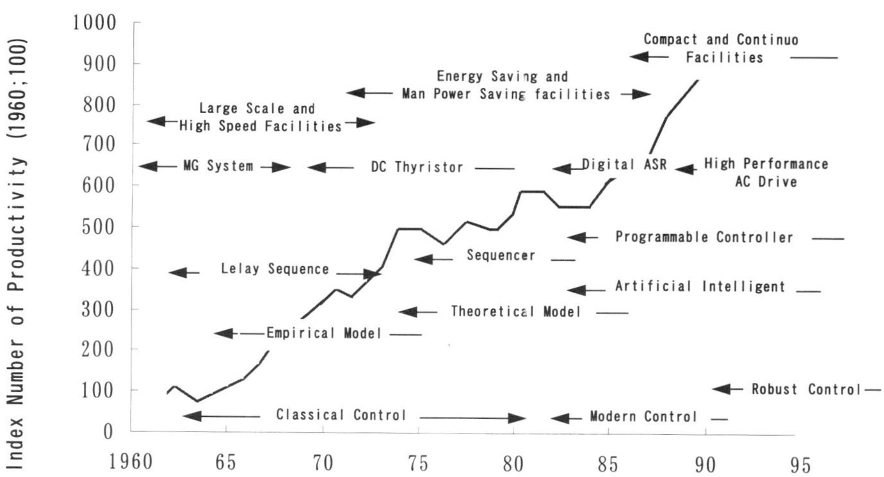

Fig. 4 Productivity and control systems.

want to introduce these specific matters now, but would like to present the results of his study for solving the identified issues.

3. STUDY FOR SOLVING THE IDENTIFIED ISSUES

3.1 Options to Make Liquid Steel

The blast furnace and basic oxygen furnace complex is a stable, cheap and high quality supply source of liquid steel. In contrast to these advantages, such a complex requires a high investment and relining cost, high quality of sintered ore or pellets, and also sophisticated operational know-how. Also, it generates large amounts of emissions and wastes, and produces a large and constant quantity of steel but is inflexible for responding to changes in market conditions. Under such circumstances, it is natural that the various iron- and steel-making processes other than the blast furnace and basic oxygen furnace complex have emerged in response to local conditions. This is not the case only in developing countries but also in developed countries. As shown in Fig. 5, there are a lot of limiting conditions for making liquid steel. Therefore, the most appropriate process which meets the specific local conditions has to be selected in consideration of these limiting conditions as shown in Fig. 6. Figure 7 illustrates several theoretical routes for making liquid steel and Fig. 8 indicates the position in which reported processes are placed in relation to reductants and raw materials. Figure 9 shows estimated investment costs in relation to the unit

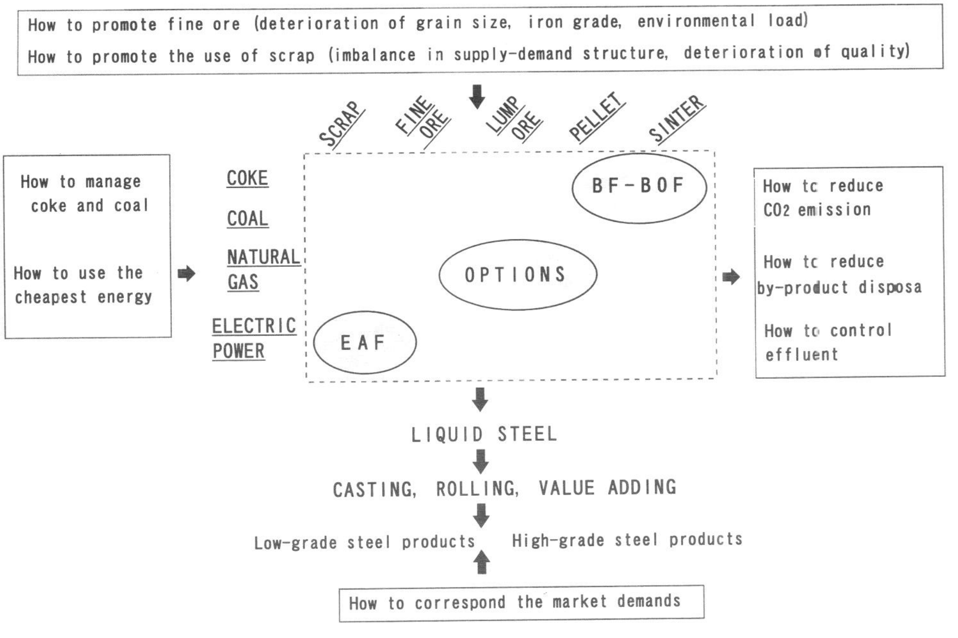

Fig. 5 Limiting conditions for making liquid steel.

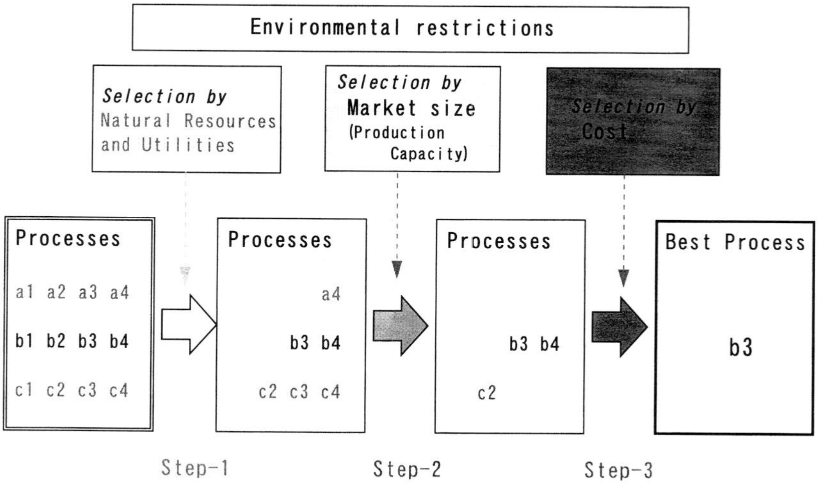

Fig. 6 Logical steps to decide the most optimum process.

capacity of each process. As for operational costs, it is extremely difficult to obtain reliable information on this because of its confidential nature. In accordance with a bold assumption, however, the production costs of liquid steel in each process was estimated, taking into account the local conditions as shown in Fig. 10.[10]

Although various observations can be made based on this analysis, it is now sufficient for us to say, that the competitiveness of the electric arc furnace strongly depends on the cost of scrap and its supply conditions which will be

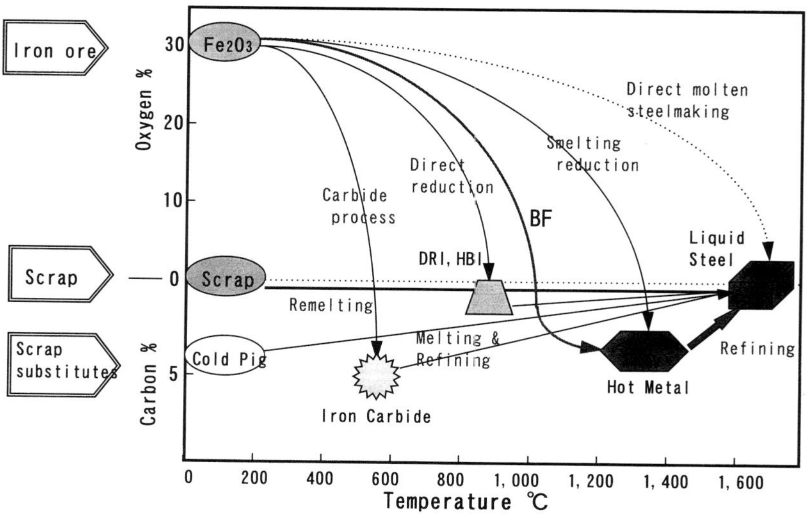

Fig. 7 Theoretical routes for making liquid steel from various materials.

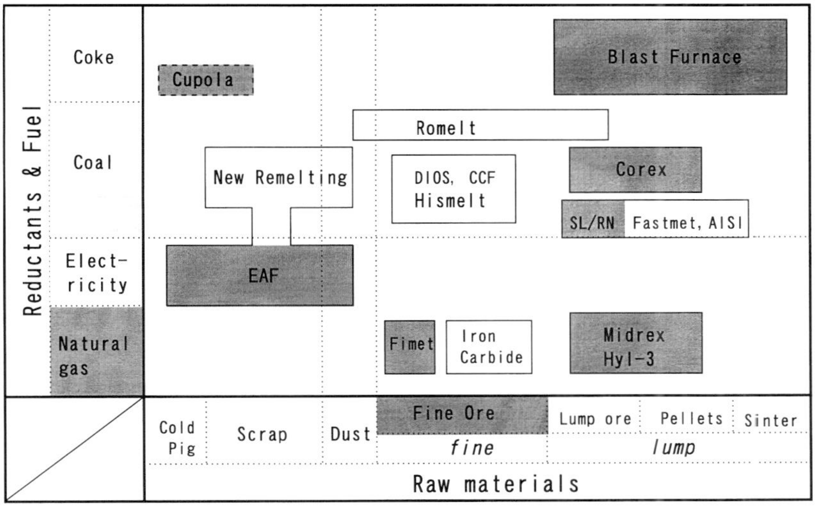

Fig. 8 Position of processes in terms of raw materials and energy.

discussed later and that some processes have competitiveness in limited local conditions.

3.2 Scrap Processing

With the rapid growth of production by the electric arc furnace, scrap shortage has become a key issue. There have been a lot of discussions and forecasts on

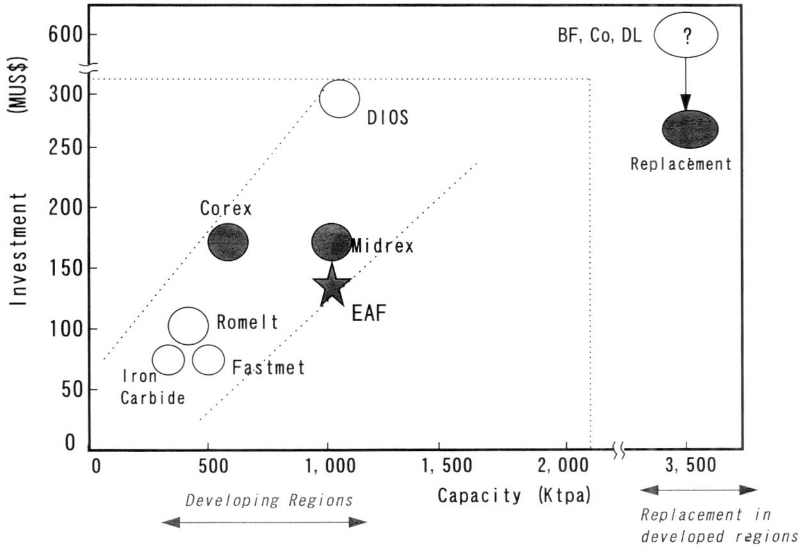

Fig. 9 Relationship of initial investment and capacity.

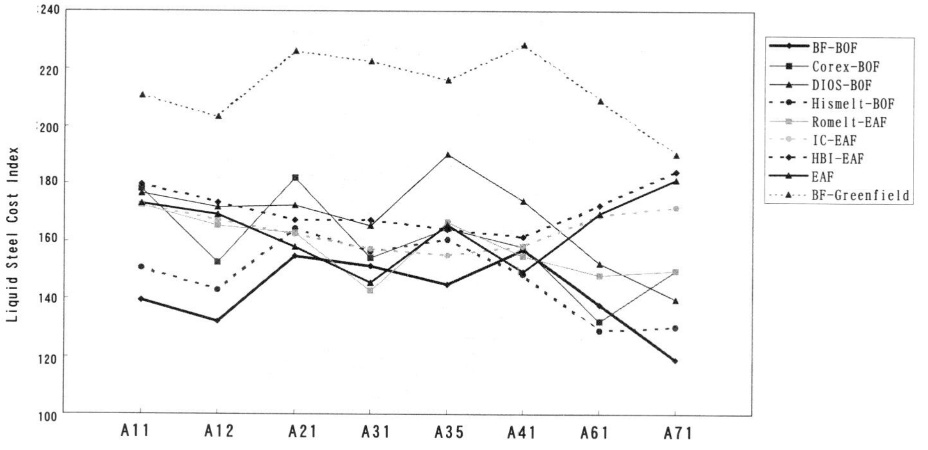

Fig. 10 Comparison of liquid steel cost by processes and regions.

demand and supply of scrap, while there have been few discussions on scrap processing. Steel is an intrinsically recyclable material as shown in Fig. 11. However, with the increase in the use of recycled scrap, incidental elements such as copper, tin, nickel and molybdenum that cannot be removed during refining have accumulated in steel. How much copper and tin are accumulated in steel after the cyclic use of scrap is illustrated in Fig. 12, based on a simulation

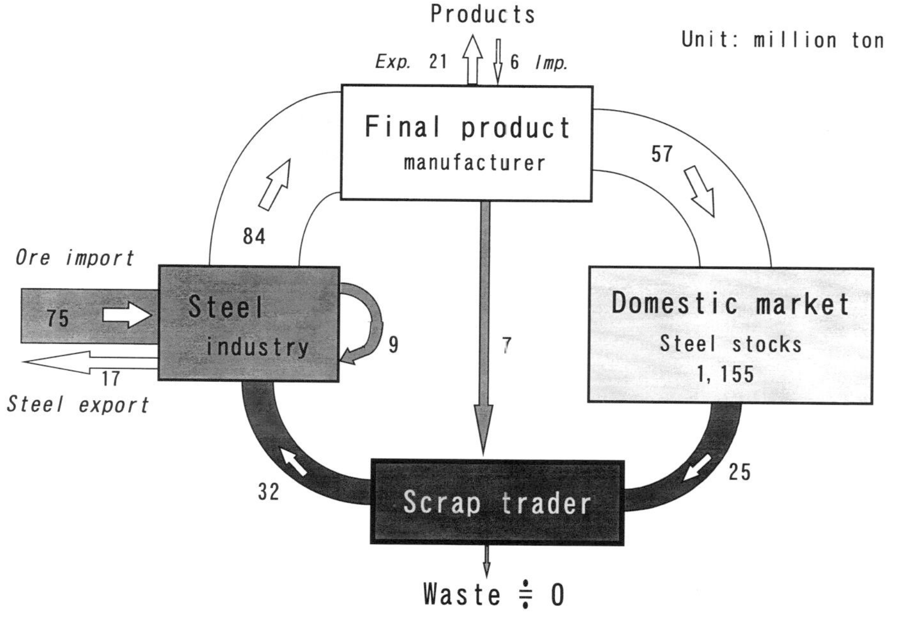

Fig. 11 Material loop of steel in Japan ('94).

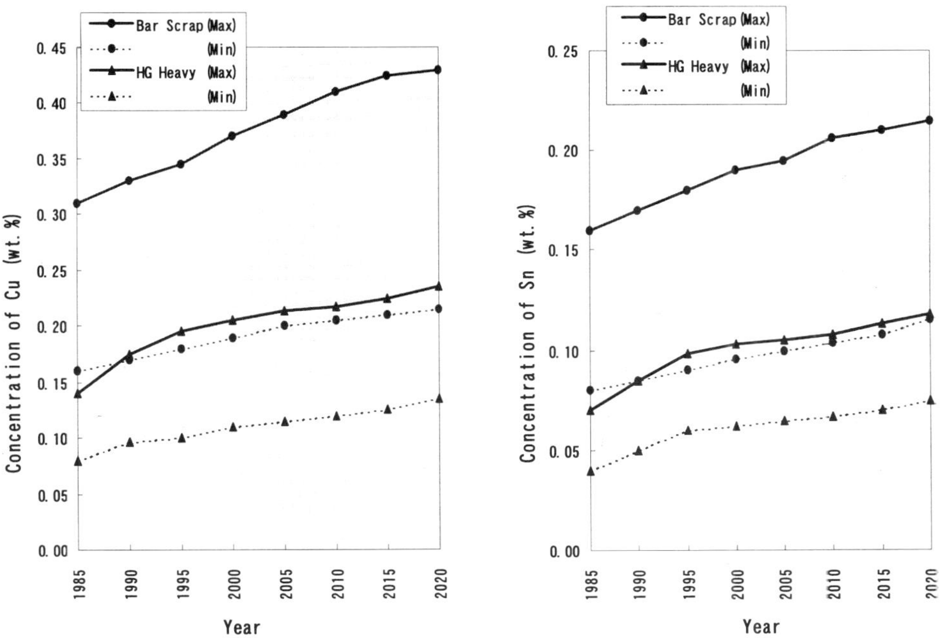

Fig. 12 Simulation of Cu and Sn concentration in obsolete scrap.

calculation, and the permissible concentration of incidental elements is shown in Fig. 13. From these figures, it is apparent that technologies to remove these elements from scrap have to be developed. In the past, copper was removed manually during scrap processing, but today nobody wants to do such a dirty job and no profit can be expected from selling copper thus recovered. In 1991, twelve Japanese companies and one French company formed a research consortium for developing new steelmaking technologies incorporating scrap processing. As illustrated in Fig. 14, shredded scrap is cooled down by liquefied nitrogen, and fed into a crushing machine, then separated by a screen. Copper is removed by this process, and the estimated process cost may not be high taking into account the cost of disposing of used cars and the premium for upgraded scrap thus recovered.

The steel industry has never paid attention to scrap processing, instead it has solely depended on scrap traders for scrap processing. But this is not someone else's problem but our own problem.[8] We should thus develop various technologies to enhance scrap recycling and upgrading.

3.3 Design of Eco-Materials and Eco-Processes

A material or process which is compatible with the environment is called an eco-material or eco-process. It is well known that some substances have an unfavourable effect on the environment. Such substances are said to give an eco-load to the environment. However, there has been no quantitative analysis

(wt. %)	Cu	Ni	Cr	Mo	Sn
Deep Drawing	0. 06	0. 10	0. 07	0. 02	0. 01
Sheet & Coil	0. 10	0. 10	0. 07	0. 03	0. 015
Wire Rods (low carbon)	0. 12	0. 08	0. 07	0. 02	0. 01
Springs	0. 20	0. 10	0. 10	0. 03	0. 02
Wire Rods (special steel)	0. 25	0. 12	0. 12	0. 03	0. 03
Bar & Shape	0. 35	0. 15	0. 15	0. 04	0. 03

Fig. 13 Allowable tramp elements in steel.

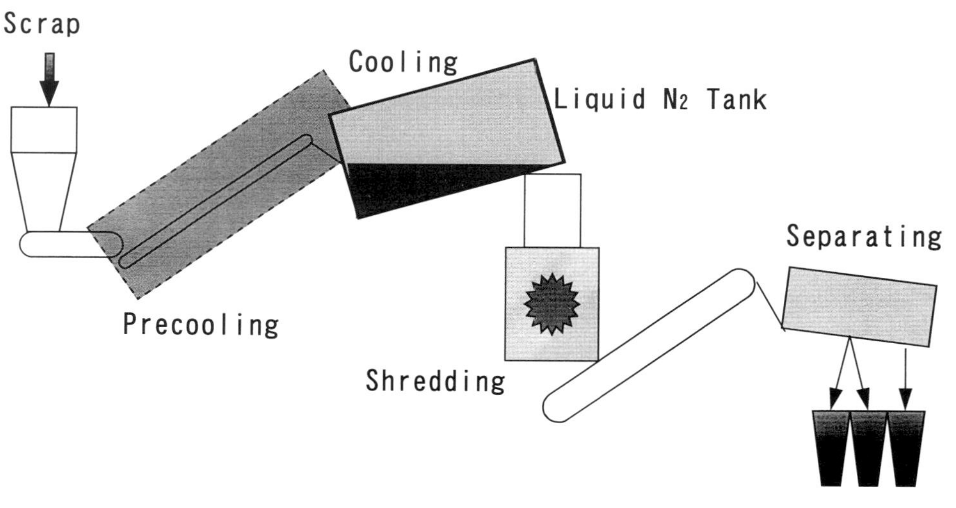

Fig. 14 Schematic diagram of the process.

made about which substance influences which eco-load. Since many substances which give an eco-load are generated in the course of energy consumption, energy consumption can be selected as a criterion alternative to eco-load.

Now, if energy consumption in the whole process of material processing is E_s, energy consumed in the service life of a product is E_p, and energy consumed in the process of recycling is E_r, then the total energy consumed in the whole life cycle E_t is represented as eqn (1)

$$E_t = E_s + E_p + E_r \tag{1}$$

Since E_r is net energy, which the potential energy of recycled material is subtracted from total energy consumed in the recycling process ($E_{r(gross)}$.), E_r can be represented as eqn (2).

$$E_r = E_{r(gross)} - r{\cdot}\alpha E_s \tag{2}$$

where r; recycling rate.

α; coefficient to adjust E_s in relation to potential energy of recycled material.

Given the combination of E_s, E_p and E_r which makes the smallest E_t, the most favourable product and process design could be possible.[9]

According to these criteria, the products made of steel can be categorised into four groups as shown in Table 1.

Table 1. Energy consumption and steel product.

	Type of products	Category
Products relevant to energy consumption	Directly energy consuming products	I
	Indirectly energy consuming products	II
Products irrelevant to energy consumption	Short life-cycle products	III
	Long life-cycle products	IV

In the case of the mid-size passenger car, it is estimated that E_t is 91 GJ, while E_p is 81 GJ.[10] This means that the reduction of E_p has the most substantial effect as regards the improvement of E_t. Since E_p strongly depends on fuel consumption, which depends on the weight of a car, reduction of the weight of the car is now being strongly promoted. The ULSAB project is a good example of this.

Good examples of category II are a transformer and a boiler from a power plant. Neither of them consume energy but are closely connected to power loss and thermal efficiency. Development of low watt-loss silicon steel and higher heat resisting boiler tubes are important goals.

The products in categories III and IV such as beverage cans or steel structures like bridges and buildings, are not connected to energy consumption. In this case E_p is almost negligible.

In the case of the beverage can in category III, the uni-steel can is better than the hybrid can because E_s of aluminium is larger than that of steel. Furthermore, aluminium used for the lid cannot be recovered as aluminium metal because it is oxidised during refining. This means that maintenance-free and long-life structures are most favorable.

This concept will lead us to develop various types of eco-products and eco-processes.

CONCLUSION

Steelmaking technologies have progressed by fusing together science, technology, experience and even intuition. As the paradigm has now shifted, however, we must harmonise technologies focused on natural science with technologies compatible with man and society. In the past, technology addressed only what were visible and measurable, and required that the unit of the left-hand side of the equation be the same as that of the right-hand side. Today, however, there are a number of problems technology has to address that are not visible, measurable and quantifiable in a unit element. As I mentioned, the diversification of an individual's desire, which is invisible, unmeasurable and unquantifiable, has already mirrored so clearly the change of products and processes, and this change will be more rapidly accelerated with complicated objects involving heterogeneous elements having of technologically unfamiliar functions.

Especially, these anthropo-technological matters which bridge man and technology will increase their importance and will not be able to be solved without international cooperation involving cultural and social scientists.

REFERENCES

1. Aulis, V. A. Saarinen and Azusa Tomiura: 'Prospect for New Technology Towards the 21st Century', *IISI-26*, pp. IISE/E/2614/0.
2. Ad-hoc Committee on International Affairs Committee on Policy Matters, Japanese Prime Minister's Council for Science and Technology, May 1996.
3. *IISI Yearbook 1994*.
4. A. Tomiura: 'New Direction of Steel Technology Towards the 21st Century', *TETSU-TO-HAGANÉ*, 1995, **81**(4), 249.
5. Paine Webber: *World Steel Dynamics 1996*.
6. A. Tomiura: *IPEC-95*.
7. A. Tomiura: *IISI-30*, 103–117.
8. *Scrap Demand, Supply and Quality in Japan and South East Asia*, IISI Metallics for Steelmaking.
9. A. Tomiura: *Monthly Eco-Industry*, **1**, 9–11.
10. *Competition Between Steel and Aluminium for the Passenger Car*, IISI, 142.

Innovations in Production Technologies

SUMMARY: TECHNOLOGICAL INNOVATION

F. BONOMO

c/o Centro Sviluppo Materiali, Via di Costel Romano 100/102, I-00129 Rome, Italy

The iron and steel industry is facing great structural changes. This is due to many important factors:

- free world borders and free markets open to international trade;
- stringent environmental regulations;
- modifications of people's expectations in areas such as shorter working time, more comfortable working conditions and desire for tailor made and more sophisticated products;
- growth of world population, especially in developing countries where more steel production will be required, despite capital shortages and small local markets;
- rigidity of the traditional megamills where huge amounts of capital have been invested;
- availability of new technologies to make mini mills more flexible and enable them to compete with the integrated megamills.

Strong competition is taking place between the traditional concept of steelmaking based mainly on integrated works and the present trend towards mini mill production. This competition has significantly accelerated manpower reductions in the iron and steel industry that were already taking place due to technical improvements. For example, the labour required to make cold rolled sheet has gone down from about 10 man hours per tonne at the end of the 1970s to the current 3–4 man hours per tonne at an integrated works, and less than one man hour per tonne at the most recent mini mills.

Competition from mini mills is very active due to the following issues:

- Lower specific investments and much lower total investment needed, allowing newcomers to enter the business. This has helped bring about the collapse of the past oligopolistic structure of the industry;
- The mini mills advantage of lower man hours per tonne of product;
- The opportunity to locate production units near steel users;
- Greater flexibility of the mini mill route;
- Lower environmental impact.

A further incentive to the spread of mini mills is the recent development of important new technologies more suitable for application in mini mills as compared with conventional integrated plants. In particular, the following technologies have to be mentioned:

- Smelting reduction using the Corex process from VAI. By avoiding the use of coke, Corex makes it possible to save the investment costs of coke ovens.
- Thin slab technology for hot strip production. Examples discussed here include CSP from SMS and ISP by MDH/Arvedi. By using direct rolling and lower thickness slabs, reheating furnaces can be saved. A temperature homogenisation furnace can be used in place of a reheat furnace. The length of the rolling train is reduced as less reduction is required from a thin slab.
- Continuous linking of downstream pickling and cold rolling allow the production cycle to be further compacted.

Important technological steps are expected in the near future. For example, second generation smelting reduction should allow, not only the substitution of coke with coal, but also the use of fine ore in place of lump ore. In this way it will be possible to eliminate sinter plants used to agglomerate fine ores in existing integrated works. Successful development of strip casting can be anticipated. Casting steel strip directly will get rid of the rolling process. This would allow a great saving on the transformation cost. Above all it would save on the huge capital investment in hot strip mills.

The electric arc furnace is at the heart of the competition between megamills and mini mills. The EAF is more suitable than other liquid steel sources for the '*simple, flexible and compact*' concept, at least in the short to medium term. In this light it is reasonable to forecast a strong increase in electric arc steelmaking from the present share of about 33 per cent of world steel output. EAF steelmaking might reach a share nearer to 50 per cent by the year 2010.

The growing importance of the electric arc furnace has given rise to rapid technological progress. In the past decade there have been remarkable savings in energy, refractory use and electrode consumption coupled with significant improvements in productivity. Recent major innovations in electric steelmaking have aimed to reduce environmental impact, to further improve energy savings and productivity, and to diversify the charge material. The following technologies deserve mention:

- Twin shell furnaces;
- continuous charging and preheating of scrap;
- Oxygen and carbon injection and post-combustion technologies, allowing substitution of electrical energy with less expensive chemical energy.

It is recognised that increased electric arc furnace utilisation is linked to availability of ferrous charge material, at the moment mainly scrap. This focuses attention on the scrap problem: either there is enough scrap to meet the likely higher demand, or its price will increase to the point where electric arc steel will become far less competitive relative to integrated steel production. If scrap prices increase, electric arc steel output will not reach the high level forecast for the year 2010. Scrap substitutes play an important role in the overall picture. These include directly reduced iron (DRI) and ordinary iron (either hot metal or cold pig iron). Development of new directly reduced ironmaking technologies, like FINMET and others, may significantly decrease the cost of DRI to the point where it could become competitive with the scrap price, even in places where energy costs are currently too high for economic production. In this event, the forecast for future electric arc production could well be realised. Substitutes for scrap will play an important role as a supplementary charge material in any event. Some high purity charge material is required to dilute residuals in scrap. The spread of mini mills for flat products increases the importance of scrap substitutes as a pure charge material. Even if scrap supplies are readily available, some alternative charge material will be required for quality reasons.

This general scenario does not apply with equal force in different geographical regions. Perhaps scrap will not be a problem in the older industrialised areas such as the USA, the European Union and Japan. Whereas a shortage can be envisaged in the emerging industrialised economies, especially in Asia. For example, South Korea, China, Taiwan and India are likely to experience scrap deficits. Scrap substitutes are likely to be widely used in these countries. Current examples include production of hot iron from COREX in South Korea, pig iron from small blast furnaces in China and coal based DRI in India.

The impact of the emerging *alternative iron-modern electric arc steelmaking* technologies on existing plants is hard to assess. These new technologies compete with the traditional blast furnace-basic oxygen steelmaking route and with scrap based electric arc based plants using conventional continuous casting and rolling technologies. No doubt the most modern or recently upgraded works will continue producing in the usual way for many years. It is not economic to replace still efficient plants with new plants since the extra capital costs cannot be compensated by the flow of lower operating costs. A few years ago it was thought that coke shortages arising from coke ovens reaching the end of their lives would result in closure of integrated works. Reconstruction of coke ovens is expensive. Instead, pulverised coal injection (PCI) is now routinely used to replace coke in the blast furnace. The technique is likely to be further used and expanded in the near future. In the longer term blast furnaces might be replaced by smelting reduction plants. The possibility of installing more environmentally friendly and technologically advanced coke ovens cannot be excluded.

Remarkable plant engineering activity can be anticipated over the next few years. Some forecasts indicate that world steel demand might reach 1000 million tonnes a year by the year 2010. Rising demand coupled with a high propensity towards innovation are expected to trigger a high level of plant construction. Looking forward, estimates suggest by the year 2010, new plant of about 250–300 million tonnes annual capacity, and upgrading of existing plants with a capacity of around 170 million tonnes a year will be required.

To sum up, we can draw the following conclusions from our analysis:

- The steel industry is facing structural evolution using new technologies.
- The present trend is to abandon the concept of new megamills of 5–10 million tonnes capacity a year and pursue smaller, more profitable production units of 1–2 million tonnes capacity or even less. These plants are characterised by more compact production cycles, lower specific investment, greater flexibility and lower environmental impact.
- Smaller plants have been helped by recent technological developments such as smelting reduction and thin slab casting. Further improvements to these technologies and developments such as generation 2 smelting reduction and direct reduction processes will give an additional impulse to small plants.
- Electric arc furnaces are more suitable for these smaller plants and are expected to increase their share of steel production.
- An increase in the electric arc furnace share may be restrained by scrap shortages, although local shortages may be offset by substitute materials like pig iron or directly reduced iron.
- Scrap availability will vary across regions. Surpluses are expected in the older industrialised countries and shortages in the developing economies.
- Overall world steel production is expected to increase, with most of the growth taking place in the developing countries and only steady output among the traditional net exporters.
- Existing conventional plants will be upgraded and it will not be economic to replace efficient plant with new equipment. Continuous improvement in plant efficiency will be required to match competition from new technology in new locations.
- The new technologies will be applied mainly on 'greenfield' sites. Replacement of conventional mills will come later as these become obsolete.
- The main location for new technologies will be developing countries where an increase in production is required and in the USA where there is scope for replacing imports.
- Despite the gradual application of these new technologies, a major upsurge in plant engineering can be anticipated before the year 2010. Construction of 300 million tonnes of new capacity and upgrading of 170 million tonnes of existing capacity seems likely.

In spite of the increase in world steel output, the overall effect on employment is likely to be negative. This is particularly likely in regions like the European Union and Japan where output will stagnate but productivity will further improve. Thus steel displays a similar pattern to other mature sectors, such as agriculture, where the very success of innovation has dramatically reduced manpower.

VAI TECHNOLOGIES FOR SCRAP SUBSTITUTES

A. F. EBERLE, W. KEPPLINGER, and S. ZELLER
Mineral and Reduction Technologies, VAI, Turm Strasse 4, P.O. Box 4, Ö-4031 Linz, Austria

PRESENT MARKET SCENARIO FOR SCRAP SUBSTITUTION TECHNOLOGIES

Annual world steel production is currently about 750 Mt. Liquid steel is still mainly produced via the classical route of the 'integrated metallurgical plant' (Fig. 1). An integrated mill basically features hot metal production in the blast furnace and further processing of the hot metal in a converter. At present, about 500 Mt of hot metal are produced.

Apart from this conventional route, the so called minimills have been gaining ever increasing importance over the past decades. In a minimill (Fig. 2), steel is produced by electric arc furnaces, which are normally based on 100% scrap as input material. Figure 3, a survey of global steel production from 1960 to 2010 by steelmaking processes, shows that in1994 electric arc furnaces accounted for 30% of total steel production. According to a forecast for the year 2010, 45% of all steel products are expected to be produced via the electric arc furnace route. The great success of the electric arc furnace is primarily attributable to the following facts:

- at 100% scrap input into the electric arc furnace, there is only 10% energy consumption compared with the blast furnace-LD route
- lower specific investment expenses for minimills compared with integrated mills
- higher ecological compatibility of minimills because they need neither coking plants nor iron ore agglomeration plants.

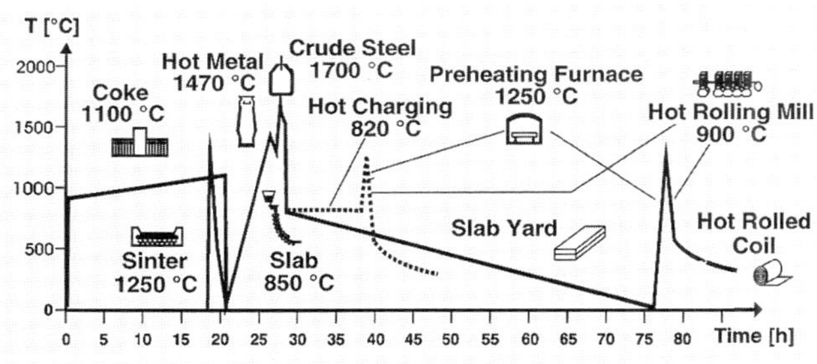

Fig. 1 Main technologies of integrated steel mill.

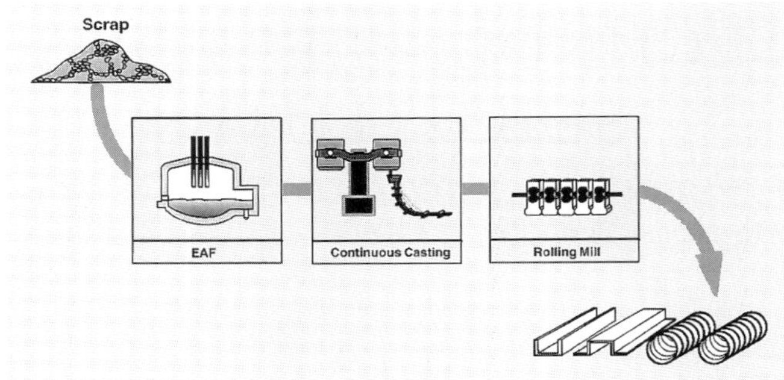

Fig. 2 Minimill route.

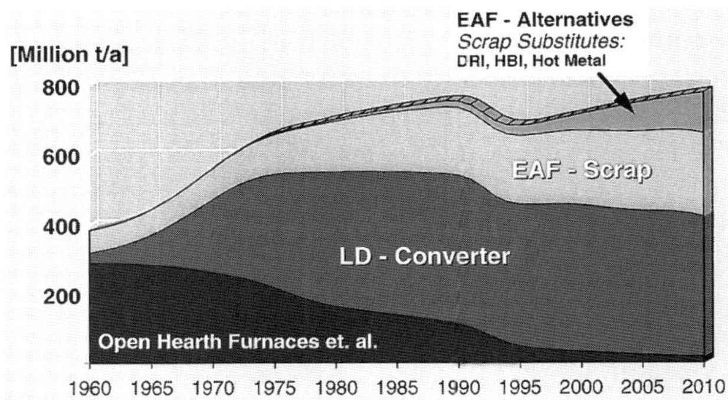

Fig. 3 Global crude steel production.

Initially, the success of the electric arc furnace route in the case of carbon steels was linked with the production of steel grades with low quality requirements. Based on the adoption of new technologies such as thin slab casting and rolling technology, the operators of electric arc furnaces have increasingly been striving to produce clean steel grades meeting high quality requirements. Scrap, with its undesirable tramp elements such as copper, tin, chromium, and nickel, is not perfectly suitable as a raw material for the production of clean steel grades. As a result, the demand for native raw materials (scrap substitutes), such as sponge iron, HBI (hot briquetted iron), iron carbide, and hot metal, is rapidly increasing.

There are various forecasts on this topic; for the year 2010, for example, a demand for scrap substitutes of 60–110 Mt/year is forecast, while for 1997 a global production of 36.2 Mt of sponge iron and HBI is stated.

DIRECT REDUCTION AND SMELTING REDUCTION: VAI'S CONTRIBUTIONS

The concept of ironmaking without a blast furnace (without a liquid phase or without coke) is based on the idea of obtaining iron (sponge iron) of a maximum degree of cleanness by reducing the Fe oxides using pit coal or lignite, petroleum, or natural gas.

The first reduction processes were developed in Sweden in the first half of this century (Wiberg and Höganäs process). The late 1950s and early 1960s saw the first bunch of innovations in direct reduction technology. In this period, the Purofer, Midland-Ross (subsequently Midrex), HYL I, and Fior processes were developed.

In the 1970s, the R&D activities in the field of smelting and direct reduction were further enhanced. As already noted, there is a market need for technologies to enable EAF operators to produce highest quality products, utilising the cost advantage of non-agglomerated fine ore for the production of HBI or sponge iron. There must also be a long term assurance that substitute technologies are available for the blast furnace as coking and sinter plants are put under increasing pressure.

Figure 4 gives a survey of smelting reduction and direct reduction processes currently on the market or under development as well as of the established blast furnace process, arranged according to products, reducing agents, and Fe bearing raw materials.

Currently, the only industrially proven smelting reduction process on the market is the Corex® process. All other smelting reduction processes are still under development and not yet industrially proven. Among those processes using fine ore directly, without preceding agglomeration, the CCF, Dios, Finex®, HIsmelt®, and Romelt processes are the most advanced ones.

In the field of direct reduction, the Midrex process and the HYL III process (shaft furnace processes) have assumed market leadership, holding a total share of 82%. Other processes worth noting are operated on the basis of rotary kiln technology (SL/RN, Krupp Codir, etc.). In recent years, fine ore process developments have been increasingly penetrating the market. (The Fior process

Product	Basic iron		Virgin iron units		
	Hot metal / Pig iron		DRI / HBI		Iron carbide
Reductant / Iron carriers	Coke	Coal	Natural gas	Coal	Natural gas
Lump Pellet Sinter	BF {TECNORED}	COREX® [ITmk3] {Redsmelt}	[AREX] MIDREX® HYL III GHAEM	COREX®-DR [FASTMET] {INMETCO} SL/RN GRATE CAR	
Fines		(CCF) [DIOS] (FINEX®) [HIsmelt®] [ROMELT] [AUSMELT]	{FINMET®} FIOR {CIRCORED} (SPIREX)	(CIRCOFER) (COMET)	IRON CARBIDE

(...) component tests [...] pilot stage {...} under implementation

Fig. 4 Scrap substitution technologies.

is the first technology of this kind that has been implemented. Since the 1970s, Fior de Venezuela has been operating such a plant in Puerto Ordaz, Venezuela.) Since 1994, Nucor, Trinidad, has been running an iron carbide plant with a nominal capacity of 300 kt/year, which is still struggling with operational troubles. The year 1994 also saw VAI's introduction of Finmet technology on the market. Recently, Lurgi has been awarded the first contract for Circored technology.

VAI'S CONTRIBUTIONS TO DEVELOPMENT OF DR PROCESSES

Research work in the field of direct reduction within VA(I) dates back to the early 1970s. In 1975, following market studies and technological investigations, a licence agreement with Armco was signed. As a result of lack of commitment on the part of Armco in the marketing of this process, VAI took out a construction licence for Midrex technology from Korf Engineering. By 1994, VAI had erected Midrex DR plants in Venezuela, India, Malaysia, and Libya.

Since the early 1990s, VAI has been making intensive efforts to develop fine ore alternatives for Midrex. Technological research work was carried out, and on 3 November 1992, a contract for joint development of the Finmet process was signed with Fior SA in Venezuela. After 3 years' development work, the first contract was signed with the Australian company BHP for the construction of a Finmet plant with a nominal capacity of 2 Mt.

DEVELOPMENT OF THE COREX® PROCESS

The preliminary stage of Corex® technology was the KR process (coal reduction process), which was based on the idea of melting down prereduced ore into hot metal in a fluidised bed coal melter gasifier and the generation of reduction gas. In 1977–78, Ralph Weber carried out tests on a 0.5 th^{-1} pilot plant in Pains, Brazil, to corroborate his basic concept. In 1978, Ralph Weber, Walter Maschlanka, and Gerhard Sanders filed a basic patent application for the KR process.

In 1979, Willy Korf acquired the patent rights and the knowhow from Ralph Weber and invited Voest-Alpine AG to develop the KR process to reach industrial maturity jointly with Korf Stahl AG.

In 1979 and 1980, VA and Korf Stahl AG engineered a 60 kt/year pilot plant, which was erected in Kehl, Germany, in 1980–81. As early as November 1981, hot metal was tapped at the plant for the first time. During the initial test campaigns, lignite, anthracite, and sponge iron were used as raw materials. These trials were intended to prove the process principle for the buildup of the fluidised bed as well as the possibility of melting down hot metal and slag.

In the second campaign, which was run in March 1982, the overall process concept was tested on the basis of oxidic pellets and coal. For the first time, hot metal was successfully produced from non-coking coal and ore according to the KR process. Further campaigns followed in August and September 1982, the results of which, however, fell short of the expectations of the ambitious development team. Hot metal was produced at a comparatively low tapping temperature, resulting in high sulphur contents, and featured low carbon and silicon content as compared to blast furnace hot metal. This was mainly attributable to the short dwelling time of sponge iron in the fluidised bed.

Finally, in 1984, high quality hot metal was successfully produced after the following essential measures had been taken:

- instead of the fluidised bed stated in the original patent, a mode of operation was adopted in the melter gasifier which provided for an active dead man and a fixed bed above the O_2 tuyeres
- originally, the dust recycling system was only designed as a material transport system. In 1984, to avoid excessive dust accumulation in the process, a technology for dust recycling providing for the gasification of dust by means of dust burners was implemented, and a patent application was filed.

Based on the described alterations, the process was renamed, from KR to Corex®. A total of 11 campaigns were carried out in Kehl, in the course of which the process concept reached industrial maturity.

In 1985, Iscor ordered a Corex® 300 kt/year (C–1000) plant for the Pretoria works. This plant was commissioned in December 1989. In November 1995, Posco started up the first plant of the C–2000 type with a capacity of 600 kt/year on a 100% lump ore basis for its iron- and steelmaking plant in Pohang.

DESCRIPTION OF COREX® PROCESS AND COREX®-DR INTEGRATION

The metallurgical reactions in the Corex® process (Fig. 5) take place in two process reactors, i.e. in the reduction shaft and in the melter gasifier.

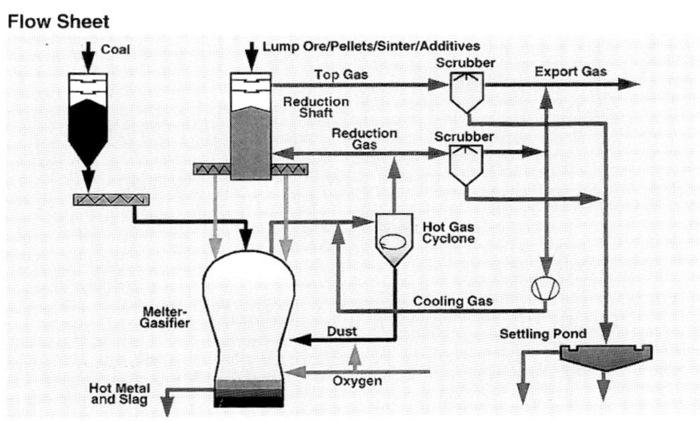

Fig. 5 Corex® process.

Lump ore, pellets, or a mix of these input materials, are charged into the reduction shaft together with additives via a hopper system. In the reduction shaft, the input materials are reduced to sponge iron with a degree of metallisation of over 90% by the reduction gas on the counterflow principle.

The sponge iron is extracted from the reduction shafts via specially designed screw conveyors and drops into the melter gasifier, where it is completely reduced, and hot metal and slag are melted. Tapping of hot metal and slag is exactly the same as with the blast furnace route.

Coal enters the melter gasifier via a lock hopper system and is converted to char at temperatures in the range of 1000–1050°C. Oxygen is blown into the fixed bed formed by the degassed coal in the melter gasifier. Apart from the melting heat, an excellent reduction gas is generated, consisting mainly of CO, H_2, as well as coal and iron dusts. After exiting the melter gasifier, the gas is cooled to the desired reduction temperature of approx. 850°C, and dedusted. The dust is largely recycled to the melter gasifier via the dust recycling line. The gas is further used in the reduction shaft to reduce the input materials as described above.

Both the tapping temperature and the hot metal composition are the same as with the blast furnace. The top gas from the reduction shaft has a net calorific value of approx. 7500 kJ m^{-3} (STP) and a low sulphur content. Following scrubbing, this export gas represents a high grade byproduct.

Export gas can be used for heating purposes in the iron- and steelmaking plant, for the production of sponge iron, for electric power generation, or for the generation of synthesis gas.

Figure 6 gives a schematic of sponge iron production using Corex® export gas in a reduction shaft. The CO_2 contained in the export gas is removed by CO_2 scrubbing. Subsequently, the gas is heated up to approx. 800°C. Analogous to the Corex® reduction shaft, the reduction of pellets and lump ore is accomplished by the reduction gas flowing through the shaft in counterflow.

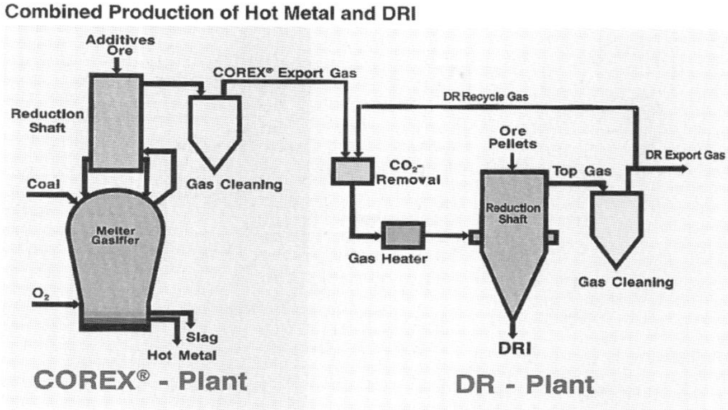

Fig. 6 Use of export gas in Corex® system.

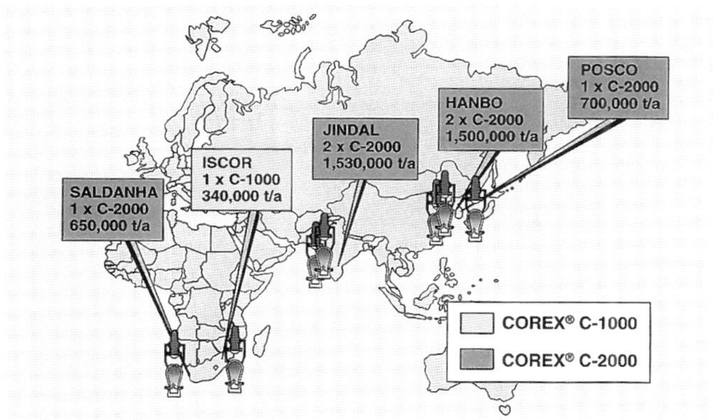

Fig. 7 Corex® plants in operation or implementation.

SURVEY OF COREX® PLANTS AND ORDERS

Figure 7 gives a survey of Corex® plants currently operated or under construction.

Iscor, Pretoria, South Africa
The plant, featuring a nominal capacity of 300 kt/year of hot metal, was started up in 1989. To date, Iscor has produced more than 2.3 Mt of hot metal in Pretoria after the last blast furnace was shut down in 1992. The hot metal produced is sold on the international pig iron market.

Posco, Pohang, South Korea
Posco's Corex® plant was started up on 14 November 1995. It has a nominal capacity of 600 kt/year of hot metal based on 100% lump ore input (see Fig. 8), which corresponds to an hourly capacity of 80 t. Since its startup, the plant has

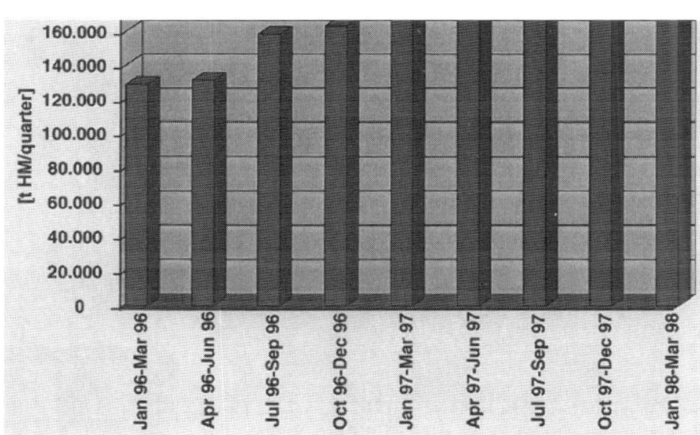

Fig. 8 Corex® Posco plant production figures.

Table 1. Corex® Posco: ores used
Analysis of used ores

		Algarrobo Pellets	CVRD Pellets	LKAB Pellets	Peru Pellets	Mt. Newman Lump Ore	Sishen Lump Ore	IHG Lump Ore
Fe	(%)	65.2	65.75	66.32	65.50	64.3	66.56	65.6
SiO_2	(%)	2.1	2.52	2.02	3.60	3.04	3.10	2.1
AL_2O_3	(%)	0.53	0.56	0.32	0.47	1.04	0.97	1.53
CaO	(%)	2.63	2.5	0.25	0.47	1.3	0.12	0.09
MgO	(%)	0.41	0.04	1.62	0.82	0.11	0.07	0.03
K_2O	(%)	<0.02	<0.02	0.04	0.1	<0.02	0.12	0.02
P_2O_5	(%)	0.06	0.06	0.06	0.02	0.12	<0.1	0.09
SO_3	(%)	0.02	0.01	<0.01	0.02	<0.02	0.02	0.01

Table 2. Corex® Posco: coals used
Analysis of used ores

		Mount Thorley	South Blackwater	Coalex	Optimum	Bays Water	Ensham
Ash	(%)	13.8	9.5	10.7	10.6	7.4	8.2
Volatiles	(%)	32.6	26.7	29.8	37.7	38.6	26.6
C_{fix}	(%)	53.4	63.7	59.6	56.7	56.0	65.2
S_{total}	(%)	0.35	0.36	0.38	0.6	0.6	0.39

been operated with several types of lump ores, pellets, and coals aimed at optimising the operating mode and the operational parameters of the plant.

The chemical analyses of the ores, pellets and coals used are given in Tables 1 and 2. Good operating results were achieved in conformity with the scheduled startup sequence and the acceptance of the plant, comprising the contractual performance tests, was completed in summer 1996.

A survey of the Corex® hot metal quality produced at Posco gives the following hot metal analysis at temperatures of 1500–1530°C: 4.4–4.6%C, 0.01–0.02%S.

The plant availability already exceeds the contractual value of 93%, and the C_{fix} and oxygen consumption figures/t of hot metal most of the time are below the contractual values. To date, over 1.8 Mt of hot metal have already been produced.

Hanbo, Dangjin, South Korea

In 1994, Hanbo commissioned VAI to construct two Corex® C-2000 plants for the production of 1.5 Mt/year of hot metal. The Corex® export gas is used for the production of 800 kt/year of sponge iron in a direct reduction shaft (to be constructed by the consortium VAI/Linde/Kobe) as well as in a conventional power plant. The hot metal and the sponge iron are charged into electric arc furnaces.

The complete steel plant including the COREX® plants at Asan Bay went bankrupt in March 1997. Due to that a revised start-up plan is not available at the begining of 1998.

JINDAL Toranagallu Works

JINDAL, one of the largest Indian flat steel producers in the private sector, is in the process of realising a 'greenfield' steel works near the settlement of Toranagallu, Bellary Hospect District, in the State of Karnataka, India. Towards realising this goal JINDAL VIJAYANAGAR Steel signed contracts in July 1994 and December 1995 for the supply of two COREX® Plants C-2000. The first COREX® plant is scheduled for start-up in 1998, the second plant approximately six months later. Figure 9 shows the status of the COREX®

Fig. 9 COREX® tower and casthouse A & B, Module 1, JINDAL.

plant under construction as of June 1998. The two COREX® plants will be operated with a mixture of 30% lump ore and 70% pellets, which will allow for a design capacity of 1 500 000 t/a hot metal.

The export gas from both COREX® plants will be used for internal applications witin the Toranagallu Works, for the generation of electrical energy in two adjacent power plants, as well as for the production of pellets in a pelletising plant with a nominal capacity of 3 000 000 t/a (Fig. 10). About 50% of these pellets will be processed in the COREX® plants and the rest will be sold to third parties.

Saldanha, Saldanha Bay, South Africa

In late 1995, VAI was awarded a contract for the erection of a Corex® C–2000 plant. In early 1996, the consortium VAI/Linde/Midrex was awarded a contract by Saldanha for the erection of a DR plant using Corex® export gas. The startup of both plants is scheduled for 1998.

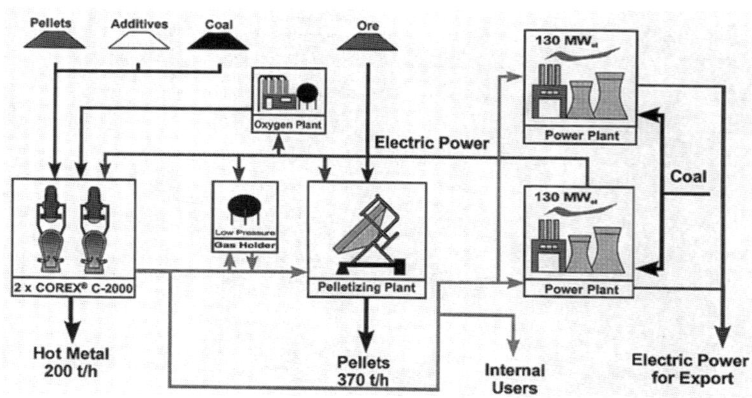

Fig. 10 Combined production of hot metal, pellets and electric power.

DESCRIPTION OF FINMET TECHNOLOGY

The Finmet process uses fine ore with a high Fe content and a maximum grain size of 12 mm as the raw material. The share of grain sizes less than 0.15 mm is limited at 30%.

The fine ore is dried, stored in bins, and subsequently conveyed to a pressure vessel system (11 bar system pressure) via a belt type bucket elevator. From there, the ore is charged into the first reduction reactor, in which fluidisation, heating, and preliminary reduction take place.

The fine ore flows from the upper reactor to the next one by gravity. In the fourth reactor, a degree of metallisation of more than 90% is achieved at the reduced fine ore. The fluidised condition is maintained by reduction gas which is injected by counterflow.

From the pressurised fourth (last) reactor, the reduced iron is fed into a briquetting bin via a pneumatic conveyor system. From this bin, the reduced hot fine ore flows into the briquetting machines. The briquetting strands produced are crushed into single briquettes in drums connected downstream. The material discharged from the drums is screened, and the fines are returned to the briquetting machines by bucket elevators. The hot briquettes are conveyed to an air cooler, where they are cooled and passivated (= prevention of oxygen absorption) before they are transported to the storage yard. The briquettes have a very high density and feature excellent storage properties.

The reduction gas required for ore reduction is generated from natural gas by steam reforming in a conventional reformer. The reduction gas from the upper reactor is cooled and cleaned in a scrubber system in order to separate the ore particles carried over and the water formed during reduction from the gas.

The reduction gas of the steam reformer, which consists of about 62% hydrogen, 6% carbon monoxide, and a residue of carbon dioxide, methane, and water, is mixed with the cleaned and compressed gas from the upper reduction reactor. The CO_2 content of the gas is adjusted in a chemical scrubber; subsequently, the gas is preheated to approx. 840°C and fed into the reduction gas cycle.

BHP FINMET PLANT, PORT HEDLAND, AUSTRALIA

In 1995, the Australian company BHP ordered a Finmet plant with a nominal capacity of 2 Mt/year for Port Hedland in Australia. The plant is of modular design; each production unit comprises four reduction reactors and has a nominal production capacity of 500 kt/year of HBI. The Fe content of the Mt Whaleback iron ore that will be used is specifically adjusted in an ore beneficiation plant according to requirement. The startup of the plant is scheduled for early 1999.

Besides the plant for BHP, Australia, another Finmet plant with the same capacity has been ordered by Orinoco Iron to be implemented in Puerto Ordaz, Venezuela.

OUTLOOK

A further related development of the VAI Group in smelting reduction technology is Finex®, where liquid hot metal can be directly produced from iron ore fines and non-coking coals. VAI submitted the general patents for the Finex technology in the mid 1980s. In 1992 a joint development cooperation agreement was signed between Posco/RIST, Korea and VAI. Process development is expected to be completed by the end of this decade.

The estimated expert global hot metal projections and the experience which VAI has in the field of smelting and direct reduction suggest promising potential for the Corex® and Finmet processes.

EVOLUTION OF EAF STEELMAKING ROUTE

E. REPETTO, G. BRASCUGLI* AND G. PERNI*

**CSM, *Acciai Speciali Terni; *Acciai Speciali Terni*

ABSTRACT

The share of steel produced via Electric Arc Furnaces (EAF) increases steadily at international level. A new generation of minimills for long and flat products will provide the means for coping with this production increase.

The EAF is the core of a minimill and the innovations which have been introduced in the field of electric arc furnace technology are reviewed in this report.

A number of different technical solutions, some already in operation and others soon to be introduced, are described and compared. The following features are compared taking into consideration the charge materials presently available and those foreseen in the future: manufacturing practice, electric energy consumption, electrode and refractory consumption, productivity level, degree of automation, organisational structure.

The environmental impact is also discussed and some EAF dust treatment technologies are described.

1. INTRODUCTION

Steel manufacture in Electric Arc Furnaces (EAF) is gaining new impetus both in emerging countries where the overall capacity is increasing and in industrialised countries.

The share of the EAF in the production of raw steel (31.8% in 1994) increases steadily and is expected to exceed 50% by the year 2010.

New-generation minimills will mainly contribute to this production increase. While in the past their capacity was 0.3 to 0.5 Mt/y, they are now in the position to produce as much as 1 to 1.5 Mt/y. Additional growth is expected to occur in the next few years up to 2 Mt/y and more, mainly for flat products.

A significant contribution to the technological modernisation as the basis of this expansion has certainly come from Nucor whose Hickman plant is equipped with DC furnaces and thin-slab casters for the manufacture of flat rolled steel. Its example has been followed worldwide and several other minimills have been constructed to produce flat products and to a lesser extent long products.

At present the introduction of minimills is spreading, particularly in the following frameworks:

- green-field construction of minimills for the manufacture of flat products (Nucor, Steel Dynamics, Gallatin, etc.)
- replacement of existing blast furnace/converter plants. In Europe the

following instances can be mentioned: Georgsmarienhütte, Preussag Stahl, ProfilArbed at Differdange and Unimétal.[1,2]
- construction of new minimills in emerging countries where the poor availability of scrap is compensated for by locally-produced substitutes.[3]
- introduction of new EAFs in existing electric plants.

Minimills have undergone a real revolution as far as continuous casting (thin slab technology) and particularly the EAF are concerned.

The result of this revolution is a minimill with the following characteristics:

- line compactness
- semi-continuous flow of materials
- product specialisation
- high productivity using a small number of workers

The EAF is the core of a minimill. As a function of different requirements it has specialised and differentiated also thanks to the introduction of the DC furnace. However, owing to some residual discontinuities (e.g. tapping) the EAF is still an obstacle to be overcome until a line with continuous flow of materials and increasingly uninterrupted process can be obtained.

Various alternatives are now available or expected in the near future. However, before going into detail it is first necessary to examine which raw materials can be used to ensure the aimed production growth worldwide. As an example, if 30 Mt/y of flat products from the new minimills add to the overall production as expected by the year 2005, then the demand for low-impurity materials will dramatically increase and scrap will be able to meet this requirement only to a minimum extent.

2. CHARGE MATERIALS

EAF production of raw steel is expected to increase by 4.6% per year as an average. This means growing from 240 Mt/y up to 340 and 390 Mt/y by 2000 and 2005 respectively.

To cope with this situation the increase of the availability of both scrap and alternative metallic materials will be necessary.

The following types of scrap are presently available: 'home scrap' generated by the steelmaking process and the associated primary processes; 'prompt scrap' generated in the starting stages of steel consumption by the main manufacturing companies and 'obsolete scrap', i.e. what is recovered from used or dismantled products. Only the latter is likely to be available in strongly increasing quantity, but the quality of scrap will be progressively worsening unless cleaning treatments are provided which are justified only at low cost. The other two types of scrap are expected to be slightly decreasing (home scrap) or maintaining steady levels (prompt scrap).

The resulting effect will be a growing scrap availability, although quantities will not be enough to cover steelmaking requirements. As a consequence, it is estimated that the scrap shortage by the year 2005 will be amounting to 46 Mt/y (in a mid case scenario).[4]

To balance this shortage, non-integrated minimills will have to purchase substitutes and new integrated minimills incorporating direct reduction (DR) and/or smelting reduction (SR) processes will have to produce substitutes locally.

Direct reduced iron (DRI) production is growing fast. In 1995 it totalled 30.7 Mt/y (64.8% of which was obtained with Midrex process) with a 12% increase over 1994.

Other substitutes such as iron carbide and hot metal by COREX process are progressively coming into stream. As regards the COREX process, eight plants are either operating, next to be run at steady state or just on order for an overall production capacity of approx. 4.8 Mt/y.

Considering all the above, the EAF technology is called to face a double challenge:

- it will have to evolve towards technological solutions that are able to increase productivity, decrease conversion costs and conform to ever stricter environmental laws; and
- it will have to ensure a high inherent flexibility in the use of strongly diversified charge materials.

Hot metal, pig iron, direct-reduced iron, hot briquetted iron (HBI) and iron carbide, real or anticipated shifts in EAF charge materials, are driving evolutionary changes in the anatomy of the arc melter.[5]

3. ELECTRIC-ARC FURNACE TECHNOLOGY

In the past years the EAF technological evolution has been directed by the following driving forces:

- increase in productivity and power-on time with regard to tap-to-tap time;
- reduction in electric energy consumption;
- reduction in electrode and refractory consumption;
- decrease in labour cost.

Two additional factors have recently played a significant role, namely:

- reduced environmental impact;
- use of strongly diversified charge materials.

Pressure from the above forces during the 1970s and 1980s has led the AC furnace towards an intensive technological evolution (UHP operation, water cooled panels, EBT, oxy-fuel burners, oxygen lancing, carbon injection, foamy slag practices, etc.) which is still under way. For an extensive study on these practices, literature will provide a wide source of material.

Since the late 1980s when the DC furnace was introduced on industrial level, a new phase has opened that is still at the evolutionary stage. This led to radical

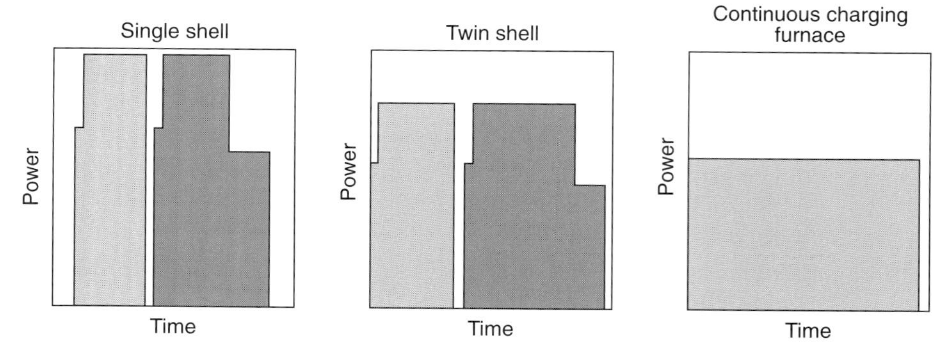

Fig. 1 Tap-to-tap period for different phase structuring, for three different furnace types.

changes in the EAF technology and, together with technical developments in continuous casting, it changed the operation of the whole manufacturing cycle from molten steel to semi-finished products.

In particular, under the pressure of the above-mentioned driving forces and as a response to the specific requirements of steel producers or network and location requirements, the EAF has strongly differentiated. This has resulted in three main furnace types:

(a) single-shell furnace (top charging mode);
(b) double-shell furnace (top charging mode);
(c) continuous charging furnace.

Each furnace type has got a different phase structuring within the tap-to-tap period, as shown in Fig. 1. In practice, the power-off time is higher in type (a) and it progressively lowers in types (b) and (c). Consequently, the Power Utilisation Ratio is approx. 70–75% for type (a) and more than 90% for type (c).

The AC mode is widely used for type (a) furnaces while the DC mode – which was initially tested on type (a) – is now also applied to type (b) and (c) furnaces.

3.1 Single-Shell Furnace (top charging mode)

The single-shell furnace is charged from above by introducing baskets through a roof opening.

The tap-to-tap time in the two-basket practice is composed of the following main steps:

– *first charging, melting, second charging, melting and refining, tapping and repairing*

which stresses how significant the incidence of the power-off time may be.

The single-shell furnace with the above features is now available in the DC and AC versions. The DC furnace has been spreading so rapidly since the late 1980s that in 1996 there are at least 90 DC furnaces.[6,7] Out of these, single-shell furnaces are still the majority over twin-shell and continuous charging furnaces, and in all they provide a total capacity of 60 Mt/y.

For a compared analysis of DC furnaces versus AC furnaces, the following parameters are to be measured:

- electrode consumption
- refractory consumption
- flicker and harmonics
- bath stirring

Although the following indications on DC vs. AC furnaces refer to single-shell furnaces, they can also apply to twin-shell furnaces.

Continuous charge furnaces and scrap preheating furnaces will require a separate analysis.

Regardless of any comparison of AC and DC furnaces, some recent innovations like *post-combustion* which were initially tested on single-shell furnaces can now be applied generally. Other innovations, such as *air-tight operation*, are still restricted to AC single-shell furnaces.

Graphite electrode consumption A modern DC furnace consumes 1 to 1.5 kg/t of electrodes. New-generation high-impedance AC furnaces consume 2 to 3 kg/t and in some applications even less than 2 kg/t of electrodes.

The gap between AC and DC furnaces that existed when the latter came on stream is now partially filled.

To achieve this result, the secondary voltage of AC furnaces, presently around 1000 V, was increased with consequent reduction in current values. However, at the same time it was necessary to improve the arc stability with the introduction of series reactors.

The present high-power DC furnaces (80–100 MW) install 700 to 750 mm electrodes with a current carrying capacity up to 120 kA, for which electrode manufacturers apply a 20% extra charge over the price of 600 mm electrodes that are used on AC furnaces of equal powers.

Unless the electrode market situation changes, the comparison of AC and DC furnaces with regard to specific costs is basically balanced. DC furnaces still have an advantage when electrode diameters are equal to or lower than 600 mm.

Refractory consumption. The presence of a bottom electrode in the DC furnace implies a construction and maintenance burden. However, it is counterbalanced by the fact that refractory wall consumption in DC furnaces is lower than in AC furnaces.

The DC furnace is currently available in four electrode designs:

(a) conductive refractory bottom
(b) water-cooled billets
(c) multipins
(d) multifins

To ensure a homogeneous refractory wear on the DC furnace walls it is necessary to avoid arc deviation which is responsible for hot spots on the furnace walls. This requirement mainly arises in the long-arc mode also when foamy slags are applied.

Furnace manufacturers have developed various arc deviation control systems which are based on the eccentric position of the graphite electrode, optimisation of the external cabling and control of the DC current distribution through each billet (water-cooled billet design).

Arc deviation control also results in lower flicker. Based on the latest progress, the useful life of a bottom electrode can be as long as 2000–3000 heats.[8,9]

Flicker and harmonics. Mini-mills are big sources of noise in the electric network owing to the flicker and harmonics produced above all by the electric arc furnaces.

Compared to AC furnaces, DC furnaces generate about 50–60% less flicker than AC furnace. This is because to generate DC the electricity must run through capacitors, rectifiers and inductors.[10] In AC furnaces, the gap has been reduced in part by increasing the secondary voltage and by using long electric arcs which operate at lower current values; the resulting benefits have been less electric noise in the network as well as reduced electrode consumption. However, it was necessary at the same time to increase arc stability and this was done by increasing the system reactance, since the transformer reactance was insufficient for the purpose. Series reactors with on-load regulation were installed (or incorporated in the transformer) dimensioned for producing the necessary reactance values to ensure arc stability.

Recently, a new system has become operative which utilises a saturable reactor system that enables current limiting during the melting process and therefore a reduction of flicker. A reduction in current values also leads to a reduction of the electrodynamic stress factors arising from the current. Hence, it is possible to reduce both the vibration and rate of breakage of the electrodes.[11] On the other hand, DC furnaces are indirect sources of harmonics owing to their thyristor rectifiers; harmonics from DC furnaces can be worse than those from AC units but, in any case, can be coped with by using suitable filters.[10]

Bath stirring. Bath stirring is gaining increasing importance in high-production furnaces as it promotes better temperature distribution in the steel bath, especially in peripheral furnace areas. When pre-reduced material is charged continuously, it prevents the formation of 'icebergs'.

DC furnaces provide for a natural electromagnetic stirring under the action of the current passing through the bath. The stirring intensity can be increased with busbars of suitable geometry coming out from the anode or using special coils.

AC furnaces require bottom gas stirring systems that use argon or nitrogen, resulting in additional costs. Measurements have shown that the stirring effect of either systems is basically the same.[2]

With the use of multi-media submerged nozzles for the introduction of oxygen and carbon carriers, an efficient bath stirring results as a so-called waste product so there are no additional costs.[1]

Post combustion. Post combustion (PC) is an emerging technology whose aim is to lower the energy cost and increase the EAF productivity. It is a direct evolution of foamy slag practice.

Several PC installations are supplied by Praxair, Air Liquide (ALARC system), Danieli, More, American Combustions Inc. just to mention a few, each one running its own PC package. Initially they were applied on AC single-shell furnaces and more recently they have been introduced in twin-shell furnaces.

The main product of foamy slags is CO. This technology is based on blowing oxygen above the bath with a PC lance. The aim is to post-combust most of the CO in the bath which in conventional practice is post-combusted in the exhaust gas system without taking advantage of any potential oxidation energy.

Together with CO also H_2 is post-combusted into H_2O which increases the energy gain.

In principle, every Nm^3 of PC oxygen generates 5.8 kWh when burning CO at 1600°C. In practice, as much as 3–4 kWh/Nm^3 are likely to be saved.[12]

In order to optimise the efficiency of post-combustion, Praxair during the startup determines the pattern of CO evolution over several heats. At the end of startup, the oxygen flow through the PC lance is controlled by pre-defined blow profiles, with no need for a permanent off-gas analysis system.

Other PC manufacturers prefer to install an on-line regulation of oxygen injection based on a continuous off-gas analysis system which measures CO, CO_2, O_2 and H_2 concentrations.

Air-tight system. Reducing the air intake into the EAF is likely to provide several potential advantages, such as: lower energy consumption as there is less nitrogen to be heated, lower power at the direct evacuation system and improved tapping yield due to reduced oxidation of charged material. Reduction in NO_X and dust content in off-gases is also expected.

Special sealing devices are required near the furnace door, the gap between the furnace wall and roof and the electrode openings. At the same time, a dynamic control system for the exhaust rate needs to be developed.

This makes furnace operation more sophisticated. That is why the system is still restricted to a small number of steelworks (e.g. DDS-Frederiksværk; Van Moos Stahl AG – Emmenbrücke, etc.).[13]

Operating results. The results obtained with DC and AC single-shell furnaces operating with conventional and post-combustion practices are listed respectively in Tables 1, 2 and 3.

Table 1. DC furnaces.

Plant		SMI, Birmingham	BOËL	L.M.E.	Daiwa Steel	Essar Steel		North Star Steel	Okayama, Tokyo Steel	[20] Typical 100 t
Transformer rating	MVA		95	84	100	160	160	80/88		100
Electric power	MW		72 (max)							
Max secondary voltage	V		720					790		
Charge mix			100% scrap	1 bucket + cont. charging	1 bucket + hot metal	100% scrap	20% scrap + 80% sponge iron			
Tapping weight	t		85	85	120	150	150	93 short tons	150	
Tap-to-tap time	min		52	48	53	60	90		65.68	55
Power-on time	min		n.a.							
Electric energy	kWh/t	380	480					<400		420
Total O_2 consumption	Nm^3/t		20					17		25
Burners fuel	Nm^3/t		n.a.							
Electrode consumption	kg/t	0.95?	1.52					1.2		1.1
Electrode diameter	mm	700	700	700					750	
Production per year	Mt/y		0.5	0.55						
Productivity	t/h			106	130	150	100			

Table 2. AC furnaces.

Plant		MJS	Chaparral Steel 'A' furnace	Chaparral Steel 'B' furnace	ISPAT	Basauri, Sidenor	AST (Stainless steel)
Transformer rating	MVA	44–48	81	138	105	85	110
Electric power	MW		57	90	78		60
Max secondary voltage	V		755	1050	960	1050	763.5
Charge mix		35% CPI 65% scrap	100% scrap	100% scrap	100% scrap	100% scrap	100% scrap
Tapping weight	t	90	130	140	115	105	125
Tap-to-tap time	min	60					95
Power-on time	min	54				53	55
Electric energy	kWh/t	390	460	420	430		425
Total O_2 consumption	Nm^3/t	33	28	33	25	21	15.5
Burners fuel	Nm^3/t				5		3.8
Electrode consumption	kg/t	2.2	2.9	1.9	1.8	2	3
Electrode diameter	mm	550	600	600	600	600	600
Production per year	Mt/y	0.64	0.54	0.86	0.895		0.65
Productivity	t/h	89	68	110	112		79

Table 3. AC furnaces with post combustion.

Plant		Dalmine	Ferriera Valsabbia	BSW	Nucor Steel	Von Roll	Vallourec	Cascade Steel
Transformer rating	MVA	100	50	57/68	33	60	85	85
Electric power	MW	70	36	50		45		
Max secondary voltage	V	1000	800	660			835	
Charge mix		100% scrap	100% scrap	100% scrap				
Tapping weight	t	93	70	78	60	70	86	95
Tap-to-tap time	min	46	57.7	47.8	54	58	54	76
Power-on time	min	29	49.9	36.8	37	40		47
Electric energy	kWh/t	380	384	347	327	403	398	456
Total O_2 consumption	Nm^3/t	37	32.7	45.6	36.1	32	30.3	
Burners fuel	Nm^3/t	8	5.1	3.6	7.7	n.a.		
Electrode consumption	kg/t	2	1.9	1.8	2.5	2.32		
Electrode diameter	mm	600	500	550		600	600	
Production per year	Mt/y	0.850	0.466	n.a.		0.355	0.550	
Carbon consumption	kg/h	n.a.	10	11.8	11.5	7	16.5	

Despite the limited data available, the results substantially confirm electrode consumptions of 1 ÷ 1.5 kg/t for the DC furnace and of 1.8 ÷ 3 kg/t for the AC furnace, in line with the average data of literature.

For the same operating practice, electric energy consumption is more or less the same for either type of furnace. Adoption of the post-combustion practice has permitted an appreciable reduction of power consumption to values even below 400 kWh/t, with tap-to-tap times often lower than 60 minutes. Correspondingly, the overall consumption of O_2 increased and was always greater than 30 Nm^3/t. These data support the present tendency of increased use of O_2 by EAFs reaching, it is expected, consumption around 50 Nm^3/t (of which 11 Nm^3/t each for post-combustion and for slag-foaming) by the year 2000.[5]

Productive capacity remains generally below 1 Mt/y but it is expected that this limit will be exceeded in the near future.

3.2 Double-Shell Furnace

3.2.1 Twin-shell furnace. The Twin-shell furnace is the technological response to the need of increasing electric power efficiency by reducing the power off-time.

At present 15 furnaces have been counted, some of which have been operating since 1992 and some others are expected to start working in 1996 (Table 4). Several important manufacturers (Nippon Steel, MDH, Clecim, NKK-GHH, ABB, VAI/Fuchs) have designed and implemented furnaces of this type.

The furnace is composed of two vessels (Fig. 2) which are fed by one power supply, an electrode-holder system with rotating movement which is common to both shafts or alternatively two separate systems.[14] Usually the power supply

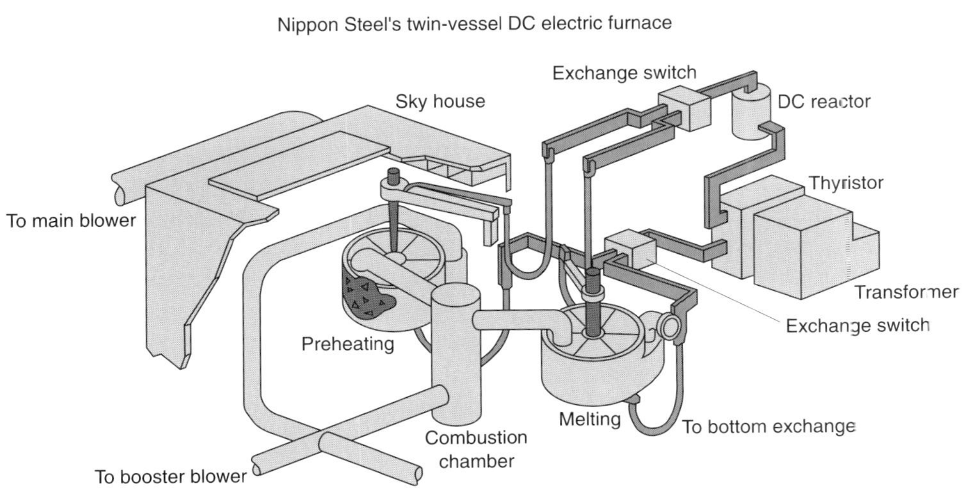

Fig. 2 Twin-furnace system.

Table 4. Twin shell furnaces.

Plant	Tap weight t	Start-up y	Transf. rating MVA	Electric power MW	Prod. per year (*) Mt/y	Charge mix	Tap-to-tap time min	Electric energy KWh/t	Total oxygen Nm^3/t	Electrode consump. Kg/t	Electrode diameter mm	Types of products	Supplier
Kansai Billet Center, Nippon Steel, Japan	2.120 DC	1992											Nippon Steel
Profil Arbed, Differdange, Luxembourg	2.155 DC	1994	140	100	1.4	scrap	59	401	37.5	14	700	long products	MDH
Takuman Steel, Japan	2.40 DC	1994											Nippon Steel
Unimetal, Gandrange, France	2.160 DC	1994	150	110	1.2	75% scrap 25% hot metal	85	490	21.5	1.95	750	long products	CLECIM
Sumitomo Metal, Osaka, Japan	2.40 DC	1995											Nippon Steel
Gallatin Steel, KY, USA	2.196 DC	1995		90	1–1.2 (1996)	scrap + 17–20% pig iron	85				700/750	flat rolled steel	NKK, GHH
Baosteel, Shanghai, China	2.150 DC	1996											CLECIM
Kwangyang Steel, Posco, Korea	2.130 DC	1996									700		Nippon Steel
Ipsco Inc, Iowa, USA	2.135 DC	1996	140		1–1.5							flat rolled steel	MDH
Southern Steel Malaysia	2.85 DC	1997	85	60	0.75	scrap + 20% HBI		350 (exp.)		<1 (exp.)	700	long products	ABB
Dofasco (Ont.), USA	2.165 AC	1996	134		1.35	100% scrap or scrap + 30% h.m.	50				600	flat products	VAI/ Fuchs
Steel Dynamics (SDI), USA	2.165 AC	1996	120		1.2	scrap						flat rolled steel	Fuchs system
Inchon, subsidiary of Hyundai, S. Korea	?	1997?			1.5								NKK
Dongbu, S. Korea	?	1997			1.5								Nippon Steel
Hybrid twin shell furnace: combined Converter/Electric Arc Furnace													
Nippon Denro Ispat, India	2.180 AC	1996	137	up to 81.4	1–1.6	Scrap, hot m. pig iron DR/HBI	51–75	187–244	34–42		700	flat products	MDH

(*) Designed capacity.

is in the DC mode with the only exception of VAI/Fuchs that supplies an AC three-electrode system.

While the charge material is molten in one shell, the first basket is charged into the other shell. When the first shell is ready to tap, the electrode is positioned in the other shell where the power-on phase is started. This results in a reduction of 6–10 min in the power-off time and a remarkable increase in furnace productivity.

Some designs provide that, while the furnace is melting, hot off-gases are passed through the second shell to preheat the scrap. Otherwise, when VOCs (Volatile Organic Compounds) and NO_X problems arise, it is preferred to preheat scrap in the stand by shell by means of burners.

The benefit of charge preheating is to decrease the power-on time and electric energy consumption, a saving which Nippon Steel has assessed at 35 kWh/t for a preheating time of 25 min.[14]

The twin-shell technology is bound to undergo a further differentiation in terms of high flexibility of the charge composition (scrap, hot metal, iron carbide, pig iron, HBI and DRI), that is often characterised by high carbon contents.

In particular, direct charging of hot metal into the EAF together with scrap or directly reduced iron is an economical solution for the production of high-quality flat products, as it involves a drastic reduction in electric energy consumption.

The need of high decarburisation rates close to 0.1% of carbon per minute has led to the design of a hybrid vessel which combines the essential functions of a converter and AC or DC furnace.

In the twin-shell design, one vessel works in DC or AC mode being fed with DRI or HBI and the other works in converter mode by decarburising the hot metal with a top oxygen lance.

ABB and MDH have provided different technological responses to the new concept (named Arcon and Con Arc respectively).[5,15]

In particular, MDH is constructing the first hybrid AC Twin-shell furnace at Nippon Denro Ispat in India, which will become operative in 1996.

As shown in Fig. 3, this furnace is characterised by its high flexibility in the use of energy and raw materials. Part of the O_2 (up to 85%) will be blown in using a vertical multinozzle supersonic lance capable of flow rates up to 300 Nm^3/min, depending on the type of charge, electric energy consumption will range from 177 to 244 kWh/t (see Table 4); electrode consumption too is expected to be very low given the reduced power-on times.

Saldanha Steel in South Africa is considering installing this type of furnace downstream of the two Corex C 2000 plants (capacity 0.65 Mt/y of hot metal) that are already on order.[5]

3.2.2 Shaft furnace. The Shaft Furnace was developed by Fuchs System Inc. and traded since 1992. It is equipped with a preheating system and delivered with either a single or double shaft, either in DC or AC mode.

Seven single-shaft furnaces and two double-shaft furnaces are already operat-

CONARC – combined **Con**verter/Electric **Arc** Furnace

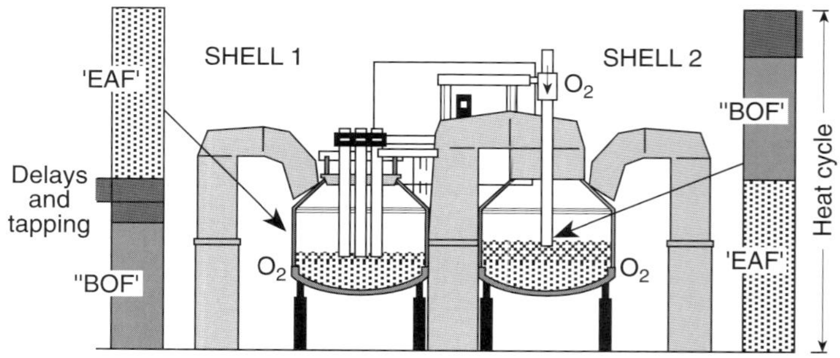

High flexibility in raw material

Scrap	0–100%
DRI; hot/cold	0–100%
Pig Iron	0–50%
Iron Carbide (Fe_3C)	0–30%
Pre-reduced Iron Ore	0–60%
'Garbage scrap'*	0–25%
Hot metal	0–70%

* depending on residual leval

High flexibility in use of energy

Secondary energy	
Electricity	15–100%
Primary energy	
O2/C – door (Arcing process)	0–30%
O2 – Top blowing (Converter process)	0–85%
O2 – Bottom/side blowing (Demag Korfarc process)	0–65%
Burners	0–40%
Post combustion	0–10%

Fig. 3 Diagram of CONARC – combined converter/Electric Arc Furnace.

ing or soon to be operating. However, one double-shaft furnace at Arbed, Esch-Schifflange, only made a few trials with double-shaft operation and is now working as a single-shaft.

Figure 4 shows a single-shaft furnace. It utilises the sensible and chemical heat of the off-gases with the additional heat from oxy-fuel burners at the shaft bottom to preheat a column of scrap positioned in a water-cooled shaft on one side of the furnace roof. The thermal efficiency of the burners is higher compared to a conventional furnace, as burners are submerged in the scrap for much longer and at a deeper level.

The shaft can hold at least 40% of the scrap charged. The remaining scrap is charged directly in the furnace before starting to melt. In the double-shaft furnace where 100% of scrap is preheated, the presence of two vessels reduces the power-off time significantly.[13]

Table 5 gives some operational data of single and double shaft furnaces in operation which are reported in the literature.

3.3 Continuous-charge Furnaces

A new generation of furnaces has been developed, which operate with con-

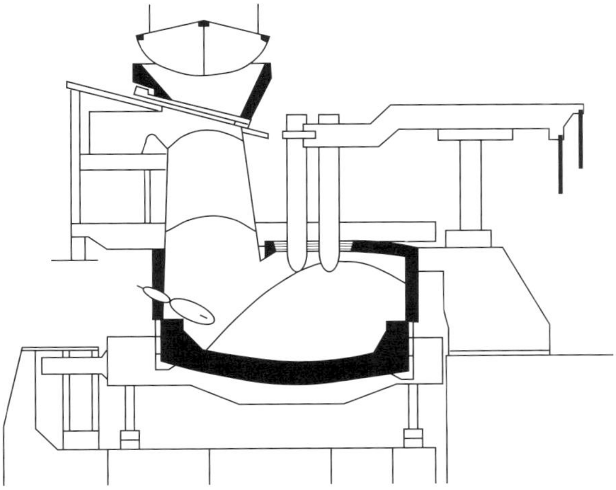

Fig. 4 Scheme of a Single Shaft Furnace.

tinuous charging of the scrap and substitutes and which incorporate a section that preheats the scrap according to various different designs.

The advantages of this technology are:

- reduced electric power consumption even if offset in part by other energy consumption;
- power utilisation factor from 80% to values close to 100%, since there is no need to open the furnace roof periodically for loading the charge;
- very smooth electrical operation with limited voltage flicker and harmonic distortion;
- possibility of operating in a continuous controlled environment.

For the future, it is possible to envisage a continuous tapping mode and a manufacturing line featuring a continuous uninterrupted flow of material from scrap charge to final product.

3.3.1 The Consteel® process. The first example dates back to 1989 by Florida Steel Corp. From 1992 to 1994 three other Consteel systems were supplied, all currently in operation. One more unit is under construction. The Consteel Process is a system developed and patented by Intersteel Technology Inc. (now belonging to the Techint Group).

Figures 5 and 6 show the operating principle and one construction example of the straight-line configuration. The scrap is put on the charge conveyor which carries it through a dynamic seal into the preheating section of the system.[15]

As the scrap moves through the preheater, it is heated to an average 550–600°C only with the chemical and sensible heat of the off-gases from the furnace. The scrap is then dropped into the furnace bath.

A continuous metal heel is kept in the furnace to melt the incoming scrap

Table 5. Shaft furnaces.

Plant	Tap weight t	Start-up y	Transf. rating MVA	Electric power MW	Prod. per year (*) Mt/y	Charge mix	Tap-to-tap time min	Electric energy KWh/t	Total oxygen Nm^3/t	Natural gas Nm^3/t	Electrode consump. Kg/t	Electrode diameter	Types of products
SAM, Montereau, France	2.90 AC	1993	96	50–60	0.75	100% scrap	62	380	33.3	4.9	1.60		long products
ARES, Esch-Schifflange, Luxembourg	2.95	1993	105				n.a.	328	24.2	6.5	1.35?		
Co-Steel, Sheerness, UK	90 AC	1992	80	55			50	360	33	11	2.0		rod and bar
Hylsa, Monterrey, Mexico	135 DC	1994	156	80	0.75	55% scrap + 45% DRI	55–70					710–810	flat rolled steel
Ege Metal, Aliağa, Turkey	90 AC	1994											
ZSS, Zhangjiagang City, P.R. China	90 AC	1995											
Von Roll Stahl, Gerlafingen, Switzerland	70 AC	1995											
North Star Steel, Kingman, USA	95 DC	1995											flat products
Cockerill-Sambre, Charleroi, Belgium	140 DC	1996	100		0.7	scrap + hot metal							

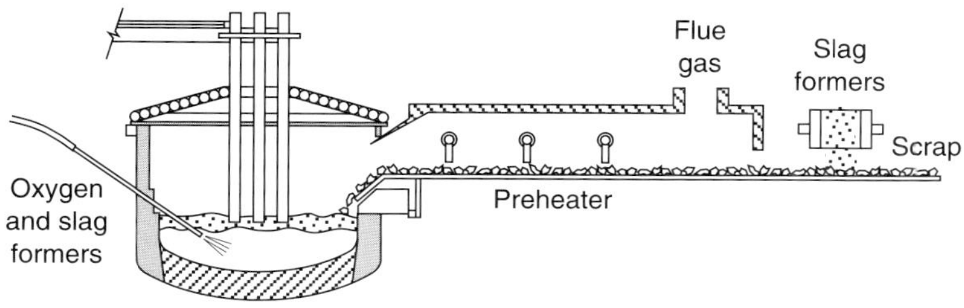

Fig. 5 Basic concept of the Consteel process.

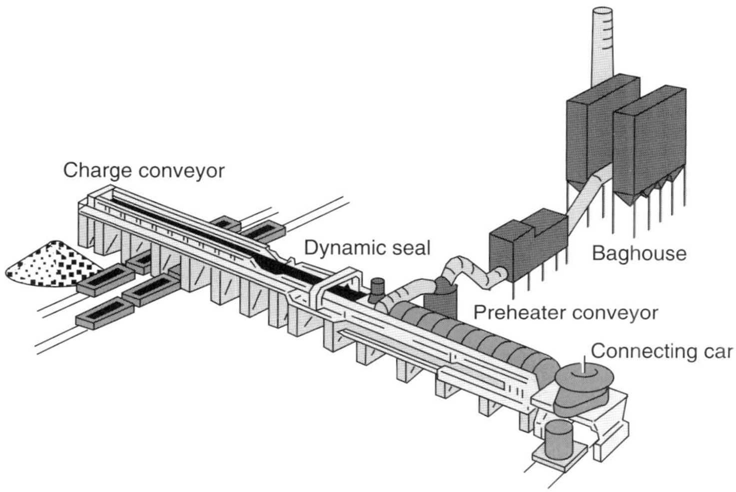

Fig. 6 Three dimensional rendering of a Consteel® system.

while the arc is used to keep the bath hot. This provides the furnace with a constant flat bath condition which makes it possible to continuously refine the bath while scrap is being melted.

The charge materials that can be used in the Consteel Process are: scrap (electric furnace grades under specification of Scrap Iron and Steel, Inc. (ISISI) of USA; pig iron and HBI added to the ferrous scrap charge mainly for the production of flat products and hot and cold DRI charged preferably in a selected part of the preheater.

Hot metal (from blast furnace, Corex, etc.) can be used in a proportion from 30 to 40% together with scrap. In this design, the hot metal is charged continuously through a channel in the vessel.

The following Table 6 reports some operating data of five Consteel plants, the last one being under construction.

Efficient management of the scrap preheated, maintaining foamy slags during the whole process and continuous charging of scrap ensuring constant thickness

Table 6. Consteel furnace.

Plant		Florida, Charlotte AC	Kyoei, Nagoya DC	Nucor, Darlington DC	New Jersey, Sayreville AC	Nakornthai, Thailand AC
Transformer rating	MVA	30	55	50	50	120
Electric power	MW	24	46	42	40	95
Max secondary voltage	V					
Charge mix		100% scrap	85% scrap 15% pig iron	100% scrap	100% scrap	80% scrap 20% DRI
Tapping weight	t	36	110	91	75	180
Tap-to-tap time	min	45	53	54	55	47
Power-on time	min	42	50	51	51	44
Electric energy	kWh/t	370	315	325	390	340
Total O_2 consumption	Nm^3/t	22.2	35	33	23	35
Burner fuel	Nm^3/t					
Electrode consumption	kg/t	1.75	1.15	1	1.85	1.7
Electrode diameter	mm	500	700	700	600	600
Production per year	Mt/y	0.45	0.45	0.75	0.6	1.5
Productivity	t/h	54.5	120	92	82	229

of the scrap bed are the prerequisites to the full exploitation of the Consteel® system strengths.

Less noise and flicker, shorter power-off time, lower dust content in the off-gases and full heat incineration of CO and dioxins are, according to Intersteel, other benefits to be added to the above.

3.3.2 Double electrode DC furnace. This new-generation furnace was developed by IHI (Ishikawajima–Harima Heavy Industries Ltd.) and ABB and became operative in 1996 at Tokyo Steel's Utsonomiya plant. With a tapping weight of 140 t, the expected production is 0.8 Mt/y of long products.

The furnace is completely sealed and includes two main parts: a scrap charging system and a DC oval-shaped furnace, as shown in Fig. 7.

The continuous charging system consists of a shaft-type preheating chamber and a charging unit.

The furnace exhaust gas flows upward in the chamber to preheat the scrap to about 800°C, while the temperature of the off-gas at the chamber outlet is less than 200°C. A two-stage scrap pusher delivers the charge into a space between

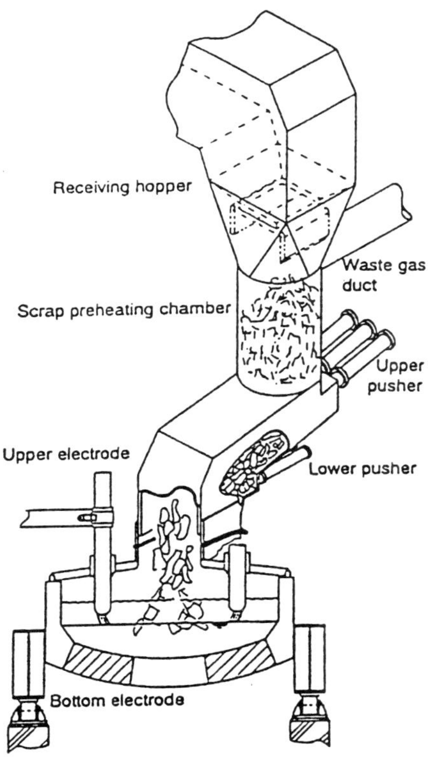

Fig. 7 Double electrode DC furnace.

the two electrodes, where the energy of the arcs is concentrated. As a consequence, there is no need for water-cooled walls.

Benefits attributed to this innovative design are:

- electric energy consumption: 260 kWh/t (carbon injection: 25 kg/t; oxygen injection: 33 Nm^3/t)
- lower noise (15 to 20 dB less than batch-type DC furnaces)
- reduced flicker (50 to 60% less than batch-type DC furnaces)
- reduced dust emission.[13,17]

3.3.3 Contiarc. Although tested by MDH just on a pilot scale, the Contiarc furnace is to be mentioned among the emerging technologies for its potential benefits. The Contiarc furnace (Fig. 8) is a DC arc-heated shaft furnace with an annular shaft formed by an outer and inner vessel that surrounds and protects a central graphite electrode.

Scrap is fed continuously with an appropriate system in the upper part of the annular shaft at a rate corresponding to the melt-down rate in the lower section

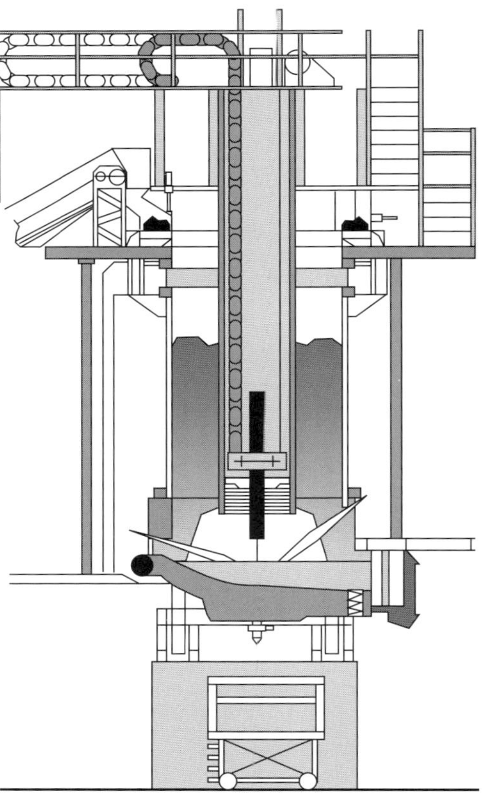

Fig. 8 Contiarc electric melting furnace.

of the furnace. During its descent, scrap is preheated by the ascending hot gases. When these gases leave at low temperature from the top of the scrap column, they are captured in a ring duct and conveyed away for waste gas treatment.

Apart from the slag door and the passage of the scrap conveyor in the upper section, the furnace is completely sealed. Thus the amount of inflowing air is very low as is the off-gas volume.

Since there is always scrap to protect the furnace walls, and no arc deflection, an arc of maximum power can be permanently maintained at flat bath conditions with no problems for the wall refractory. As this untiltable furnace can be tapped during operation, the resulting power-on time is close to 100%.

It has been calculated that the Contiarc furnace as a result of its low energy losses and its integrated scrap preheater arrangement has an expected electric energy consumption of only 260 kWh/t. The Contiarc furnace is also characterised by lower volume of dust emission through off-gases, owing to the filtering effect of the scrap column. The next step will be to check the operational suitability of the Contiarc furnace in a demonstration plant.[18,19]

3.3.4 Comelt. A new DC furnace, named Comelt, has been developed and tested on a pilot scale by VAI (Voest-Alpine Industrieanlagenbau), Fig. 9.

This is a multi-electrode furnace composed of two separate and integrated parts:

- a tiltable melting vessel
- a superimposed vertical scrap preheating shaft.

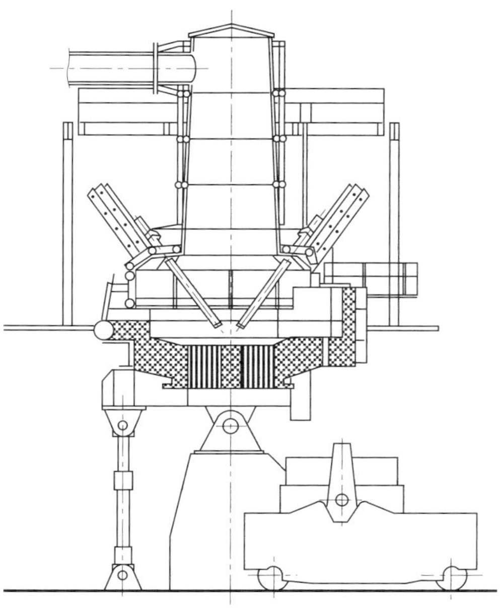

Fig. 9 Side view of Comelt furnace.

Scrap is carried by a conveyor belt and charged into the furnace through a side charging door located near the top of the shaft. Scrap is melted by sloped graphite electrodes which are inserted laterally into the melting furnace. The electrodes can be moved inwards and outwards according to the current arc conditions.

Arcs are directed towards the furnace center where they melt the charge. The hot gases generated flow to the middle of the scrap pile and then upwards through the whole scrap column in the preheating shaft.

Jet burners and lances for CO post-combustion are planned to be installed in order to reduce electric energy consumption. Expected results should be: electric energy consumption around 335 kWh/t (with 30 Nm^3/t of O_2 and 6.7 Nm^3/t of natural gas), tap-to-tap time lower than 45 min and electrode consumption lower than 1 kg with conventional diameter electrodes (i.e. 400 mm).

Other expected benefits would be: reduced dust emission, lower noise level and smaller off-gas volume.[21]

4. REVIEW OF TECHNOLOGIES

4.1 Electric Power Consumption

Table 7 lists the electric energy consumption data reported in literature for the types of furnace previously discussed.

Taking top-charge furnaces as a reference, no significant difference can be noted between the consumption rates of the single-shell and twin-shell designs whether AC or DC powered. For the same operating practice, the addition of post-combustion obviously further increases the energy gain in both cases.

Comparatively lower rates are reported instead for furnaces equipped with

Table 7.

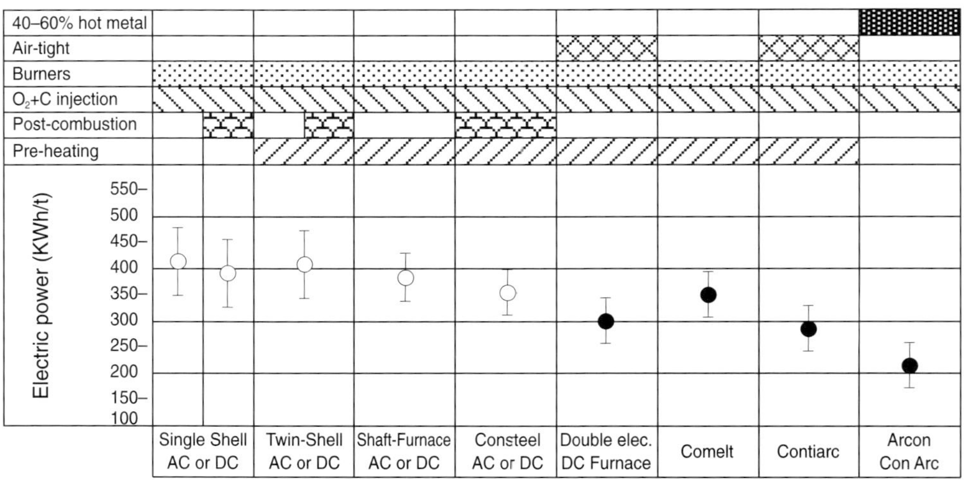

preheating systems. Within this group, the lowest values (less than 300 kWh/t) are given for the Double Electrode DC furnace and for the Contiarc furnace due, probably, to the improved thermal efficiency of their electric arcs, to lower heat losses and to the fact that both are air-tight furnaces.

Rates of even less than 200 kWh/t are given for the Hybrid Twin-Shell furnace, but this must be considered a particular case since the charge includes a high percentage of hot metal.

4.2 Electrodes

The diameter of the electrodes is one of the factors limiting the power input to DC furnaces. At present, the largest single-electrode DC furnaces install 700 to 750 mm diameter electrodes with power inputs up to 100 MW (transformer rating, 140–150 MVA) and current carrying capacities up to 125–130 kA. In the future, use of 800 mm electrodes is expected.

Owing to the fact that the market for large electrodes is fairly limited, suppliers apply an extra 15–20% charge for 700 mm electrodes which partly offsets the cost advantage of an electrode consumption lower than that of AC furnaces of equal power.

An alternative to installing larger-diameter electrodes is to reduce electric energy consumption (while ensuring the same productivity) by undersizing the power input (lower transformer rating), either by introducing scrap preheating (e.g. Consteel, Contiarc) and/or by increasing the number of electrodes to two (Double Electrode DC furnace) and even up to four (Comelt).

Further progress is expected in reducing electrode consumption with the introduction of new systems for operating and controlling the process (long arc operation, foaming slag) but also, and especially in the case of single-electrode furnaces, by injecting liquid/gaseous substances into the plasma-arc for the purpose of cooling the tip of the electrode (tip consumption) or of creating a neutral or reducing atmosphere around the electrode.

4.3 Automation

New ways of process automation have just been designed or already installed on the EAF. They range from automated scrap charging in continuous-charging furnaces based on the scrap level in the furnace (Contiarc, Tokyo Steel's Double electrode DC furnace, etc.) to automated oxygen and carbon blowing by lances mounted on robot, according to sensor signals correlated with slag foaming.[22]

Other types of process automation concern electric-arc furnace control by means of neural networks in order to optimize the power input and provide for maximum arc power by maintaining a stable long arc (Birmingham Steel), automatic handling of electrodes, steel and slag sampling, bath temperature measurement and tapping and deslagging operations.

These innovations immediately result in a decrease in the number of workers needed and they also dictate the future steps that should be taken in order to achieve a fully automated management of the whole EAF steelmaking cycle.

4.4 Personnel and Organisational Aspects

Smaller size of manufacturing lines, introduction of various continuous or semi-continuous processes and closer links among the vartious plants have led to a drastic personnel reduction in the latest minimills.

The Tokyo Steel's new minimill at Utsunomiya will be producing 0.8 Mt/y of long products with no more than 150 workers (production and office workers plus contractors' employees) corresponding to 5000 tons per worker or 0.4 man-hours per ton (mpt).[3]

As an example, only two operators are needed to manage the double electrode DC furnace and the ladle-refining furnace. All scrap-handling operations from scrap yard to furnace charging are automated.

Further to personnel reduction, new-generation minimills favour a light and flexible organisation structure with only three layers of management.[6]

At Gallatin steelworks (Dofasco/Co-Steel) with a rated melting capacity of 1.2 Mt/y of flat products (240 employees): level 1 is for senior management, level 2 includes the employees that plan the operations and level 3 is composed of hourly workforce.[23]

4.5 Environmental Impact

Harmonic, flicker. The effects of the power supply (AC or DC) on flicker and harmonics has already been discussed in Section 3.1.

The extent to which the charging mode (top charging or continuous charging) influences these two parameters should now be examined. The top charging mode creates, mainly during the boring and melt-down stages, large and sometimes erratic variations in current and voltage levels due to shortcircuit with the scrap.

Conversely, with continuous charging (Consteel, Double Electrode DC furnace and, potentially, Contiarc and Comelt) the electric arc is far more stable given the presence of a constant flat bath condition.

These aspects have been studied extensively at Florida Steel on a 45 t/24 MW AC-EAF equipped with a 54 t/h Consteel system by operating the furnace under the same transformer and electric network conditions in two different modes, i.e. with top charging and with continuous charging.

It was established that the power quality impact of the network with regard to voltage flicker and harmonic current distortion was considerably reduced when operating in the continuous charging mode. In particular, voltage fluctuations were 30–40% less in magnitude than those in the top charging mode.[24]

Waste gas treatment. Waste gas treatment and controlled emission of off-gases into the atmosphere are regulated by ever stricter laws from which no derogation will be provided in the future.

Main pollutants are:

- VOCs, NO_X, CO
- dioxins (polychlorinated dibenzo-p-dioxins: PCDDs)
- furans (polychlorinated dibenzo-furans: PCDFs)
- EAF dusts.

VOCs are composed of organic substances, coming from the scrap; NO_X are nitrogen oxides mixtures that arise at temperatures above 1000°C.

Dioxins and furans can be formed by several organic substances in the range between 200 and 480°C. Once they have formed, dioxins and furans must be fully oxidised at high temperature together with intermediate organic compounds to generate CO_2 and H_2O.

If combustion is not correct, the resulting substances must be cooled down across the thermal range between 500–600°C (depending on the authors) and 200°C with at least 300°C/s gradient, in order to prevent new dioxins from being generated.

All the above substances are hazardous for the environment, although in differing degrees. They are emitted in different ways according to the type of EAF in use and especially according to the presence of burners, post-combustion equipment or scrap preheaters.

For a standard (e.g. single-shell) EAF, the best solution is currently an oxidising combustion at 950–1000°C inside and/or outside the furnace followed by appropriate cooling and gas-treatment practices.

CO, whose content in the gas emissions must be low for environmental reasons, can act as an indicator of an efficient combustion of organic substances. When CO content is around 50 mg/Nm^3, one may be reasonably certain that VOCs, dioxins and their causes, i.e. organic reagents, are efficiently combusted.

For furnaces equipped with scrap preheaters it is worth stressing that the more we optimise scrap preheating to the detriment of off-gas temperature, the more it will be necessary to recover the heat balance for ecological reasons. A correct solution involves finding a good compromise between these two requirements.

Careful monitoring or pretreatment of scrap is however needed in order to minimise the presence of organic substances.

Scrap pre-treatment. The physical and chemical pre-treatment of scrap before entering the furnace could have some harmful effects on the environment. This treatment that is an integral part of scrap recycling generates waste for dumping with an additional cost.

The solution suggested by Intersteel is to treat this waste in the preheater of the Consteel Furnace for co-generation of hot water, steam or electric energy, depending on local situations.

Dust generation. The technologies based on scrap preheating (e.g. Consteel, Shaft Furnace, etc.) are characterised by less dusts in the waste gases.

In the Consteel process dust settles in the scrap due to the low velocity of the waste gas inside the preheater. At Florida Steel, with Consteel the dust production has decreased by 30%.

In the Shaft Furnace the scrap in the shaft acts as a filter and reduces waste gas dust load by 20%.

Similar if not better results are expected from the double electrode DC furnace and the Contiarc furnace. In Contiarc, according to the packed density of the scrap, a portion of the coarse dust is captured and it is possible to expect up to 40% less dust in the waste gas and an accumulation of volatile elements (Zn, Pb, Cd) in the filter dusts.

5. EAF DUST TREATMENT

Measures for developing EAF dust treatment technologies are intensifying worldwide. Their objective is to find an alternative to dumping and be economically profitable.

These mainly hydro- or pyrometallurgical technologies are based on the recycling/recovery of materials – first of all zinc in the form of either metal or oxide, Cr and Ni ferroalloys – from the EAF dusts that arise from stainless steel manufacture.

The Ezinex process by Engitech Impianti is an instance of a hydrometallurgical treatment based on selective extraction of heavy metals with Zn deposit on cathodes by hydrochloric electrolysis. A production unit capable of treating up to 10,000 t/y of EAF dusts has recently come into stream at Ferriere Nord at Osoppo of Pittini group.[25]

The solution by Acciai Speciali Terni is based on a pyrometallurgical plasma-arc process developed together with Multiserv of HARSCO Corporation.

5.1 Ilserv Process

Acciai Speciali Terni represents an interesting example of ecological steel making. As shown in Fig. 10, in addition to the primary flow made by scrap, liquid metal, slabs and coils, there is a secondary flow made by: EAF dusts, slurries and mill scales. These residues are treated in the Ilserv furnace that can be considered as an integral part of the Steelworks.

The Ilserv plant, based on a plasma-arc technology, has the capacity to treat 20,000 t/y of residues where EAF dusts are prevailing. These dusts coming from the production of stainless and electrical steels, contain generally iron oxides, MgO, ZnO, CdO, SiO_2, Cr_2O_3 and NiO.

The feed to the furnace consists of a mix of residues, coal and a flux material, which is introduced throug the roof (Fig. 10). The plasma arc is generated by a single graphite electrode fed by a 7 MW DC power unit, at up to 20,000 amps. The electrode is provided with an axial hole where nitrogen is introduced to produce plasma-arc.

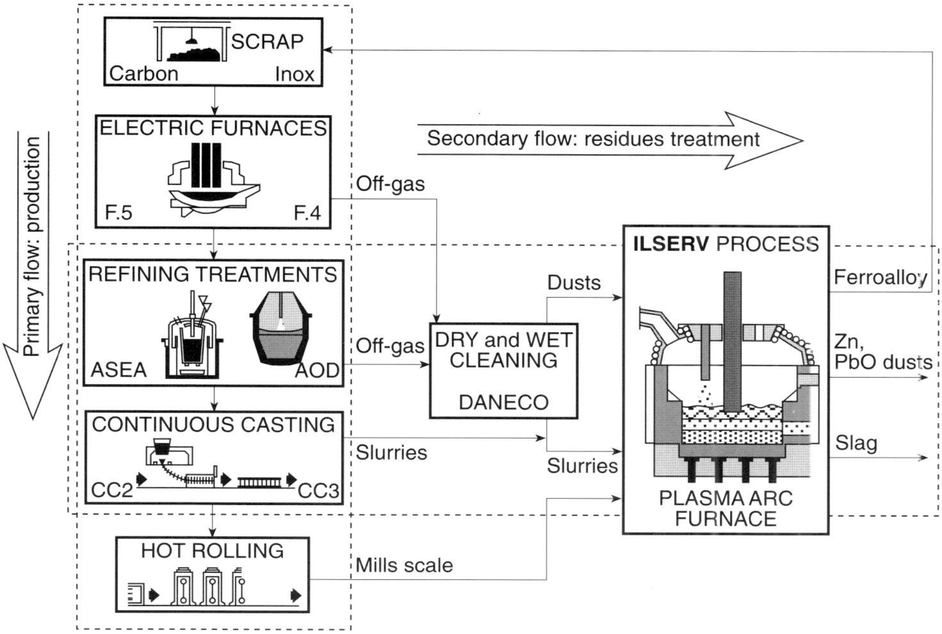

Fig. 10 Ecological Steelmaking in Acciai Speciali Terni.

An off-gas exhaust and filtration system allows the recovery of Zn, Cd and Pb oxides and other solid particles entrained in off-gases.

The products of the Ilserv process are:

- a ferroally containing Cr, Ni and Mn which is reintroduced into the production cycle of the EAF;
- an inert slag, which is sent for dumping,
- volatile metal oxides dust (Ilserv dust).

Table 8(a) reports typical Ilserv operating and analytical parameters.[26]

The Ilserv dust is subsequently treated in a water leaching plant, for the elimination of some alkaline elements, namely Cl, Na, K and S, in order to increase the commercial value of these dusts.

Table 8(b) reports the analytical composition before and after the leaching treatment.

The Ilserv process is suitable also for treating dusts that arise from carbon steel manufacturing cycles.

The treatment of Cr- and Ni-free dusts can be made economically attractive by preliminary enrichment (recharging in furnace, air-tight furnace etc.) of the dusts before loading them in the Ilserv furnace. Figure 11 shows schematically the following three stages of the complete Ilserv process:

Table 8(a). Ilserv furnace operating parameters (1995).

Total oxide: EAF's dusts (71.5%), rolling mill scale (22.1%), slurries (6.42%)	Kg	13132
Coal + lime	Kg	3545
Charge mix per heat	Kg	16677
Tap to tap time	min	287
Charge mix feeding rate (on feeding time)	t/h	3.92
Mean electric power	MW	5.28
Specific electric power consumption (on oxides)	KWh/t	1868
Productivity (charge mix on tap to tap)	t/h	3.68
Analytical parameters		
Slag chromite content	%	4.2
Alloy chromium content	%	7.58
Chromite content in the charge mix	%	6.44
Alloy carbon content	%	4.93
ZnO content in ILSERV dusts	%	71.52

Table 8(b). Chemical analysis.

	ZnO	Pb_3O_4	C1	Cr_2O_3	NiO	CdO	Fe
Dry dusts	71.5	4.3	3	1.18	0.39	0.04	5.48
Slurries after leaching	78.3	4.9	0.06	1.26	0.25	0.04	7.43

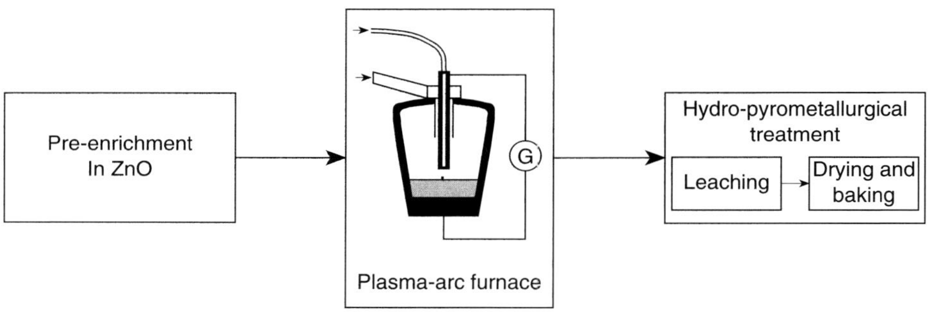

Fig. 11 Treatment of EAF dust to obtain high-ZnO products.

- enrichment of EAF dusts;
- treatment of dusts in an innovative Ilserv furnace;
- final hydro-pyrometallurgical treatment for conferring high commercial value to the dusts.

6. CONCLUDING REMARKS

As the amount of steel produced via the EAF increases worldwide, rapid evolution is taking place in EAF technology and in the way steel is manufactured.

Since the DC EAF was introduced in the late 1980s as an alternative to the prevailing AC EAF, new furnace technologies have been launched: twin-shell furnaces, scrap continuous charging, scrap preheating systems, etc.

While the confrontation DC EAF vs. AC EAF is fought on equal ground for single-shell furnaces, assessing the situation of the other known furnaces is more complex, although the DC power supply seems to provide more technological potential.

As a consequence of this evolution, the consumption of electric energy has been reduced and in several steelworks it is now far below 400 kWh/t. A further reduction down to 300 kWh/t is expected with the new Tokyo Steel furnace which has recently come into stream.

Various factors have generated these results: reduction in heat losses and power-off time, scrap preheating, increase in thermo-chemical energy by larger use of O_2 that in some cases is already as high as 50 Nm^3t.

Lower electric energy consumption allows for a smaller power supply both in AC and DC mode, while ensuring the same productivity.

The above actions, together with the use of high arc voltages with foamy slags, have resulted in significant electrode savings. As a matter of fact, electrode consumption is now 1 to 1.5 kg/t for DC furnaces and 1.8 to 3 kg/t for AC furnaces.

Further evolution and diversification of the EAF are due to the use of increasingly differentiated charge materials, which often requires furnaces to be tailored according to customer requirements.

The EAF is reaching a highly automated level to cover a number of operations, from scrap preparation to scrap transportation and furnace charging and process automation in general. The recourse to robots is also increasing for electrode preparation, O_2 and C injection, sampling and temperature measurement.

Existing situations (e.g. Nucor, Tokyo Steel, etc.) highlight that the EAF is becoming increasingly integrated with a manufacturing line featuring a continuous, uninterrupted flow of material operated by a small number of workers.

In new minimills personnel are organised into highly flexible work structures on few hierarchical levels.

Environmental impact is playing an increasingly important role. It incorporates the aspects associated with flicker, noise, waste gas treatment, controlled gas emissions, EAF dust treatment, disposal of wastes from steelmaking and the provision of a pleasant working environment.

REFERENCES

1. P. Heinrich: 'Minimills and new developments in DC electric arc furnaces', *MPT International*, February 1995, 34–45.
2. R. L. W. Holmes *et al.*: 'Replacement of BOF Steelmaking with a Twin Hearth DC EAF at Unimétal', *I&SM*, September 1995, 63–67.

3. B. Berry: 'The Nucors of Asia: Hanbo and Tokyo Steel', *New Steel*, January 1996, 60–69.
4. P. F. Marcus and K. M. Kirsis: 'Scrap is King', *Steel Survival Strategies Conference*, New York, June 1995.
5. Jo Isenberg-O'Longhlin: 'Morphing The Melt Shop', *33 Metal Producing*, March 1996, 29–90.
6. G. Dollè: 'L'evolution du four électrique à arc', *La Revue de Métallurgie – CIT*, Octobre-Novembre 1995, 1177–1186.
7. J. Vervacke: 'Development of DC – electric furnaces with arc control', *Metallurgical Plant and Technology International*, 1994, **4**, 48–53.
8. M. A. Pedroza *et al.*: 'Steelmaking Design for Flat Products at Hylsa', *I&SM*, November 1995, 33–36.
9. S. C. Beason: 'Operational Experience of Charter Steel's Conductive Refractory Bottom DC Arc Furnace', *I&SM*, June 1994, 36–38.
10. B. Read: 'Reducing flicker and harmonics from EAFs', *New Steel*, January 1996, 30–34.
11. M. Kratz: 'Effect of Adding Reactance in AC Electric Arc Furnace', *5th European Electric Steel Congress*, June 1995, 273–282.
12. A. E. S. Pengelly: 'Conference report on 5th European Electric Steel Congress', *Ironmaking and Steelmaking*, 1995, **22**(4), 273.
13. S. Köhle and B. Kleimpt: 'Scrap preheating: technical, economic and ecological assessment', *B.F.I.*, 1996.
14. T. Furukawa: 'Nippon Steel sells Posco a Twin-vessel electric furnace', *New Steel*, May 1994, 34–35.
15. 'On the DC learning curve in Turkey', *Steel Times International*, November 1995, 29–30.
16. J. A. Vallomy and A. Manenti: 'Implication of Continuous Steelmaking in EAF on Charge Materials Handling and Feeding', *XXVII Seminar on Melting, Refining and Steel Solidification*, Belo Horizonte, Brazil, May 1996.
17. T. Furukawa: '5,000 annual tons per worker at Tokyo Steel', *New Steel*, May 1995, 40–41.
18. W. Hofmann and W. Reichelt: 'Contiarc – A new scrap melting technology', *Steel Times*, March 1996, 103–106.
19. W. Reichelt and W. Hofmann: 'Contiarc – A new scrap melting technology', *MTP International*, February 1996, 56–60.
20. Battelle Europe: *Iron & Steel Industry*, Multiclient Study, Geneva, 1993.
21. H. G. Müller *et al.*: 'New Concepts in Electric Steelmaking', *5th European Electric Steel Congress*, June 1995, 491–501.
22. O. Takeuchi *et al.*: 'New Scrap Melting Furnace with continuous preheating system', *I&SM*, April 1995, 13–14.
23. A. Ritt: 'Out of the gate at Gallatin', *New Steel*, July 1995, 20–24.
24. Intersteel Technology, 'Power Quality Impact of the Consteel Process'.
25. C. M. Beckovich: 'Treating and Managing Electric Arc Furnace Dust', *I&SM*, April 1996, 39–42.

26. ECSC Multinational Project, 'Ecological and Economical Electric Arc Furnace Steelmaking'.
27. *The Steel Industry in the year 2005*, International Iron & Steel Institute, Brussels (1995).
28. F. Bonomo: 'Le Materie Prime nell'Acciaieria Elettrica', *CSM*, January 1996.
29. W. D. Huskonen: 'The New Look at Hysla', *33 Metal Producing*, October 1995, 27–32.
30. M. Beirer *et al.*: 'Operational Results of a Double Shaft Furnace at SAM Montereau', *I&SM*, May 1995, 49–54.
31. J. P. François *et al.*: 'The new DC electric steel plant at Unimétal Gandrange', *MPT International*, 1996, **1**, 40–44.
32. S. R. Mendis *et al.*: 'Investigation of Transmission System Voltage Flicker Due to Multiple AC and DC Furnace Operation', *I&SM*, October 1995, 77–83.
33. R. Scott: 'The debate between AC and DC', *New Steel*, May 1995, 44–48.
34. H. J. Heine: 'Notes on AC vs. DC for EAF Steelmaking', *33 Metal Producing*, October 1995, 71–73.
35. Jo Isenberg-O'Loughlin: 'Thinking Exothermically', *33 Metal Producing*, March 1995, 29–62.
36. J. Wypych *et al.*: 'Commissioning and 1st Year of Operation the Number one Slab Caster in Poland', *CCC '96*, Linz, May 1996, 1–9.
37. A. E. S. Pengelly: 'Perspective for steel at the millennium', *Iron Steel International*, 1996, 5–8.

CSP TECHNOLOGY: AN UPDATE

W. HENNIG, F. HOFMANN, and B. KRÜGER

SMS Schloemann-Siemag AG, E. Schloemann Strasse 4, Postfach 230229, D-40088 Düsseldorf, Germany

Some 7 years ago the first compact strip production (CSP) plant was commissioned at Nucor Steel, Crawfordsville, Indiana. Now, after years of production experience, the technology has achieved worldwide recognition (Fig. 1).

There are now six production plants and one CSP caster for special trials in operation. Nine further plants are being built (Fig. 2). The combined total capacity of these plants is 22 Mt/year (Fig. 3).

The reference list (Table 1) as well as showing the main technical data also indicates some trends. Due to the more economical production with the two strand solution three plants are already running with two strands (Nucor, Crawfordsville; Nucor, Hickman; and Hanbo). For two further plants the second strand is under negotiation (Hylsa and NDIL).

In the cases of ASM, Malaysia, and Baotou, China, the plants will run from start up with two separate strands. Another trend is to achieve thinner final gauges (h_{min} ~1–1.2 mm) and to use newly developed equipment like the liquid core reduction (LCR), electromagnetic brake (EMBr), hydraulic oscillator, and others.

Up to the present, the total production of hot rolled strip of all the CSP plants in operation is about 12 Mt/year (Fig. 4). The production curves of the new CSP plants in operation are steeper compared to Nucor, Crawfordsville as a result of experience and technological improvements. The highest production achieved in the first production month was by Steel Dynamics, with 27 kt in January 1996, which by April 1996 had already risen to 62 kt.

We shall now examine the reasons behind the success of CSP and where this novel technology will go in the future.

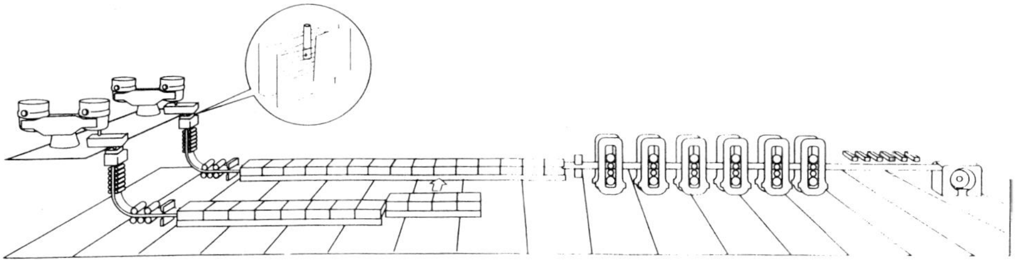

Fig. 1 Basic principle of CSP technology.

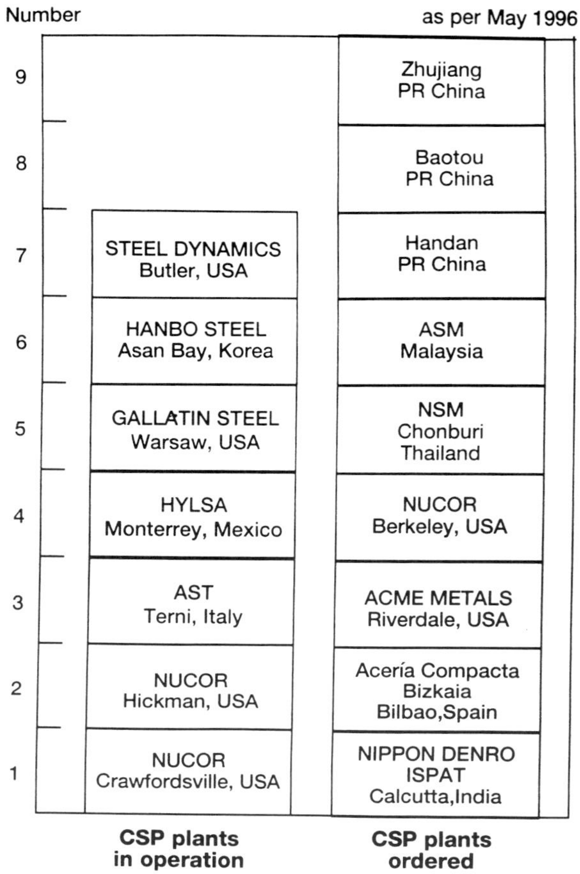

Fig. 2 The evolution of the CSP technology.

BASIC PRINCIPLES OF CSP TECHNOLOGY

Principle no. 1 is the absolute minimisation of the number of process stages (Fig. 1). The route from liquid steel to finished coil includes only those process stages which are physically necessary, such as:

- casting process in the CSP caster to produce thin slab
- temperature equalisation in the roller hearth furnace to achieve a very uniform slab temperature
- rolling process in the rolling mill to achieve the desired final dimensions and tolerances
- cooling on the run out roller table to obtain the desired metallurgical properties
- coiling process on the coiler to ensure a readily transportable product.

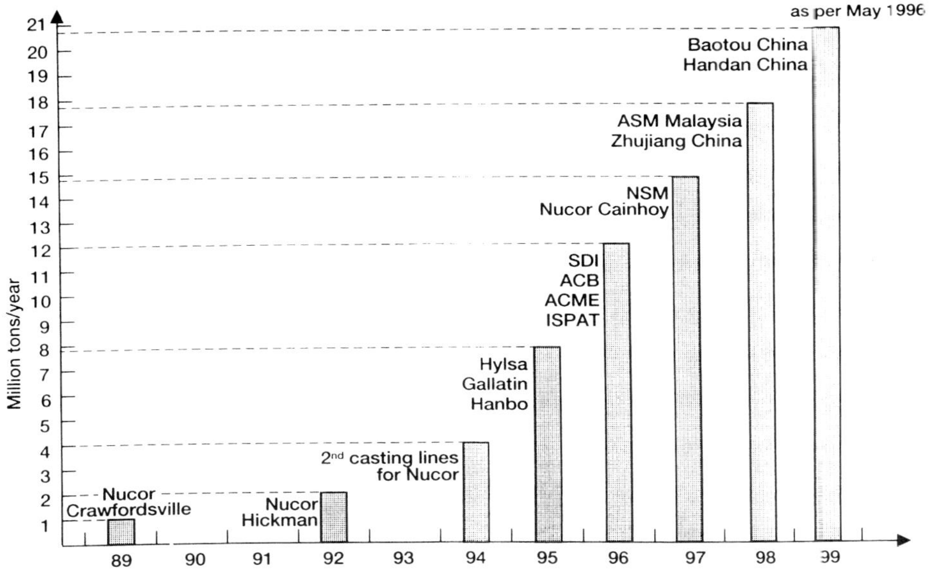

Fig. 3 CSP plants: Installed capacity (cumulative).

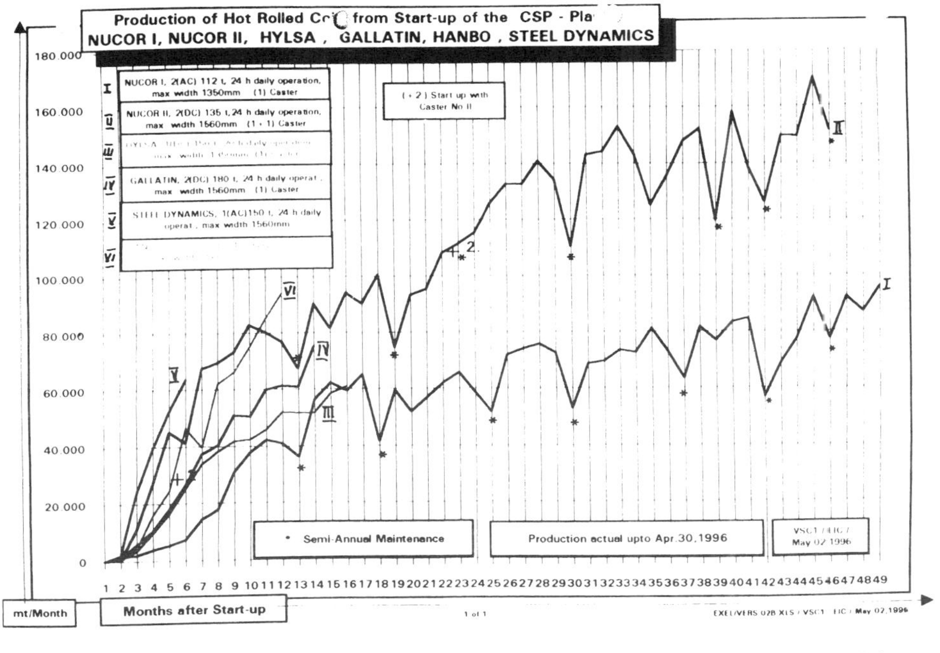

Fig. 4 Production of hot rolled coils from startup of CSP at Nvcor I, Nvcor II, Hylsa, Gallatin, Hanbo, Steel Dynamics.

Table 1. Reference list of CSP plants

Customer (date of order)	Caster (no. of strands)	Rolling mill, no. of strands	Heat size, t	Rated capacity, t/year	Slab dimensions, mm	Min, strip thickness, mm	Startup
Nucor Steel Crawfordsville, IN, USA (Dec. 1986)	1	6 (sixth stand ordered July 1995)	112	900,000	40–50 × 900–1350	1.76 (1.43 at 1000 mm)	July 1989
Nucor Steel Crawfordsville, In, USA (May 1993)	1	–	112	900,000	40–50 × 900–1350	–	April 1994
Nucor Steel Hickman, AR, USA (Nov. 1990)	1	6	135	1,000,000	50 × 1220–1260	1.95 (1.5 at 1150 mm)	Aug. 1992
Nucor Steel Hickman, AR, USA (June 1993)	1	–	135	1,000,000	50 × 1220–1560	–	May 1994
AST Terni, Italy (Sept. 1991)	1	–	120	Alloyed, stainless, and electrical steel	50 × 1000–1560	–	Dec. 1992
Hylsa SA Monterrey, Mexico (Dec. 1992)	1	6	150	900,000	50 × 790–1350	1.7 (1.2 at 1140 mm)	Feb. 1995
Hanbo Steel Asan Bay, Korea (Feb. 1993)	1	6	150	1,000,000	50 × 900–1560	1.7 (1.5 at 1200 mm)	June 1995
Hanbo Steel Asan Bay, Korea (Dec. 1993)	1	–	150	1,000,000	50 × 900–1560	–	Sept. 1995
Gallatin Steel Co. Warsaw, KY, USA (March 1993)	1	5 6 (1996)	175	1,000,000	50 × 1000–1560	2.1 at 1250	April 1995
Steel Dynamics Inc. Butler, IN, USA (March 1994)	1	6	150	1,200,000	40–70 × 990–1560	1.35 (1.20 at 1240 mm)	Dec. 1995

Table 1. Reference list of CSP plants – *continued*

Customer (date of order)	Caster (no. of strands)	Rolling mill, no. of strands	Heat size, t	Rated capacity, t/year	Slab dimensions, mm	Min, strip thickness, mm	Startup
Nippon Denro Ispat Calcutta, India (March 1994)	1	6	220	1,200,000	50 × 900–1560	1.6 (1.2 at 1150 mm)	Aug. 1996
Acería Compacta Bizkaia SA Bilbao, Spain (June 1994)	1	6	132	910,000	53 × 790–1560	1.5 (1.3 at 900 mm)	July 1996
ACME Metals Inc. Riverdale, IL, USA (July 1994)	1	7	86	1,000,000	50 × 900–1560	1.55 (1.25 at 900 mm)	Oct. 1996
Nucor Steel Berkeley, SC, USA (March 1995)	1	6	150	1,500,000	50 × 1220–1680	1.45	Oct. 1996
NSM Chronburi Thailand (August 1995)	1	6	180	1,200,000	45–70 × 1220–1600	1.35 (1.20 at 1250 mm)	Aug. 1997
ASM Malaysia (January 1996)	2	6	160	2,000,000	50 × 900–1560	1.6 (1.2 at 1150 mm)	March 1998
Zhujiang Steel China (April 1996)	1	5	150	792,000	50 × 1000–1300 (future 40 with LCR)	1.27 (1.90 at 1300 mm)	Nov. 1998
Baotou Iron and Steel China (April 1996 LOI)	2	6	230	2,000,000	50–70 × 980–1560	1.20 at 1000 mm 2.10 at 1560 mm	Still open
Handan Iron and Steel China (April 1996 LOI)	1	1 rougher (5 finishing stands)	101	1,230,000	70 × 900–1680	1.20 at 1500 mm 2.43 at 1680 mm	Still open
Total	21	90		20,732,000	–	–	–

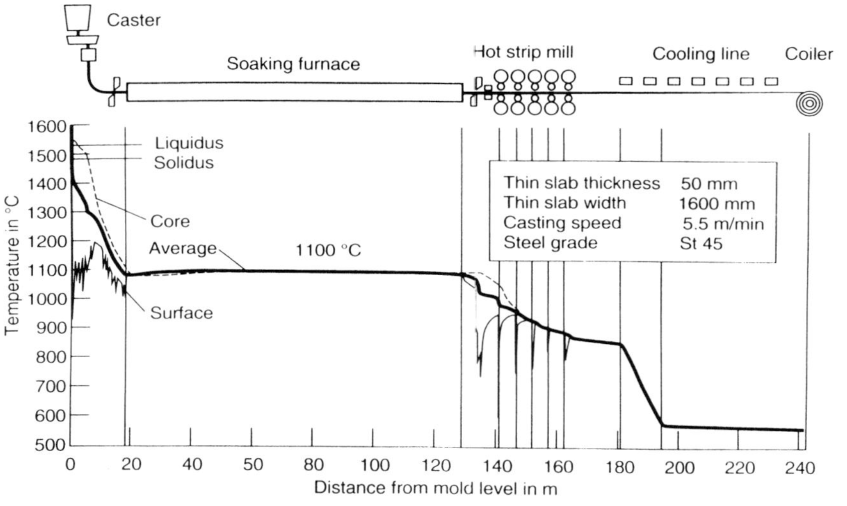

Fig. 5 Temperature levels in a CSP plant.

Each of these process stages forms a technical unit in itself and hence permits the number and costs of operating and maintenance staff to be minimised. Investment requirements are coming close to the achievable minimum and profit loss due to yield losses cannot be cut any further.

Principle no. 2 is the minimisation of energy consumption (Fig. 5). Thanks to the minimisation of the process stages, the energy demand for the entire production process can be covered by the original melting heat without any need to reheat the material. Apart from a desired temperature difference of 20 K on furnace entry side, all that needs to be compensated for are the no load operation losses of the roller hearth furnace, which are in the order of 25 kWh t^{-1}.

Principle no. 3 is to keep the thin slab at an absolute constant temperature. The holding time of the thin slab in the soaking furnace, 15–20 min, ensures an almost perfect temperature uniformity over the width, thickness, and length. Therefore, the subsequent process stages, rolling and cooling, can be performed at constant speed. There is no need for any corrective measures such as strip edge heating, temperature speedup, or interstand cooling. The product consequently has a microstructure of very high homogeneity. The geometrical precision with respect to thickness and profile easily reaches minimum values, much better than with strips rolled in conventional hot strip mills.

The level of success that the CSP concept has achieved in practice is attributable to the specifically developed and patented mould (Fig. 6), the main characteristic being the funnel shaped expanded opening of the mould in the upper part to provide sufficient space for the installation of a large diameter submerged nozzle.

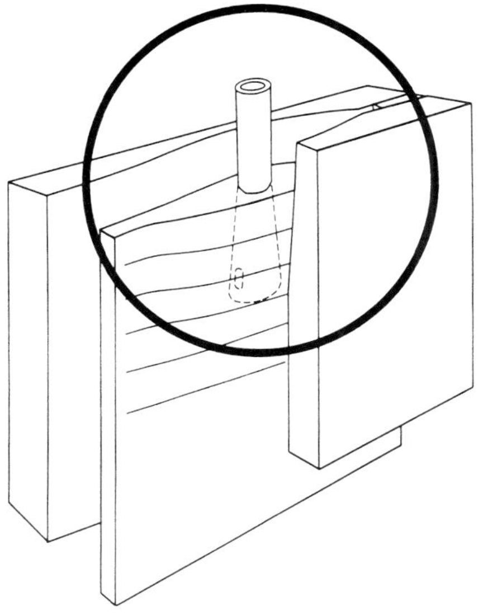

Fig. 6 CSP mold.

With regard to the shape of the mould, the dimensions and flux conditions in the submerged nozzle, and optimisation of the entire system (namely the mould and submerged nozzle), extensive model calculations have been made and a water model test carried out, in both of which constant optimisation steps are performed in line with further development of the CSP technology.

This not only relates to the development of optimal mould and submerged nozzle shapes, but also to the overall design and use of novel and improved materials to increase the mould and submerged nozzle service lives, and hence to reduce the specific production costs and augment the productivity of the process by casting long sequences.

The funnel shape of the mould is proportional to the shrink of the strand shell whereby stresses in the strand shell in the horizontal direction are avoided or minimised. Moreover, the distance between submerged nozzle and copper wall is sufficiently long to reliably avert the formation of bridges and to have a sufficient amount of heat available in the bath for casting powder melting. This shows that the SMS developed mould type offers decisive benefits over a parallel walled mould as also used for the production of thin slabs.

The currently used CSP moulds are made from a copper-silver alloy and feature a long service life. The mould surface is not plated.

As already mentioned, the submerged nozzle in particular has been undergoing constant development and optimisation since the commissioning of the first CSP plant, attention being directed on two focal points:

- optimisation of outflux velocities, outflux pulse, and flow pattern in the mould
- achievement of as high a submerged nozzle service life as possible.

Thanks to the use of corresponding materials and a constant variation of the submerged nozzle immersion depth during the casting process, the sequence length could be further increased up to 18 heats/sequence (some 2 kt) which no longer necessitates flying tundish or submerged nozzle changes (Fig. 7).

Figure 8 illustrates the benefits of the CSP mould-submerged nozzle combination as used today, compared to the parallel mould. The large outlet opening also permits a sufficient steel flux in the case of cloggings and the bath level area in the submerged nozzle zone is large enough for melting a sufficient amount of casting powder and avoiding the formation of bridges.

The narrowest gap between submerged nozzle and mould wall is 26 mm and, at a casting speed of 5 m min^{-1}, the strand shell grows to *c.* 5 mm in the immersion depth area of the submerged nozzle. The geometry of the SEN is the subject of permanent improvements, to optimise the flow pattern inside the SEN and inside the mould.

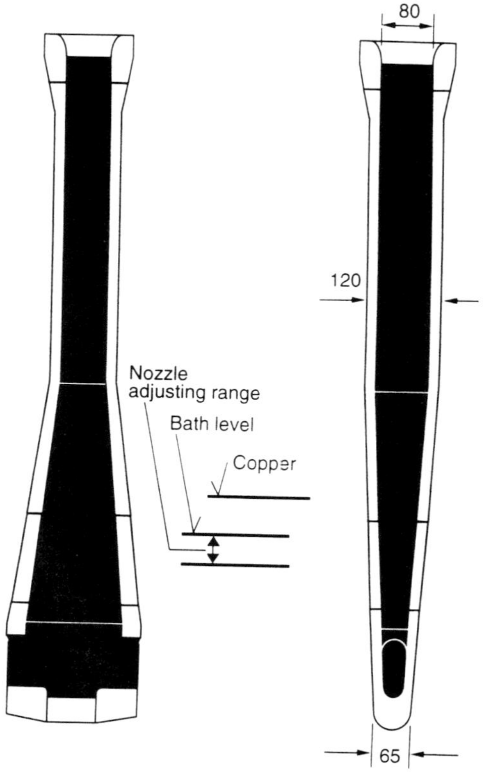

Fig. 7 CSP mould SEN – Submerged Entry Nozzle.

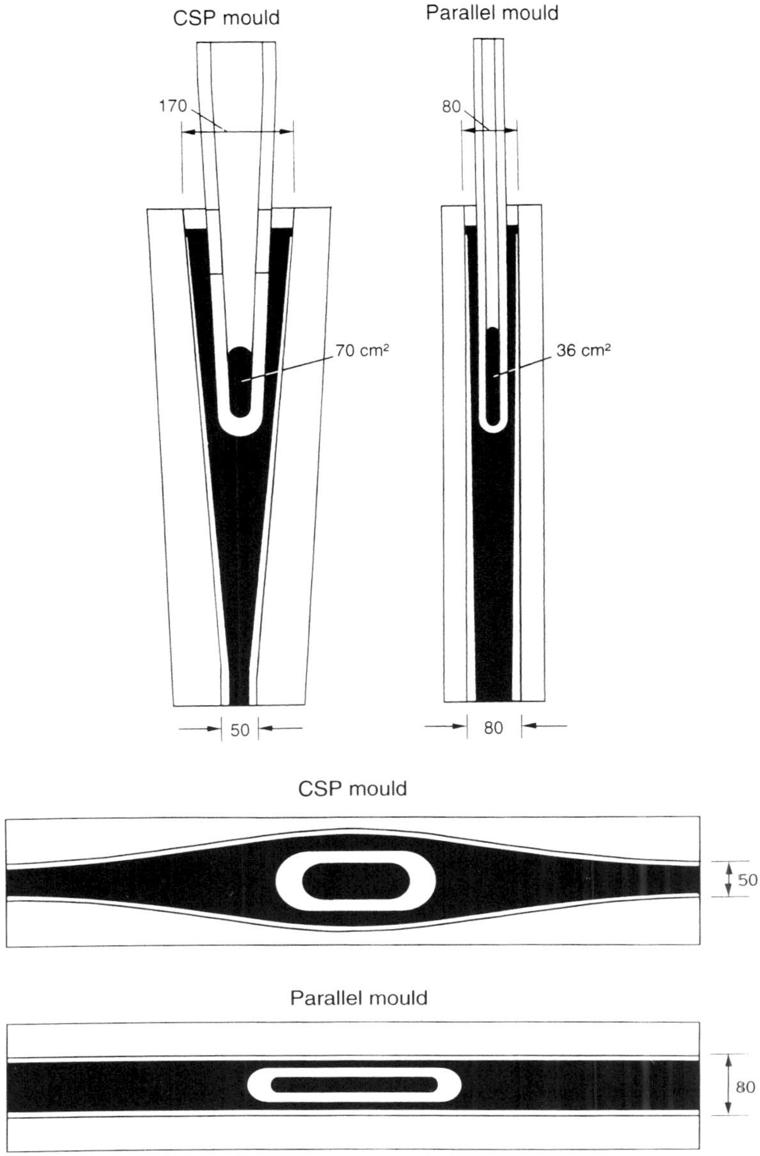

Fig. 8 Comparison between CSP mould and parallel mould.

ENHANCEMENTS IN CSP CASTER

Electromagnetic Brake (EMBr)

It is quite natural that the possibilities of steadying the flows inside the mould via the configuration of the submerged nozzle are somewhat restricted (Fig. 9).

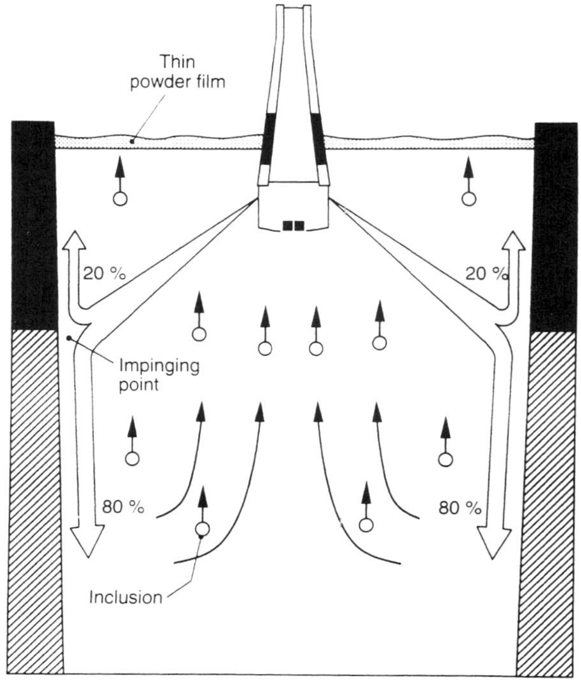

Fig. 9 Steel flow without electromagnetic brake.

The main stream remains directed against the narrow sides where *c.* 80% flow down and *c.* 20% up. Both part streams dissipate in a return-flow back to the centre, which even in the lower part is directed upwards due to the steel surplus in the narrow-side area. A uniformly slow steel downflow which would be desired to minimise inclusions is not practicable.

An electromagnetic brake is expected to minimise the number of inclusions (Fig. 10). Such an electromagnetic brake produces a magnetic field in the upper part of the mould, the intensity of which is variable. The inflowing liquid steel crosses this magnetic field and induces voltages and hence currents which short circuit in the liquid steel. The resultant forces slow down the steel flow velocity and, in the ideal case, produce the desired uniformly distributed downflow.

The effect and/or efficiency of the electromagnetic brake is visible from the extent to which the flatness of the bath level is influenced (Fig. 11). With the brake cut out, the upflowing part stream produces a convex bulging of the bath level in the narrow-side area. This bulge clearly reduces when the brake is cut, from which one may conclude that the downwardly directed, invisible part stream, too, is markedly steadied. Directly and clearly visible, however, is the positive impact on the flatness of the bath level and the resultant better uniformity of the slag layer on top of the bath.

The positive effects of steadying the flow inside the mould are expected to

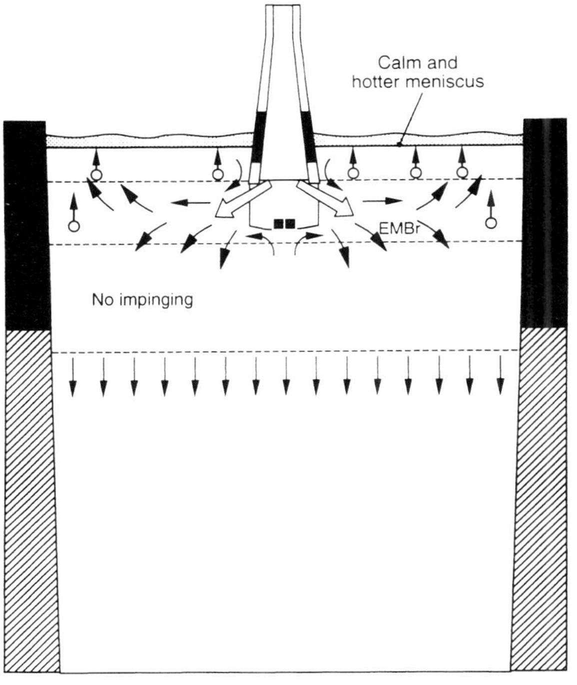

Fig. 10 Steel flow with electromagnetic brake.

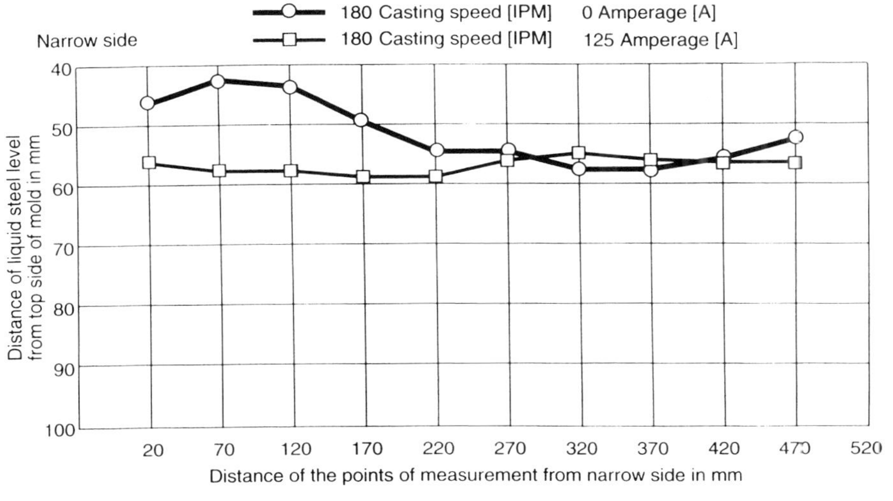

Fig. 11 Measured effect of the electromagnetic brake.

comprise not only a further minimisation of inclusions, but also a stepwise increase of the potential casting speed. Development work continues.

Liquid Core Reduction (LCR)

Another approach to steady the liquid steel inside the mould appears to be the so called liquid core reduction (LCR) (Fig. 12). This process adopts the known funnel shaped CSP mould, but with an outgoing thickness increased, for example, from 50 to 70 mm. Downstream in the mould, the strand shells are pressed together from 70 to 50 mm in proven manner with a liquid core left in between. This liquid core reduces during further strand guiding until the thin slab has fully solidified.

The positive effects of this process are expected to come from the increase of the mould volume. Opening the mould from 50 mm to, for example, 70 mm allows more steel in the mould. The flow velocities in the mould reduce while the throughput remains the same, which leads to calmer conditions.

Basically, the liquid core reduction process pursues the same goal as the optimisation of the submerged nozzle or the provision of an electromagnetic brake, to minimise the amount of inclusions and/or to increase the casting speed. Which of these measures or which combination of them will actually

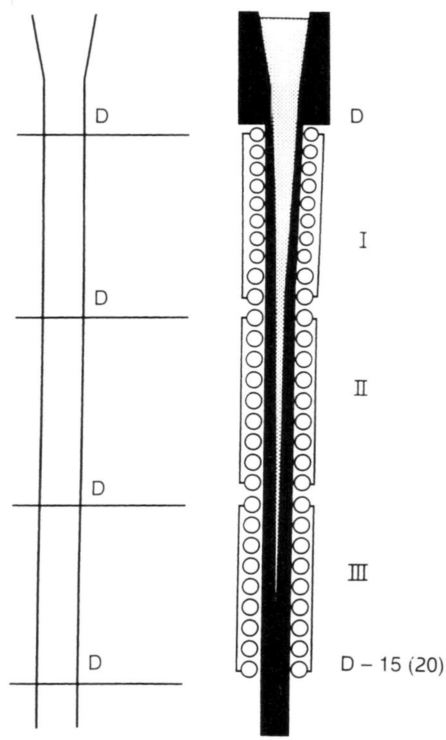

Fig. 12 Principle of liquid core reduction.

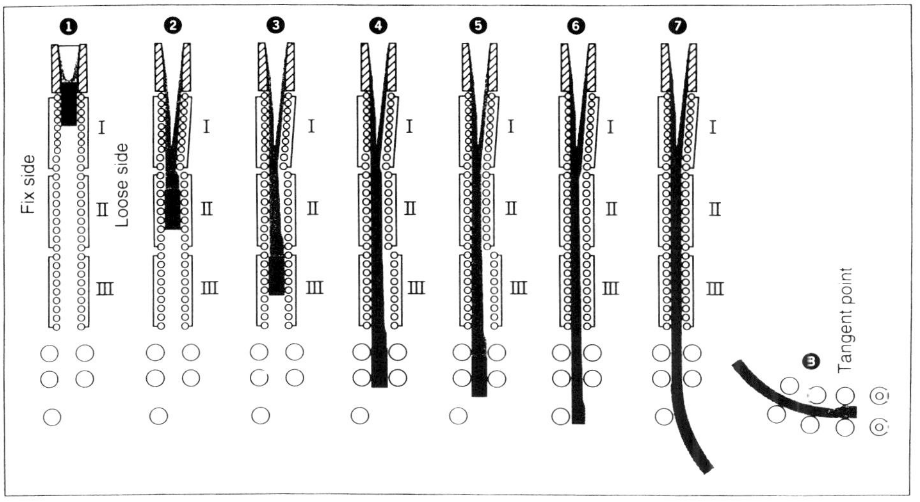

Fig. 13 Starting procedure with liquid core reduction.

produce the desired success remains to be seen from continuing development work.

For the adoption of liquid core reduction, a special casting-start process has been developed which comprises the following phases (Fig. 13):

A casting start on to the inserted dummy bar and advancing the strand until the thin slab head has passed segment 1

B swinging segment 1 into position and further reducing the strand thickness while carrying the strand on until the target thickness has been reached

C segmentwise adjustment of the strand segment pinch rolls and straightening rolls to suit the reduced strand thickness.

The goal of this process is to minimise the length of the lost head which, owing to the process, exits the machine with the exact thickness given in the parallel part of the CSP mould.

Hydraulic Mould Oscillator

A hydraulic mould oscillator is expected to produce gradual improvements in the area of contact between strand skin and mould wall. Conventional mechanical mould oscillation by means of an eccentric drive enables just a near sinusoidal mould motion. It is well known that reliable feed of molten casting powder into the area of contact between strand shell and mould wall is practicable only in the very area where the difference between mould speed and strand absolute speed is positive (Fig. 14). The balance time T_N one oscillation cycle therefore must be kept as short as possible. In the case of mechanical oscillation (upper part of figure), T_N reaches a defined value that cannot be

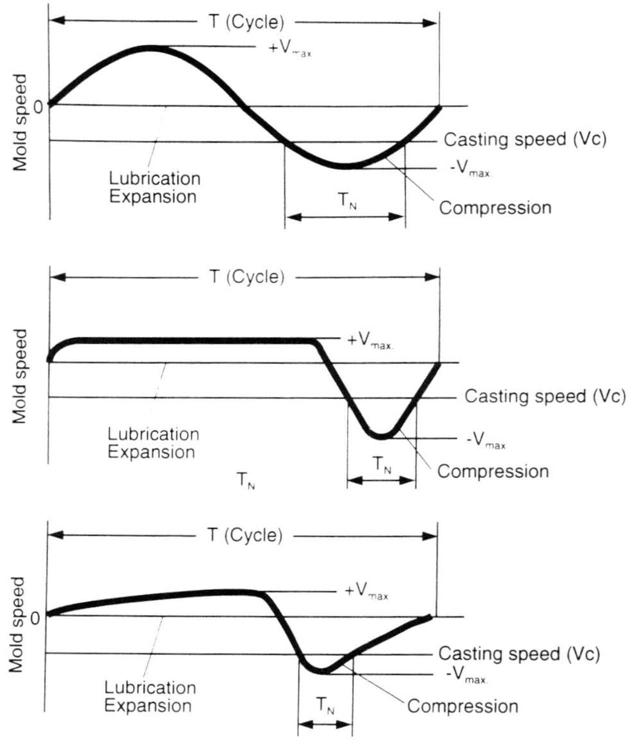

Fig. 14 Oscillation curves using hydraulic oscillation.

reduced any further. Hydraulic oscillation, on the other hand, allows this value to be reduced through an adequately selected motion function (centre of figure). The flexibility in selecting the motion curve enables a further reduction, however, by renouncing the actual desired constant speed difference in the positive part of a cycle (lower part of figure).

ENHANCEMENTS IN THE CSP MILL

High Pressure Descaler

A high performance and effect of high pressure descaling ahead of the rolling mill is of outstanding importance to the surface quality of the products. The design provided for the first production facility turned out to be inadequate. The new design being adopted today produces much better results. The distance between nozzles and thin slab surface has been reduced, which also necessitated smaller internozzle spaces. Therefore, a totally novel type of nozzle attachment was adopted. The original 250 bar high pressure pumps were replaced with max. 400 bar variable pressure pumps. In regular production, optimal results have been achieved with pressures between 300 and 320 bar.

Fig. 15 Newly developed high-pressure descaler.

The inside of the high pressure descaler features an innovative design to keep waste water away from the thin slab surface (Fig. 15).

The result of all of these measures in the course of production is a smaller number of defective areas detected on the strip surface through quality control. As an example, see Fig. 16 showing the test results from rolling steel grade AISI

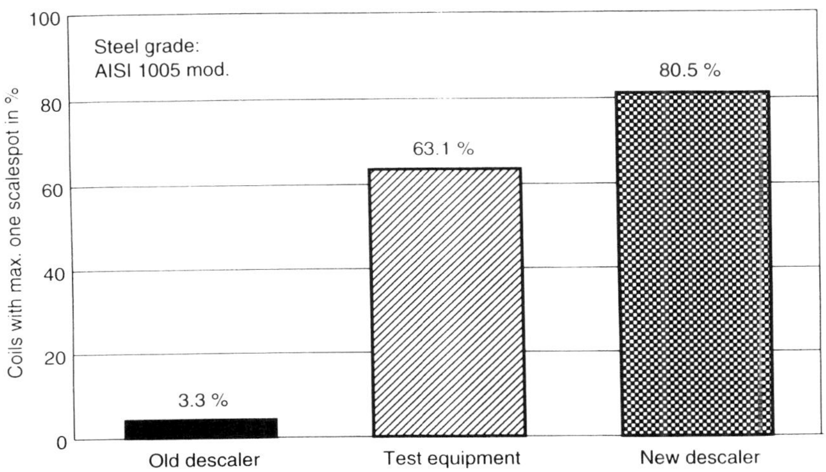

Fig. 16 Production test with the newly developed high-pressure descaling unit.

1005. By means of a test installation, the quota scored of the desired quality class, permitting a maximum of just one scale affected area in each coil, could initially be increased from 3.3% to 63.1%. By putting into practice the results and findings gained from that test installation, another increase to 80.5% could be achieved.

The results from the Hylsa plant for low carbon grades after pickling show a defect free surface (rolled in scale) in proper descaling conditions.

Roller Side Guide (Edger)
The edger is positioned in front of F1. There are several reasons to install an edger in front of the CSP rolling mill, for instance:

- higher reduction ratio in F1 due to the recrystallisation of slab edges; mainly for crack sensitive steel grades and high strength material, it is an advantage to use such an edger with a reduction of about 10 mm each side
- improvement of the slab edge quality for sensitive grades
- improved guidance of thin slab in rolling mill.

The edger can be installed, in any case at a later stage, depending on the product mix of the CSP plant.

STEEL GRADES AND QUALITY ACHIEVED

It is undisputed that the economy of a production process is fully utilised only if the products actually satisfy the quality requirements of the market by all standards. When developing a novel technology and technique, one is always facing great impatience here. Even at an early stage, while the novel technology still has to prove itself, one is expected to answer questions regarding the mechanical properties and the surface quality of all grades or materials processed in conventional hot strip mills. In particular, the casting process and its parameters need further time for optimisation, when an acceptable surface and a low inclusion microstructure can be reliably reproduced. However, results were already available in 1996 from actual production or extended trial operations, these results covering the major part of material groups customarily processed in hot strip mills (Fig. 17).

Groups 1–10 are currently produced in the CSP plants which are already in operation. The casting technology for steel grade groups 11 and 12 will be developed at AST in Terni, Italy and at ACME, Riverdale, USA after commissioning of the new CSP plant.

The significant quality features of some individual groups are presented below.

Group 1	Weldable unalloyed steel for structural use C≤0.22% Mn≤1.50%
Group 2	Medium carbon steel C≥0.22% Mn≤1.0%
Group 3	Mild carbon steel for cold forming and cold rolling C≤0.10% Mn≤0.45%
Group 4	High-strength microalloyed steel for cold forming and cold rolling C≤0.10% Mn≤1.4% Cb/V/Ti
Group 5	High-strength phosphorus alloyed steel C≤0.10% Mn≤0.70% P≤0.12%
Group 6	HSLA steel with improved atmospheric corrosion resistance C≤0.22% Mn≤1.25% Cr/Ni/Cu/P
Group 7	Microalloyed HSLA steel C≤0.22% Mn≤1.8% Nb/V/Ti
Group 8	Silicon alloyed steel for electrical application C≤0.02% Si≥1.2%
Group 9	Alloyed steel for heat treatment C≥0.25% Cr≥0.50%
Group 10	Stainless Cr-steel C≤0.10% Cr≥13%
Group 11	Stainless CrNi-steel C≤0.10% Cr≥17% Ni≥8%
Group 12	High-carbon steel C≥0.60% Mn≥0.45%

Produced on CSP plants (Groups 1–10)

Cast on pilot plants (Groups 11–12)

Fig. 17 Range of hot rolled coil made on CSP plants.

Group 1: Weldable Unalloyed Steels for Structural Use
These steels feature the following characteristics:

- a well balanced yieldpoint/tensile strength ratio
- a good elongation after fracture
- a defined notch impact strength (the structural steels produced on CSP plants feature a favourable combination of all these properties)
- the R_e/R_m is at the upper limit of the range to be expected for this steel group
- the values assessed for elongation after fracture also are within the expected range
- the notch impact strength values are equivalent to those of steel grades rolled on conventional hot strip mills.

Group 2: Medium Carbon Steels
The point that we would like to emphasise for this wide range of grades is the dependence of the elongation on the carbon equivalent. The values found are

above the applicable standards. Tests reflect a grade featuring 0.30%C, and there is a uniform distribution of pearlite with only a little banding.

Group 3: Mild Carbon Steels for Cold Forming and Cold Rolling
Plates cold rolled from CSP hot strip feature excellent deep drawing and stretch forming properties, which is confirmed by the r and n characteristic values for the deformability as well as by the experience that the processing industry has gained from the fabrication of complicated parts. The values for elongation and strength conform to those achieved in conventional hot strip production.

The point that the processing industry is particularly emphasising is the homogeneity of the fine grained microstructure over the length, width, and thickness of the strip.

Group 4: High Strength Microalloyed Steels for Cold Forming and Cold Rolling
These grades are used for drawn parts in the automobile and machine building industry. Demanded here are the following characteristic properties:

- a high yield point/tensile strength ratio
- good formability
- uniform mechanical properties in the longitudinal and transversel directions.

Group 7: Microalloyed HSLA Steels
The following are typical properties of these steels:

- high tensile strength and yield point
- good notch impact strength values at low temperatures
- very fine grained microstructure.

These properties are achieved by adding such microalloying elements as vanadium and niobium which at temperatures below 1000°C retard the recrystallisation and hence counter act the grain growth.

Results from rolling HSLA steels having yield points over 460 N mm^{-2} (65 ksi) at a finish rolling temperature of 850°C on a CSP mill with just four mill stands are as follows:

- the achieved microstructures are extremely fine grained
- the notch impact energy values found at minisamples to applicable ASTM Standards are favourable
- there is no need to set higher preheating temperatures requiring more energy, nor to add costly titanium.

It should be emphasised again that the results, especially the fine grained microstructures, were achieved with just four mill stands.

Group 8: Silicon Alloyed Steels for Electrical Application

It is well known how extremely difficult it is to roll these steels, with a Si content of up to 33, on conventional hot strip mills. The precipitations provoked by slab cooling must be dissolved by means of extremely high preheating temperatures of more than 1300°C. This gives the slab a very coarse grained microstructure so that it is susceptible to cracking during the subsequent rolling process.

This problem does not exist in a CSP mill. Rolling in one heat precludes the formation of disturbing precipitations. The well known problems during hot rolling such as edge cracks and transversal cracks do not exist. The hot strips produced feature a great uniformity of mechanical technological properties both in longitudinal and transversal directions.

ROLLING MILL

It is standard that CSP plants come with profile and flatness control systems adopting the well known CVC technology for adjustment (Fig. 18). Profile and flatness control enable defined profile heights to be set for the finished strip irrespective of the distribution of pass reductions in the mill. Elastic deformation and thermal crown and roll wear are assessed by the profile and flatness control system and compensated when presetting the roll gaps.

As a result, the specified strip profiles are kept within close tolerances. Here, too, the constant temperature over the strip length proves its positive impact. There are hardly any profile variations over the strip length.

A further development of profile and flatness control is the combination with the so called contour and flatness control, the purpose of which is to eliminate any profile anomalies which, as a result of roll wear and thermal crown, are likely to occur when rolling is done either with constant width or with changing widths over extended periods of time.

Requirements

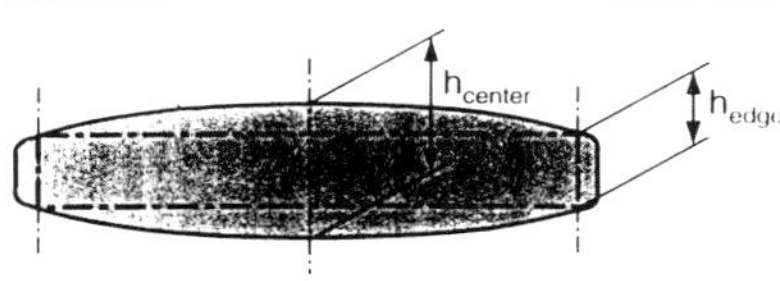

Absolute profile height (mm) = $h_{center} - h_{edge}$

Relative profile height (%) = $\frac{h_{center} - h_{edge}}{h_{center}} \cdot 100$

Apart from thickness and width tolerances, hot strip consumers are increasingly demanding the observance of specified strip profiles in narrow limits as well as the obtainment of strip flatness.

Presetting

Targeted profile setting to suit intended use of hot strip.

Obtainment of strip flatness required.

Compensation of disturbance variables

Compensation of the undesired profile and flatness changes resulting from:

- Changes in the elastic deformation of the set of rolls
- Changes in the thermal crown
- Changes in the roll crown caused by wear

Fig. 18 Profile and flatness control, using continuously variable crown (CVC).

When rolling with constant width, the coaction of wear and thermal crown in particular produces strip edge ridges, a negative phenomenon which is prevented through the PFC–CFC control. Similarly good results are achieved when rolling with different widths.

In future, the combination of PFC and CFC will be standard for CSP plants and will markedly contribute to improving the geometrical quality of the products. As well as achieving this quality improvement, an additional effect may be to extend the time between 2 roll regrinds.

FIRST OPERATING RESULTS OF HYSLA PLANT

The following figures show some examples for operating results regarding performance and strip quality achieved.

Up to 11 heats/sequence were achieved during the first 7 months of operation (Fig. 19). With these results, the funnel shaped mould and the use of a rather conventional submerged entry nozzle have proven to be reliable for productivity and quality.

The CSP rolling mill is provided with a profile and flatness control (PFC) system adopting the well known continuously variable crown (CVC) technology for adjustment.

As a result, the specified strip profiles are kept within close tolerances supported by the positive impact of the constant temperature over the entire strip length and width. The plant is currently producing mainly hot band from

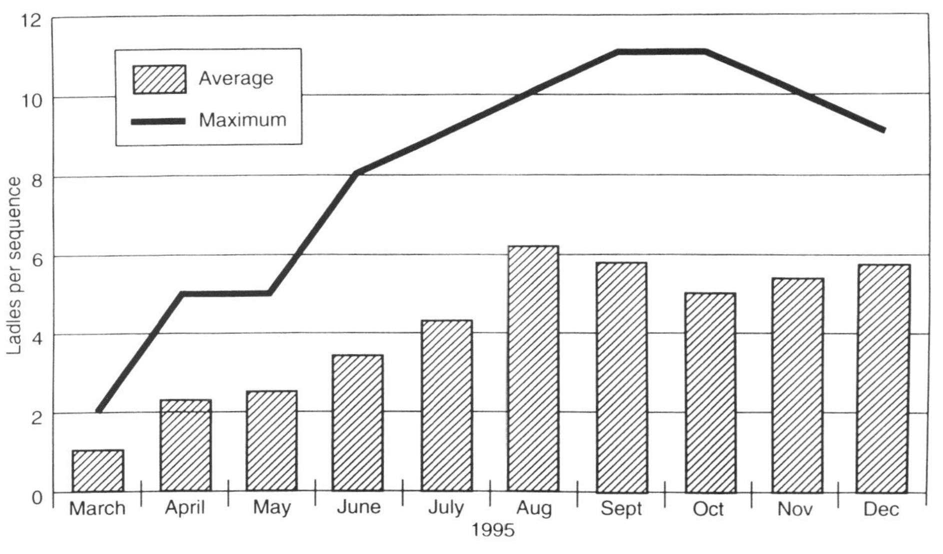

Fig. 19 Ladles per sequence.

900 to 1350 mm and from 1.4–12.7 mm thick. The 631 mm gauge was already achieved on a trial basis 6 months after start up. Prime quality reached up to 98% production within the first 4 months.

The main strategies for obtaining prime quality at Hylsa are based on steel cleanliness, meaning slag free tapping, argon bubbling, and control of sulphur, aluminium, and non-metallic inclusions.

Sulphur is controlled to a maximum of 0.012%, 0.007% being typical; aluminium is held within 0.025% and 0.045%; and non-metallic inclusions, in accordance with ASTM standards, are limited to a target value of 0.14%. Calcium injection is also very important for promoting globular inclusions.

Steel grades range from ASTM 1004 to 1035 except for peritectic grades, as well as API grade X65 and HSLA in most gauges and widths.

Stands F4 and F6 were designed with a closer spacing between the stands and smaller work and backup roll diameters. The resulting larger bite angle is required for rolling the minimum gauge of 1 mm.

The laminar cooling runout table was specifically designed for 631 mm strip gauge. A blower system underneath the measuring house will keep the strip tight to the table rollers until entering the laminar cooling section.

The coiler is provided with step control. The walking beam type coil handling system conveys the coils into the coil yard. There are coils stored by a computer controlled coil crane memorising the location for storing and removal of coils.

The total workforce consists of only 280 salaried staff and hourly wage operators, including melt shop with caster, hot rolling mill, and maintenance for the three production units. Most of the supervisory personnel have a degree in mechanical or electrical engineering. Eight employees are involved in quality control in all production units.

The worldwide demand for Hylsa products is proof of the quality oriented production: 80% of the production is being exported to countries such as Italy, France, England, and Germany, the remaining 20% to the USA and Asia. Phase II of the expansion programme for an annual production of 1.5 Mt of CSP products is under consideration.

CONCLUSIONS

The worldwide interest in CSP technology is reflected in the activities of SMS to improve the productivity of the plants, the quality of the hot rolled products, the product range, and the production cost. No fundamental modifications to the process technology are planned.

Extrapolating from the present situation and an assessment of further development, it can be expected that in the year 2000, 35–40 Mt of hot strip will be produced using thin slab technology worldwide. The CSP technology has found a firm place in this development and, 7 years after its industrial introduction, is a proven process for the production of high quality hot strip at a favourable total cost level.

REFERENCES

1. G. Flemming and W. Rohde: 'CSP technology – present state, current plans and expectations for the future', *AISE Speciality Conf.*, 11–14 June 1995, Chicago, USA.
2. W. Rohde and H. Wladika: 'Development of plant for hot strip production in the future' (translation of a paper), *Stahl Eisen*, 1991, **111**(1), 47–61.
3. G. Flemming, F. Hofmann, W. Rohde, and D. Rosenthal: 'CSP plant technology and its adaptation to an expanded production programme', *Metall. Plant Technol. Int.*, 1993, **16**(2).
4. F. Hofmann, W. Hennig, and R. Guse: 'CSP – a revolutionary technology for the hot strip production conquers the flat steel world', *Int. Symp. on Flat Steel Technology*, Cairo, Egypt, 23–25 October 1994.
5. A. Fernandez and F. Küper: 'Hylsa CSP plant, first operating results', *Technical Report CC and Rolling*, SMS Schloemann-Siemag AG, reprint of a paper represented at ATS Journées Siderurgiques 6–7 December 1995, Paris.
6. W. Hennig and F. Hofmann: 'CSP – technology the efficient way for producing high quality hot strip', *SEASI Mini Mill Conf.*, February–March 1996, Jakarta, Bangkok, Kuala Lumpur, Manila.

THE MINIMILL COMPARED WITH THE INTEGRATED CYCLE: ISP TECHNOLOGY

F. MAZZOLARI

Arvedi ISP, Via Acquaviva, 2 Zona Porto Canale, I–26100 Cremona, Italy

THE ARVEDI GROUP

The Arvedi Group's core business is represented by steelmaking activities divided into 3 business areas and 7 operating companies. In the flat rolled carbon steel area these include the companies Acciaieria ISP of Cremona and CSS. Among the companies represented in the carbon welded steel area are ATA, TL and Tubimar. Finally in stainless steelmaking the Arvedi Group includes the companies SIT, a manufacturer of stainless steel welded tubes, and Arinox, a manufacturer of ultra-thin cold rolled stainless steel strip.

The Group manufactures and markets over 1.1 million tonnes per year of which:

650 000 tonnes of carbon steel coils
340 000 tonnes of welded carbon steel tubes
27 000 tonnes of welded stainless steel tubes
16 000 tonnes of stainless steel cold rolled strip
110 000 tonnes of flat rolled products processed and marketed by the service centre.

The Arvedi Group employs 1700 staff and has a consolidated turnover of 1 200 bn. Lire. Its first two companies were established in 1963 and since then it has progressively acquired leading positions in the quality tubes sector, becoming one of Italy's most highly qualified producers. Since 1992 the Group has integrated upstream becoming a hot strip manufacturer through an innovative plant using ISP technology, jointly developed and patented with Mannesmann.

ISP TECHNOLOGY (IN-LINE STRIP PRODUCTION)

ISP (Arvedi-Mannesmann patent) is the most advanced and innovative of the near-net-shape technologies based on thin slab casting for the manufacture of flat rolled steel. The design for the new process came into being in 1988 through the successful blending of the thin slab technologies developed by Arvedi and Mannesmann, After a number of preliminary experiments conducted both at Mannesmann's works and at Arvedi's melting shop the development and fine-tuning of the technology were carried out from 1992 onwards at the prototype plant built by Arvedi at the new Acciaieria ISP di Cremona works – the first example in Europe of a mini mill for flat rolled products.

In 1994–95 the prototype plant completed the first phase of its development, reaching and in some cases exceeding its basic targets in terms of output and product quality. The Cremona plant is producing profitably and constantly improving its performance, prompting further technical development activity.

BASIC CONCEPTS OF ISP STRIP PRODUCTION

Figure 1 illustrates the compact lay out and the low investment costs of the ISP design. The plant lay out brings high energy savings and low production costs (Fig. 1). These result in a low production break-even point and a rapid return on investment without sacrificing product quality or environmental standards.

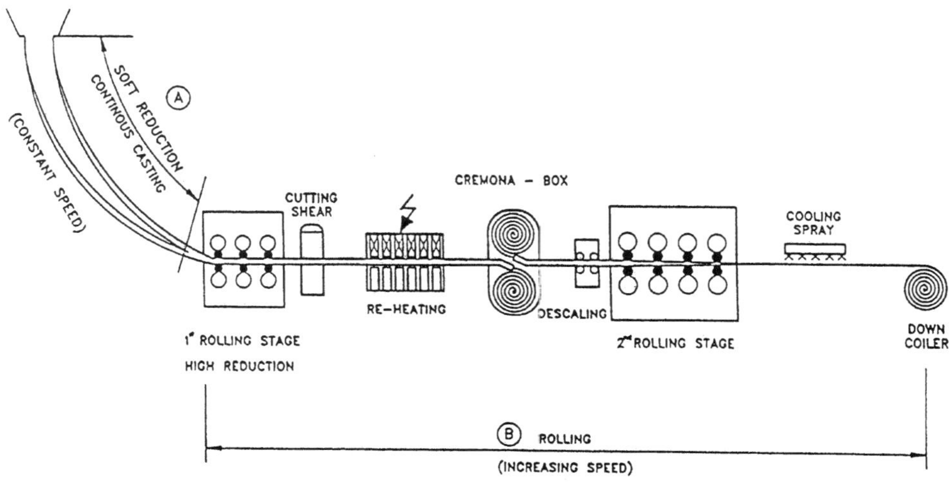

Fig. 1 ISP Technology schematic lay out.

The plant's main technical features are:

- Integration of the thin slab casting phase with the rolling stage, resulting in direct uninterrupted transformation of the liquid steel into coil;
- Cast-rolling involves liquid core thickness reduction (soft reduction) and immediate solid core rolling giving a thin slab 15–25 mm thick with the characteristics of a rolled product;
- Utilisation of the energy contained within the thin slab;
- Adjustment of the temperature prior to rolling through the finishing stands by using a high efficiency and rapid response induction furnace;
- In-line rolling through the finishing train at constant speed and temperature via the 'Cremona Box' buffer.

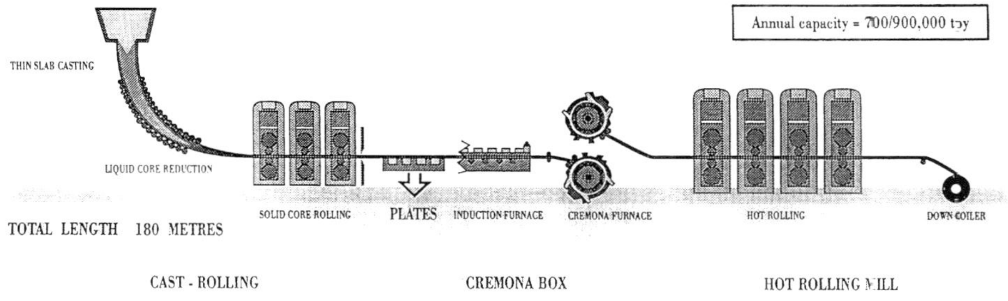

Fig. 2 ISP Process lay out.

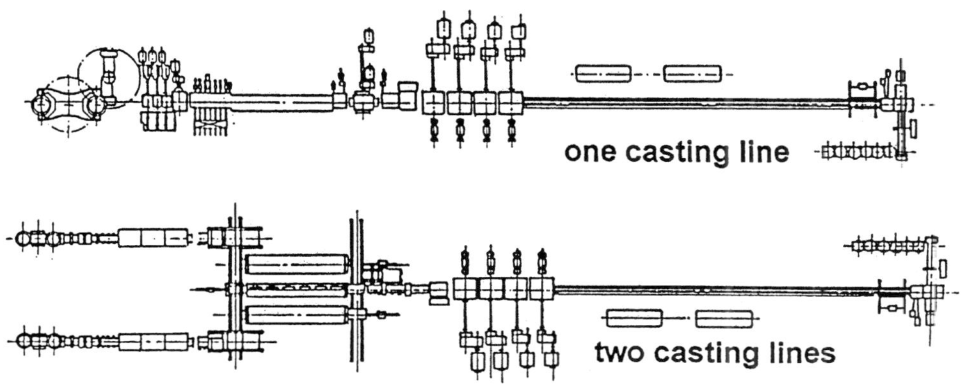

Fig. 3 Alternative lay outs for ISP plants

Figure 2 shows the sequence of rolling processes from thin slab to down-coiler.

The overall ISP plant is only 180 m. long from liquid steel to finished coil, with a production cycle of only 15 minutes. Production capacity is in the range of 700 000 to 900 000 tonnes. It is possible to double production capacity by adding a second continuous casting line at an additional investment cost of only 30% (Fig. 3). ISP technology in combination with an electric arc furnace (or new compact smelting systems) gives a slim, flexible mini mill only a tenth of the size of a traditional integrated plant. Acciaieria ISP of Cremona is among the first examples of a mini mill for flat rolled products.

QUALITY STANDARDS

ISP technology enables the production of a range of hot rolled coil products of high quality. All carbon steel grades can be produced. In addition stainless steels and high silicon electrical steels can be made. The resulting coils possess fine internal grain structure, excellent processability, improved fatigue resistance and outstanding toughness. The strip displays high dimensional accuracy, (thickness tolerance of less than 0.2%) and controlled crown profile (less than 30 microns). Thin gauges, typical of cold rolled products, can be obtained from hot rolling. The gauge is currently 1.2 mm. and expected to reach 0.7–0.8 mm. through developments now under way.

The ISP technology presents the possibility of developing products with particular mechanical characteristics. Thin slab casting results in a high casting and solidification speed, giving rise to a fine grain structure. The process of liquid core reduction gives a 'stirring' action during solidification. The 'transfer bar' has a precise and controlled profile which, in turn, gives constant thickness and profile in the final coil. Rolling at constant speed and constant temperature through the finishing train gives good dimensional accuracy and a homogeneous product. In principle it is possible to change the process parameters with every coil, given the great flexibility and good process control.

ENERGY SAVINGS AND INVESTMENT COSTS

The small size of I.S.P plant brings reductions in both energy consumption and capital outlay. The energy savings result from exploitation of the sensible heat contained in the liquid steel through successive stages of liquid and solid core reduction down to final rolling. A high efficiency induction furnace is used to normalise the temperature of the break down strip with heating confined to the passage of the strip. In-line finishing rolling takes place at constant temperature. This enables energy savings of up to 50%, compared with conventional slab casting and hot strip rolling. This is equivalent to approx. 1 tonne of oil saved for each 25 tonne coil. Similarly the investment intensity (per annual tonne of capacity) is from 2 to 3 times lower than for traditional integrated plants. Efficient production volumes of only 700 000 to 2 million tonnes are needed with ISP technology using one or two continuous casting lines. In contrast traditional integrated plants have a minimum efficient size of 3.5 to 4 million tonnes a year. As a result the total investment costs for mini mills employing ISP technology is 5 to 10 times lower than for new traditional integrated plants.

PRODUCTION COSTS AND BREAK EVEN LEVELS

Lower production costs result from high yields and lower manning levels offered by this route. ISP rolling gives better through yields (scrap and losses 30 to 50% lower than for conventional processes). It also displays better labour productivity – labour productivity is 0.5–1.1 man hours/tonne, depending on capacity compared to 2–2.5 man hours/tonne for traditional integrated plants. A compact plant, such as the ISP, functions on a lean organisational structure, yielding lower employment costs. These advantages coupled with lower capital costs give rise to reductions in conversion costs from liquid steel to finished coil of the order of 50%.

Such advantages mean production break even points are in the order of a few hundred thousand tonnes, whereas in a traditional integrated works they are in the order of millions. Consequently return on investment has fallen below ten years, whereas in integrated plants a positive return on investment may never materialise.

ENVIRONMENTAL CHARACTERISTICS

The compact dimension of the works reduces the impact on surrounding land allowing the adoption of a cost effective technology, with minimum environmental consequences. The ISP works in Cremona provides an example of clean and ergonomic steelmaking (Table 1). The fumes are completely captured, and

Table 1. ISP versus conventional systems

	ISP	Conventional
Capacity	0.8–2 M tpy	3.5–4 M tpy
Conversion liquid to coil (c.c. + rolling)		
Energy consumption	50%	100% cold charging 80% hot charging
Liquid to coil yield	1.025–1.030	1.040–1.060
Thin gauges produced without additional costs	1.2 mm (to be reduced to 0.7 mm with DSP techniques)	1.7–3 mm
Investment intensity	40–60% (according to size)	100%
Total length	180 m	800–1 400 m
Conversion cost	40–50%	100% cold charging 75% hot charging
Whole production cycle (from material to coil)		
Investment intensity	30–50% (according to size)	100%
Labour intensity (mhpt)	0.5–1.1	2–2.5

filtering enables emission of air with a residual dust content over 100 times lower than the European standard. No industrial water is discharged. Instead water is totally recycled within the plant. Noise at the works perimeter is similar to environmental background noise. Due to compact and automated lines most personnel work inside control rooms (Table 2).

IMPLICATIONS FOR THE FLAT ROLLED STEEL MARKET

The scaling down in size and investment costs that can be obtained with ISP drastically reduces entry barriers and opens up the sector to medium sized

Table 2. Summary of Arvedi ISP plant – key features

Current situation	
Plant equipment	EAF and I.S.P. process
Production (1995)	624 000 tonnes of carbon steel coils
Grades produced	• structural steel • rerolling and stamping steel • high strength steel • carbon steel for heat treatment • weather resistant steel • boron steel
Minimum gauge	1.2 mm
Thickness tolerance	≤0.02%
Employees	415 (1.1 manhours/tonne)
Total area	350 000 m^2 (60 000 m^2 covered)
Future developments	
Productivity (through increased performances)	over 800 000 tonnes/year
Extension of quality range	increase of percentages of microalloyed and hard steels; introduction of stainless and high silicon electrical steels (tests carried out have given positive results) and development of new steel grades
Development of gauges	minimum hot rolled gauges lower than 1 mm (0.7–0.8 mm) through the new DSP technique being developed in collaboration with Hoogovens. They will be able to replace a wide range of current cold rolled products

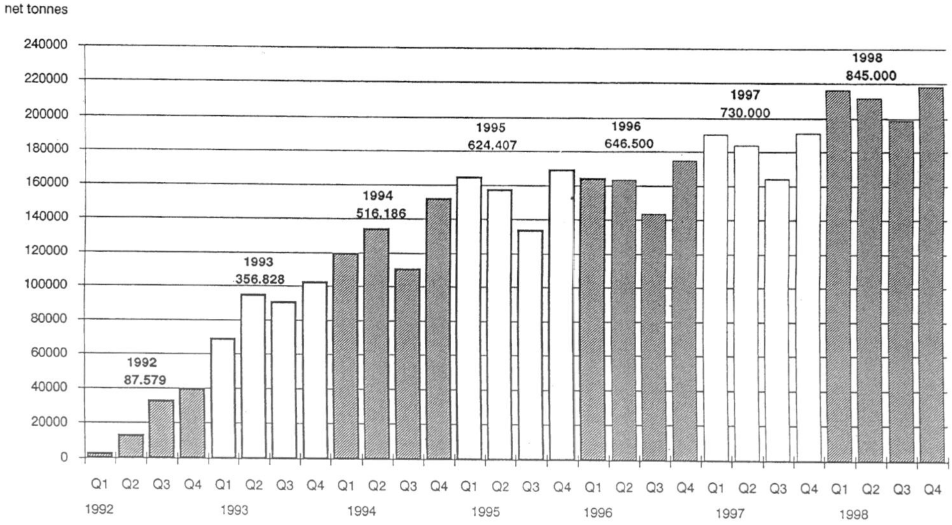

Fig. 4 Evolution of net production at the Arvedi plant: actual and forecast 1996–1998

PRODUCTION MIX 1995

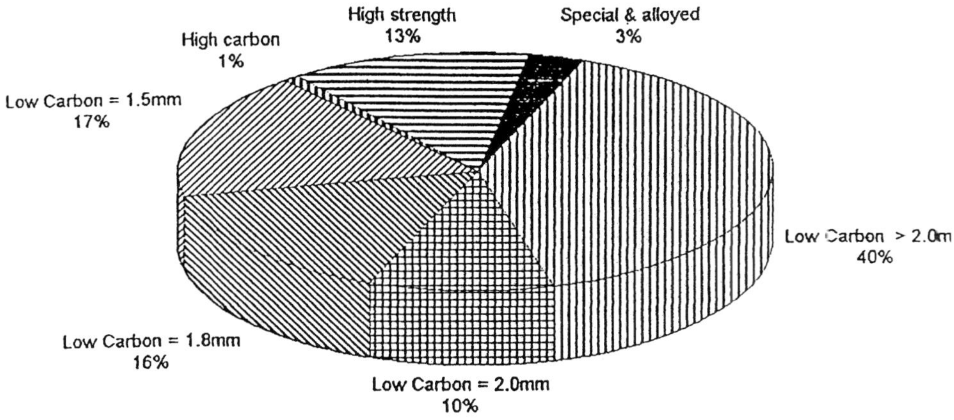

PRODUCTION MIX 1996

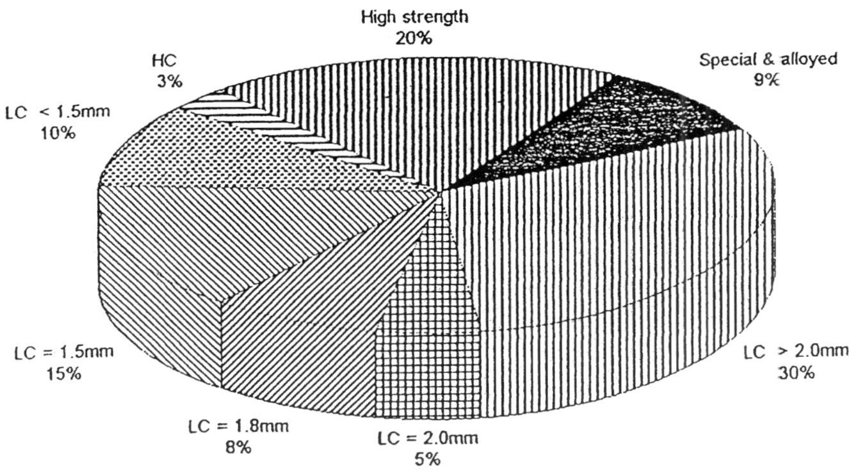

Fig. 5 Product mix at Arvedi 1995 and 1996

investors. Flat rolled products will no longer be the prerogative of the huge traditional integrated plants. Capital equipment for flat rolled steel production will be renewed with this technology, due to the rapid return of investment. Extremely flexible production units can be built, suited to small markets, the specific requirements of a particular location, or single companies. In the case of Eastern Europe, for example, it will be possible to renew plant despite limited availability of capital. The ISP process offers both flexibility and economy to these sorts of markets, particularly when making quality products, to support downstream rolling and finishing. As the ISP process develops, it should be possible to make even thinner gauges and more specialised steel grades, thus widening the range of competitive products offered by small plants in the strip market (Fig. 4). Further development could derive from the application of cast-rolling, which is at the heart of ISP technology, to plate and long products (Fig. 5). While up to now an exclusively European technology, ISP has considerable world wide potential. European companies may benefit both from its use as a replacement technology and from its sale world wide.

INNOVATIONS IN MATURE INDUSTRIES – STEEL AND BEYOND

U. WENGENROTH
Technische Universität München, D-80306 München, Germany

Mature industries are the least popular among politicians and the media alike. They are understood to be outdated, problem-ridden and, worst of all, the main sources of unemployment. Coal, steel and textiles, the pillars of European economic reconstruction after World War II, have collapsed during the sixties and seventies. Even the car industry, after a recovery from the traumatic OPEC price-shock, has run into trouble again. When it comes to industries – and not only there – 'maturity' is but a euphemism for 'decline' – a decline which first of all is measured in terms of employment. The statistical evidence of this form of decline is unequivocal, and by all means justifies the term 'deindustrialisation'.

Deindustrialisation and persistently high unemployment went hand in hand. For more than two decades European industry in total has been unable to create new jobs; instead it continued to shed labour and made recovery of labour markets increasingly difficult if not impossible. If only the process of deindustrialisation could be halted, full employment would be back within a few years because of the steady rate of expansion of the service sector.

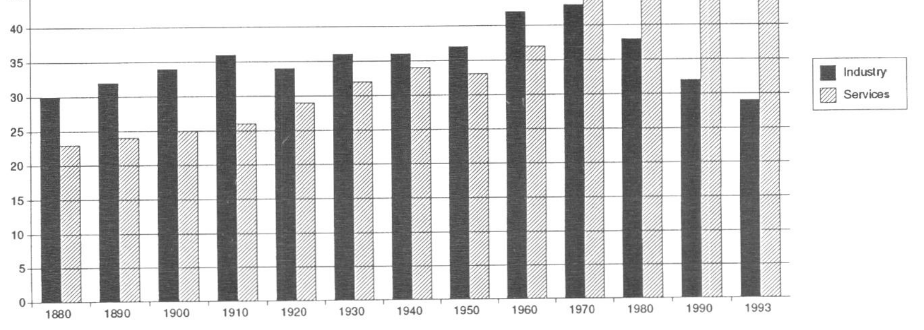

Fig. 1 Share of employment in West European industry and services.[1]

Industrial policy therefore appears to be at the core of all employment policies, and innovations are seen as the panacea not only for the labour market but for continued budget deficits as well. Forfeited taxes, unemployment schemes including early retirement, and subsidies for devastated areas have proved to be an enormous and unparalleled drain on public finances. Is deindustrialisation a solvable problem and do we have to find the way back to industrial society to recover social stability and more than minimum growth? Or is the contraction of the industrial sector irresistible and do we live through the kind of technological unemployment which has haunted sceptics of industrial development ever since the Industrial Revolution? Will 'jobless growth' be the imminent future of European economies?

The idea that technological progress will eventually destroy more jobs than it creates was a linchpin of the guilds' regulations of crafts and trades ever since early modern times and still resonated in Montesquieu's 'L'esprit de lois' (1748). In 1759 Josiah Tucker took issue with Montesquieu's technological scepticism and wrote what until today is the orthodox view of most economics textbooks: 'What is the Consequence of this Abridgement of Labour, both regarding the Price of the Goods, and the Number of Persons employed? The Answer is very short and full, viz. That the Price of Goods is thereby prodigiously lowered from what otherwise it must have been; and that a much greater Number of Hands are employed . . .'[2] This is the gospel of supply side economics: cheap mass production creates its own demand.

This gospel is also being sung in the more recent studies of the OECD, like the 'OECD Jobs Study'.[3] This study is full of complacency about the impact of technology on the labour market. Very much in the tradition of Josiah Tucker it is maintained that the labour saving effects of modern technology are being more than compensated for by increasing demand for new products. This was even more obvious in high-tech countries. The same kind of confidence in the job-creating potential of technological innovations had already pervaded a 1988-study with the programmatic title 'Industrial Revival Through Technology'.[4] In this study 'mature industries', which were seen to be characterised by a lack of innovativeness and a tendency towards standardisation of products and processes, were contrasted with technology-intensive hopefuls from high-tech branches of industry. The crux of decline was seen to be a dearth of innovations in these 'mature industries', their inability to provide for potential new outlets.

The background to this analysis is the product-cycle theory, according to which most industrial products first live through an expansion-phase to be followed by maturity and eventually contraction when they are being replaced by more recent and more attractive new products. If an entire branch of industry is lacking the necessary innovativeness to produce a continuous stream of new products, it will suffer from the product cycle and be swamped away in the same way as outdated goods. In this perspective 'old industries' are those being dominated by mostly outdated products. They have entered the stage of 'maturity' and eventually 'decline', and are being driven from the market by more dynamic new industries with a high rate of innovation. The state of 'hard-core mature industries' like chemicals, steel, aluminium, textiles &c is seen to be particularly desperate since they are being confronted with both stagnating or declining demand as well as 'saturated technology'.

The absence of or incapacity to grasp potential innovations are identified as being the main obstacles to 'industrial revival through technology'. Since almost 75% of all industrial activities in the EU merit the disquieting qualification 'maturity' the OECD-studies come to the conclusion that a shortage of innovations is the main root of disaster on the labour market and that only a most resolute innovation policy will remedy the situation. The objective of industrial policy is described as 'dematurity' through use of microelectronics, management-reforms, and a revival of innovativeness. What does history teach us about this strategy?

Old technologies and with them whole industries are stagnating and eventually declining, causing much distress among those whose livelihood depends on them and whose identities had been crafted around them – coalminers, dockers, steelworkers etc. These communities have problems indeed, often more or at least earlier than the companies they are working for. Whole regions are doomed for decline as are the old technologies they are burdened with. The centres of activity and growth go elsewhere. Old industries are those which have lost their dynamism, their innovative power and all the Schumpeterian charisma that go with it. Old industries are poor performers with labour and management blaming the slack of flexibility and foresightedness on each other and jointly demanding protection from the government for what after all was a pillar of strength in the past and should be preserved and rejuvenated for the future. Neoclassical theory has no mercy for these old industries and Schumpeterian innovation theory rejoices at their 'creative destruction' to make way for 'better' performers. Failure more than anything else is associated with old industries and the technologies underlying them. And in view of thriving new technologies it is failure to innovate. They have grown fat and complacent and obviously do not deserve better than to be pushed aside.

If one takes a closer look at these old industries and technologies, however, in many cases the presumed shortage of innovative spirit gives way to a picture of as lively and energetically driven R&D as one would expect from a hopeful new technology. The dividing line between old and new industries is certainly not one between technological stasis on one side and spurts of productivity and creativity on the other. This is not only true for the auxiliary technologies, which are often credited for the last patches of progress in sunset industries. Examples of auxiliary technologies are information technology coming to coal mines, steel works and textile mills, or cheaper transport of materials to and from these places, better packaging of product etc. It can be shown that the level of innovation was also exceptionally high for the very core technologies like spinning yarn, breaking coal and converting iron into steel.[5]

Like sailing ships came to perfection in the age of steam, and the thermal efficiency of charcoal-fired blast furnaces was much superior to that of their coke fired counterparts, or vacuum tubes never performed better than in the years when semiconductors already dominated information technology – like these and many similar examples there was never a more sophisticated and mechanised way of deep-shaft coal-mining than there is today in Belgium or at the Ruhr. Single British steel works are not only the last remnants in an industrial desert this dying industry has left, but also among the most efficient steel producers in the world – even more so than the Japanese. And individual German textile mills hold their own against any Asian competition being more afraid of their immediate neighbour's investment in the Czech Republic than anything happening east of Eden.

None of these industries, however, manages to turn inventiveness and innovative success into a medium- or let alone long-term growth perspective. Their managers show all the virtues of the ideal Schumpeterian entrepreneur but instead of creatively destroying their lazy competitors and establishing new industrial empires they hasten their own decline. Innovation alone – and technologically impressive innovation with great economic impact to be sure – does not seem to be a viable way out of industrial disaster. This is understandable in the case of sailing ships because steam navigation was so much more efficient in money-terms than wind-power, or in the case of coal with oil and natural gas catching a great share of the energy market.

In the case of steel and textiles, however, competing products are much less an issue. There is not so much pressure from alternative products. People continue to prefer textile fabrics over other forms of wrapping themselves, and steel continues to be the most popular construction material for engineers keeping both plastics and nonferrous materials at a safe distance. Although steel consumption did decline during the 1970s and early 1980s the worst may be over at least for some years and a fairly stable demand, though with some fluctuations, governs at the time and most likely into the next century. Recent trends even seem to favour these old industries: the richer people get, the more they spend on clothing, and mounting pressure on energy efficiency in the production and recyclability after use of consumer durables favours steel over its ecologically more harmful rivals.

What makes these two 'mature' industries, once the pillars of industrialisation, shrink in terms of employment and plant numbers is the very success of continued innovative activities. The capacity effect of technological progress has been out-accelerating demand for many years. In the case of steel it went through the centuries old strategy of enlarging the containers in which the transformation processes take place: blast furnaces, steel converters etc. With demand stagnating or even slightly declining since the late 1960s, the minimum efficient scale of a steel works has doubled and tripled during the last two decades with employment per plant being stable. Governments in Europe and the US paid subsidies to *slow* innovation in order to keep people employed.

In textiles, apart from multiplying output per spindle or loom up to the physical limits of the fibre, automation was driven to the extreme in achieving the 'no-touch-principle' in spinning. 'No-touch' meaning that no human being is interfering with the cotton from the bale to the finished yarn, everything is left with a self-governing set of interlocking machinery. The effect on employment is easy enough to imagine. Spinning, today, is one of the most capital intensive industries. Output per spindle has increased threefold in the 1970s alone. Similar things happened in the coal mines where slowly moving machines with several hundred meters wide shield supports are cutting and removing coal underground instead of hewers and haulers, making men and plant locations redundant at a much faster rate than the decline in demand would call for. Maintenance workers have begun to dominate these places not machine tenders or production workers.

'Technological unemployment' was one of the phrases coined for these effects of overshooting innovation. With no break on inventiveness, society's demand for goods and jobs fell apart, and the stricken industries themselves contracted, the firms cannibalising each other in a lengthy concentration process. What was meant to stem the tide of slackening demand and falling prices, more than anything else contributed to the decline of industries and regions. If there were – for some time and for a very limited number of players – individual ways out of this squeeze, the branch of industry as a whole did not escape its self-inflicted shrinking process. The inevitable outcome of continued inventiveness over a long time was the destruction of the very foundations on which the success and wealth of an industry and its underlying technologies rest. With modest increases of industrial productivity, as experienced in the earlier stages of industrialisation, we lived in a Tucker-world where growth in demand still outstripped growth of productivity. With high increases of productivity we entered a post-industrial era in the sense that industry is contracting because of over-shooting supply. Unlike the static Montesquieu world this scenario of deindustrialisation is compatible with increasing, though not sufficiently increasing demand. It is a world where industries are more competent than is good for their own survival.

They all seem to follow the path of agriculture – a much neglected example of technology related productivity gains and their effect on employment and the sector's relative importance in an economy. It was the very success of innovative farming that brought the percentage of farmers down to about 2% in spite of individually growing appetite. Agriculture for more than a century has been experiencing the fate of an old industry. Since farmers are doing their job so well – indeed too well as is shown by the huge stockpile of beef, butter etc. accumulated in the EEC's warehouses – we need so few of them. If this did happen to the production of food, there is no reason to believe that it should not happen – as is currently the case – to coal, steel, textiles, ships etc. with automobiles and other industrial products just about to follow. This is not strictly what is explained by the economists' product cycle theory – a new product coming to the market expanding its sales and eventually giving way to something new (this is the sailing ship versus steamship scenario). Steel is not disappearing, nor are textiles, but our efforts to produce them are being dramatically reduced while our demand for these goods does not grow in step with the growth of productivity.

After two hundred years of industrialisation it becomes ever more obvious that the massive allocation of human resources for the production of particular things, be they wheat or semiconductors, is of a transitory nature only, giving way to a modest (in terms of human effort) commitment to provide for a demand that eventually runs into saturation. We are about to enter phase III of the Fourastié model with industrial employment falling back to the low level of pre-industrial times.[6]

The expansion of industry into ever more new products obscured for some time the underlying principle of industrialisation, the transfer of the tools from man to machine – the way towards automation (broadly speaking) – not only automation in mechanical industries but in chemical engineering and eventually information processing as well. It is the pursuit of a technological strategy – an engineer's dream really – to reduce the input of incalculable manual (or intellectual for that matter) labour in any given production process and approach the ideal of the predictable and controllable automat.

The automat was there before the Industrial Revolution – with little use in production – as a clock, a walking or drumming figure etc. Only when the idea of the automat was taken to the workshop – when the spinning machines got their clockwork – did it become an extremely powerful economic instrument, the embodiment of the technological progress of the production function.

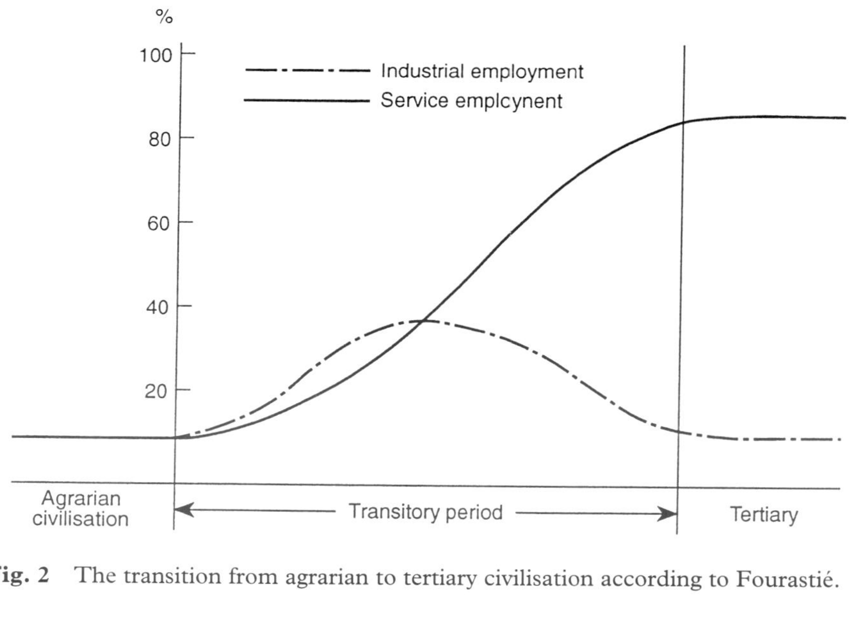

Fig. 2 The transition from agrarian to tertiary civilisation according to Fourastié.

Something that was neither an input of capital nor of labour but of an idea (dimensionless figure) how to substitute automatic elements for continuous inputs. This was technological progress that gathered its own momentum and remained a continuous challenge to generations of engineers. 'The existential pleasure of engineering' (Florman) was to make things work *alone* – for themselves. There is certainly more in technology than economic considerations.

If and when these elements of the automat proved to be cheaper than working people there was no way in a capitalistic society to prevent the stagnation and stem the decline of hardware producing industries in the long run. 'Old' stagnating industries are typically those which had a long run, the oldest being agriculture. And it was a great innovative achievement that so many automats – be they mechanical, chemical or biochemical – could be introduced in these 'old' industries. What once was consuming the efforts of the majority or of a very substantial group of the working population can now be done by a tiny fraction.

All this sounds unashamedly positivistic or whiggish, but it is difficult to see what is wrong here except the inability of our societies to peacefully benefit from this potential. To identify the problem as being one of technology, however, one inherent in the way we do things, seems to shift responsibility away from the sphere where it really originates: the political. This is not to whitewash technology altogether, since technology certainly is also a way to control people beyond their own intentions. But when it comes to declining industries, to the effect of a continued strategy of automation, the way we organise society, the way we distribute wealth and power, seems to be much more the issue, and 'technological unemployment' a very misleading concept indeed.

I suggest that in a longer history-of-technology-perspective it is more appropriate to understand decline and stagnation of certain branches of industry as a normal, foreseeable – and if I may say normatively – a welcome part of innovative processes. It is a sign of a very successful innovation process when we eventually can do with less industry or technology of a certain kind to achieve our aims.

It is a truism that it cannot be expected that all technologies and the industries built on them in the last two hundred years can go on *ad infinitum*. This is neither economically sensible nor ecologically desirable. It may well be that we have separated or even may want to separate completely from some technologies. Ecological considerations come to play here. The stagnation and gradual disappearance of industries is an eminent part of technological change, and driven by the same technological momentum as is expansion and growth. It is not failure but rather perfection. At the stage of industrial development we have entered into, the phasing-out and redimensioning of industries and technologies has become as integral a part of technological change as is the creation of novelties. And it is not only for economic reasons that we better learn how to do it peacefully.

REFERENCES

1. Data from: OECD: *Historical Statistics 1960–1993*, Paris 1995, 42 and Hartmut Kaelble: 'Was Prometheus Most Unbound in Europe? The Labour Force in Europe During the Late XIXth and XXth Centuries', in *Journal of European Economic History*, 1989, **18**, 75–80.
2. Quoted after S. Rashid, Machinery question, in: John Eatwell, Murray Milgate, and Peter Newman: *The New Palgrave. A Dictionary of Economics. Vol. 3*, MacMillan, London, 1987, 265.
3. OECD: *The OECD Jobs Study: Facts Analysis, Strategies*, Paris 1994, 27 and passim.
4. OECD: *Industrial Revival through Technology*, Paris 1988, 12–13.
5. *Innover dans la régression. Régions et industries menacées de déclin*, Paris, Stephan H. Lindner and Dominique Pestre, eds, CNRS 1996.
6. Jean Fourastié: *Le grand espoir du XXe siècle. Progrès technique – progrès économique – progrès social*, 3rd ed., Paris 1952, chapter 3.

Section III
Market Prospects
The Overview

TRENDS IN THE INTERNATIONAL STEEL MARKET

J. AYLEN
Department of Economics, University of Salford, Salford M5 4WT, UK

MARKET PROSPECTS

Introduction

The steel industry faces slow growing mature markets, yet it has been slow to adjust its industrial structure. As a result, marked cost differences persist between producers. Instead, the European steel industry is characterised by rapid product innovation as larger steelmakers develop new products not only to meet the needs of customers, but also to disadvantage weaker rivals. Product proliferation has characterised not only fast changing areas such as steel for vehicle bodies, but is also emerging in traditional areas such as sections manufacture. Meanwhile, the US steel industry is seeking adjustment to market forces through new firms entering established sectors using new process technology. The traditional producers in both North America and Europe face the dilemma of an ageing capital stock in a capital intensive industry.

Stagnant World Market for Steel

Our point of departure is the slow growth in the overall market for world steel. During the 1950s and 1960s, and up until the first energy crisis in 1974, world steel output grew at approaching 5 per cent per annum (Fig. 1) For the subsequent 23 years, total world steel production has barely grown at all. It

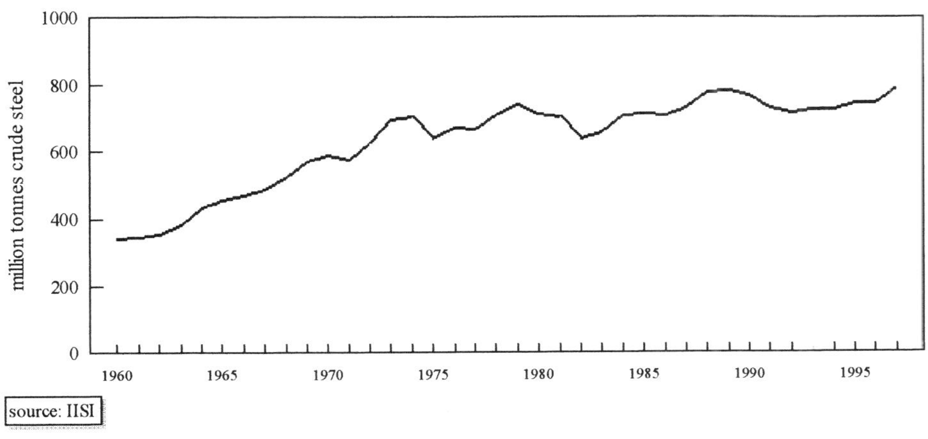

Fig. 1 World crude steel output.

follows that a competitive steel producer seeking growth and higher profitability cannot rely on overall market expansion as a source of growth.

Shifting Location of Production

The second feature of world steel demand is the shifting location of consumption and output. There is an eloquent contrast between Fig. 2 which displays steel output in the advanced industrialised economies and Fig. 3 which shows trends in four newly industrialising countries. The mature industrialised economies of the OECD area have witnessed stagnant or declining output. The notable example is Japan where steel production peaked at 119 million tonnes in 1973 and has declined since. It is possible to point to examples of sustained recovery due to improved competitiveness (the USA since 1982; the UK since

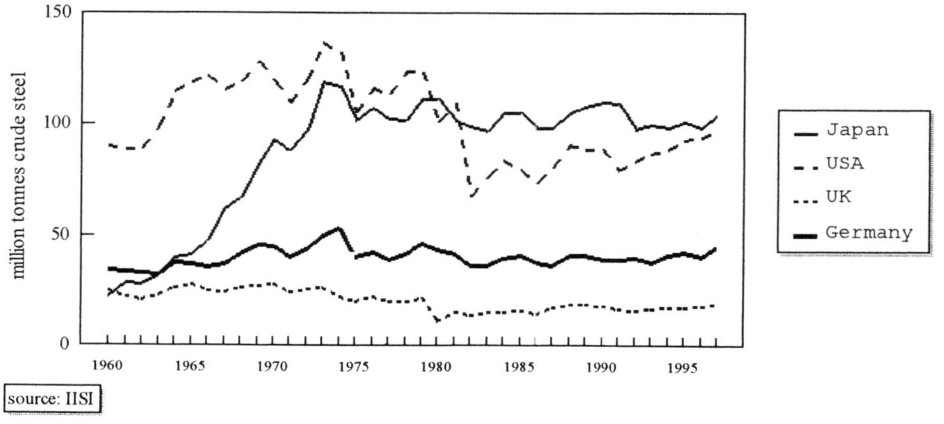

Fig. 2 OECD steel output.

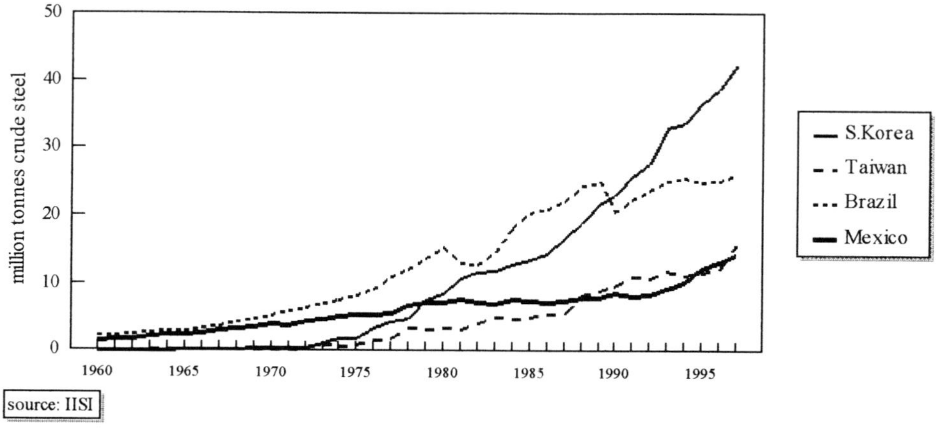

Fig. 3 NIC's steel output.

1980). But, for the most part steel output among the advanced industrialised economies has stagnated or declined over the past two decades.

In contrast, we have seen growth in steel production among the newly industrialising countries, notably South Korea where a meteoric rise in steel output since 1975 echoes expansion of Japanese steel two decades earlier.

Growth in World Trade

Overall steel output has been stagnant over the past two decades, but world trade in steel products has more than doubled between 1975 and 1995. Exports of steel products now account for about 37 per cent of steel output compared with 23 per cent twenty years ago. Successive rounds of tariff reductions, lower transport costs, development of free trade areas in Europe and North America, growing multinational procurement and greater product specialisation by individual producers are just some of the factors that have driven a growth in trade. Trade in some products has grown much faster than others: Traditional products such as rails have seen little change in export volumes in a decade. Yet trade in semi-finished slabs for further rolling has grown three-and-half-times in ten years – from around 8 million tonnes in 1985 to 28 million tonnes in 1995. There is still some way to go. Growth in world steel exports since 1980 lags far behind the overall growth rate for manufactured products as a whole. This suggests international trade and cross border specialisation has scope for further developments in this industry. Steel has been slow to globalise by comparison with other building-block industries such as petrochemicals or vehicles.

COMPETITOR TRENDS

Europe – A Failure to Adjust

European steel in particular has failed to adjust to competitive pressures brought by stagnant markets, the emergence of newly industrialising producers and changing technology. There are still marked cost differences between European steel producers (Table 1). Some larger firms, notably Usinor-Sacilor and British Steel, have pursued successful marketing policies based on product innovation. Yet government intervention in steel and the absence of a market for corporate control have inhibited rationalisation and merger.

The central problem facing the European steel industry is a failure to adjust to shifts in competitive advantage. European steel is characterised by too many firms and too much high cost, poorly located capacity. As a result, low cost producers cannot expand and new entrants are discouraged from joining the industry. Smaller producers such as Ispat of the UK and Riva of Italy have grown by purchasing privatised assets in need of rationalisation at bargain prices. But larger steelmakers have not rationalised their production facilities on a European basis. Barriers to restructuring remain. Old habits die hard. When Preussag offered the former Peine-Salzgitter steelplants for sale to Voest-Alpine of Austria or British Steel in late 1997, the government of Lower Saxony

Table 1. Comparative Steelmaking costs, 1996

US$ per tonne, CRC shipped	
United Kingdom	$445
France	$536
Germany	$598
United States of America	
– conventional integrated route	$507
– Nucor thin-slab facility	$377
Japan	$622
South Korea	$525
Taiwan	$525

Source: Peter Marcus and Donald Barnett, *World Steel Dynamics* (New York: Paine Webber, 1996)
Notes: pre-tax cost per tonne shipped at prevailing exchange rates with no allowance for differences in product quality

*re*nationalised the company in order to stave off rationalisation and politically sensitive demanning ahead of state elections.

The USA – Market Based Recovery

Conventional integrated steelmakers in the US are staging a phased withdrawal, extracting profits from depreciated assets they cannot justify renewing. Typically, they have cut costs in the face of acute competition, eschewed capital investment and made joint-venture deals with Japanese producers to gain access to strip processing technology for transplant auto-makers. The emerging mini-mill sector in the USA has grown by supplanting both imports and established domestic producers in a buoyant economy.[1–3] Ultimately, US mini-mills will have to become multinational in their search for rising profitability. Co-Steel of Canada have already pursued this route with mini-mill operations and scrap processing operations in Canada, the USA and UK.

There is a striking contrast between new entry in the US flat product mini-mill sector and slow European reaction to the same technology. In the USA, new entry to steel strip has come from entrepreneurs (Nucor, three plants; Steel Dynamics); from foreign steel companies (Gallatin, jointly owned by Dofasco and Co-Steel of Canada; Tuscaloosa now wholly owned by British Steel); and from joint ventures between established steel firms (Trico, a LTV, Sumitomo, British Steel joint venture.) In Europe, only Arvedi could be said to be a market driven equivalent venture. Ironically, much of the technology in the USA is from European plant builders who dominate the supply of DC current electric arc furnaces, thin slab casters and wide hot strip mill technology.

The central difference is that market forces have been allowed a central role in US adjustment since trade protection disappeared. Arguably, market outcomes best serve the interests of consumers and ultimately taxpayers, providing market signals are not distorted by subsidies, restrictions on ownership or trade protection.

Why has Adjustment Proved so Difficult in Europe?

There are major barriers to restructuring in European steel. Heavy sunk costs associated with the traditional blast furnace-BOS route make exit unattractive and expensive. It is easier for a marginal firm to incur continuing operating losses on a long slide towards insolvency rather than face up to the once-and-for-all costs of closure. In practice, insolvent companies have been restructured through local state aid, notably Klöckner (now Stahlwerke Bremen, with 67.68 per cent owned by Sidmar) and Maxhütte (now NMH Stahlwerke and 45 per cent owned by the Bavarian state). Steel companies lack multinational reach and fail to assess the costs of operating in different locations. Only now are larger European companies such as Arbed beginning to operate in a range of European locations (directly in Luxembourg, and through part ownership in Belgium, Germany and Spain.)

A high level of state involvement across Europe and the absence of a market for corporate control for many privately controlled assets have precluded takeovers and mergers as a route to rationalisation. The problem is diminishing as financial backers look for opportunities to shed continuing losses. Part of the Spanish steel sector has been sold to Arbed. ILVA has been sold off to various private sector interests, notably Riva. Assets in the former East Germany have come under private sector control, including Riva's acquisition of Brandenburger and Cockerill Sambre's 60 per cent share in EKO Stahl. Nevertheless, the Anglo-American style bid by Krupp for Thyssen Stahl in 1997 came up against political and trade union constraints.

The high degree of political involvement reflects the fact that large integrated plants are major sources of local employment. They have strong lobbying power in a democratic society. Ironically, overmanned plants have the greatest lobbying power as they are able to deliver more votes. Even small plants carry local clout: the Irish government bailed out their small mini-mill making sections in Cork before selling it on to UK based Ispat International in 1996.

A low degree of product diversity among steel companies makes exit via diversification difficult. European steel firms have 'stuck to the knitting'. Thyssen's strength in vehicle components and telecommunications is one exception. Hoogovens has a niche business in aluminium products for the aerospace industry. These are exceptions. Otherwise Europe's leading half-dozen steel producers have nowhere else to go.

Problems of adjustment in European steel were exacerbated by intervention from the Commission between 1980 and 1988 under the terms of the Davignon crisis cartel. In effect the cartel ossified the steel industry for eight years. The emphasis on mill capacity as a basis for allocating quotas gave steel firms a strong incentive to keep plant open. Firms were reluctant to close marginal rolling mills since the associated quota kept upstream steelplant in operation. The existence of a grey market for quotas enabled lower cost producers to 'buy' orders. Even so, the cartel delayed adjustment – all the more so because privately organised cartels persisted in some areas after the official output sharing and price fixing arrangements had come to an end.[4]

Some industries, such as the British used the 'manifest crisis' cartel of the 1980s to drastically restructure and cut costs. During the early 1980s British Steel made the remarkable transition from being among the world's highest cost steel producers to the world's lowest cost.[5, 6] The transformation under state control paved the way for privatisation in 1988. At the same time, the cartel enabled Usinor Sacilor to pursue an ambitious acquisition and investment strategy in areas such as coated strip and stainless manufacture.

The Davignon crisis cartel had unfortunate effects on German steel since it allowed their fragmented industrial structure to persist. The overall make-up of German steel barely changed during the 1980s. A renewed attempt to secure European Community intervention in 1992 and 1993 – the Braun initiative – gave further hope that change could be delayed with enervating effects.[7] In many respects, the sale of Preussag's steelmaking division; the Krupp-Hoesch merger; the eventual Krupp-Thyssen merger and the bankruptcy of Klöckner are restructuring moves which might well have taken place a decade earlier but for European Community intervention.

Marked Cost Differences Between Producers

As a result of delayed rationalisation there are dramatic and persistent cost differences between European steel producers. Taking two extremes, cost per tonne of cold rolled coil have been one third higher in Germany than in Britain throughout the 1990s. It took the depreciation of the deutschmark in 1997 to narrow the gap towards 20 per cent. These cost differences would not persist in other industries open to international competition, such as base petrochemicals.

There are two major process routes for making steel products. The blast-furnace/BOS route predominates in Europe. Costs of these integrated works operated by leading European producers are given in Table 1. Effectively, the United Kingdom is the integrated works of Llanwern and Port Talbot at British Steel and France is Usinor Sacilor, dominated by Dunkirk. The German figure is an average across leading producers. We include the world's largest producers – Japan and the USA for comparison, plus two highly efficient newly industrialising producers; POSCO of South Korea and China Steel in Taiwan. We can see Europe is an ideal location for steelmaking. Costs per tonne in the United Kingdom are lower than almost anywhere else among major steelmakers in Western market economies, a situation that has persisted since the later days of public ownership in the mid-1980s.

The second process route uses electric arc furnaces to melt scrap. This process characterises the European mini-mill sector which makes simple rod and bar products for construction and more sophisticated steels for mechanical engineering, wire for advanced applications or stainless steels. The mini-mill sector shows a similar pattern of cost differences to the integrated works.[8]

In truth, steel is an extraordinarily wide range of products with diverse end uses. The comparison in Table 1 is based on the cost of making a tonne of cold rolled coil – a standard industry product which might go to make, say, a car roof panel. Thin slab casting units fed by electric arc furnaces in the USA – such as Nucor – can make coil more cheaply still, but this route suffers from limited metallurgical quality. For instance, the thin slab route would make material for a seat back, but not a car roof.[9] Cost differences are just as marked across Europe in long products such as medium sections for the construction industry, even though they are made by a variety of process routes. Here too British Steel is the lowest cost producer.[10]

Sources of Cost Differences

These differences in steelmaking costs reflect two key factors: differences in efficiency across producers and differences in the price of inputs (Table 2). The cost differences observed here are made up of a number of small items. As so often with competitiveness, it is the cumulative effect of a large number of small cost advantages that add up to overwhelming strength.

Table 2. Differences in Steelmaking Costs, 1996

US$ per tonne, CRC shipped

Country	UK	Germany	US Intergrated	Nucor (arc/thin slab)
Labour	$108	$193	$152	$43
Materials	$308	$340	$317	$308
Capital	$32	$64	$38	$27
Total	$445	$598	$507	$377

Source: Peter Marcus and Donald Barnett, *World Steel Dynamics* (New York: Paine Webber, 1996)

Materials

As we have seen (Aylen & Albertson) the price of raw materials is much the same across Europe since iron ore, coking coal and scrap are internationally traded commodities where price is determined in a global market. There are differences due to transport costs. Spain, Italy and Turkey bear the costs of their scrap deficit in freight charges incurred trans-shipping scrap from surplus areas such as France, Germany, the UK and USA. Iron ore for Germany has to be transhipped up the Rhine from the Netherlands. But in any event a number of European works have less than ideal locations, including Llanwern and Scunthorpe in the UK where materials are hauled by train from adjacent deepwater ports and Sidmar in Belgium where the canal is too shallow to take larger vessels. German domestic coking coal is of inferior quality and high in alkalis leading to higher consumption and poorer blast furnace performance. German steelmakers appear to prefer more expensive iron ores, possibly to compensate for poor coke quality.

Rather it is the use of other materials, such as refractories, internal works transport, energy, oxygen supply or slag handling where differences in cost are found across countries. The UK's better performance reflects an accumulation of factors such as purchasing skills and wider use of sub-contractors, a slight edge in energy efficiency, cheaper transport and so forth.

Labour

Labour costs are a major advantage for the UK, which has become a low wage/high productivity producer in this sector. Labour productivity does not differ significantly across the producers quoted here. A large part of Nucor's cost advantage derives from the low manning levels inherent in a short process route. Evidence from the thin slab sector is driving integrated steel producers towards further demanning. Labour savings as well as yield gains are driving the switch from batch processing to continuous processing in cold mills and annealing operations. Further savings are achieved by combining process supervision of, say, blast furnaces or oxygen converters in single control rooms.

Finance

Financial costs are a further advantage for the UK, reflecting the combined effect of inflation, public subsidy in the 1970s and low gearing. In contrast, Usinor-Sacilor has used debt to finance ambitious overseas acquisition and new plant construction. Notice the cost of keeping capital intact in the integrated sector. Nucor's finance charges on a complete *new* thin slab strip works are lower than the necessary capital outlays required to maintain operations at an existing integrated works.

Overall

It is possible to quibble with these figures, although individual items stand independent scrutiny. Nor do they reflect subtle differences in quality which might rank Germany as the best, the UK in the middle and the USA at the bottom of the league for a product such as full finished cold rolled coil. These differences are no longer as pronounced as they once were, especially given procurement standards set by leading multinational vehicle producers. But it is evident there are dramatic cost differences across European steel which would be eliminated by market forces in more conventional industries.

AGEING STEEL

Run Down of the Capital Stock Worldwide

The existing capital stock of the world steel industry was largely installed in the 1960s and early 1970s. Adoption of basic oxygen steelmaking at a time of high growth and cheap capital sparked rapid development of large scale coastal sites fed by imported iron ore. We are now seeing modernisation of these facilities first built 25 to 30 years ago.

Recent levels of capital spending imply a reduction in future world steel capacity, especially among US integrated producers. This is less cause for concern than it appears. The productivity of capital is rising as plant is 'stretched' with the help of better process control and improved operating practice. For instance, slag blowing of BOS vessels has dramatically improved lining lives leading to sustained 1½ vessel operation in two vessel shops. Blast furnace lives are extending towards 15 years. Less surplus capacity implies more profitable operation. In turn, higher profits will generate funds for modernisation at the most advantageous sites.

Part of the potential of new technology is replacement or modernisation. In effect, the new generation of thin and medi-slab based hot strip mills in the USA are a replacement for the ageing mills of big steel of Pennsylvania and the Great Lakes. Scrap based DC arc furnaces have supplanted ageing blast furnaces fed by taconite pellets as the hot metal source for US steel production. Direct reduction facilities have been built in the Southern United States to supply iron units to the new mini-mills.

Ageing Steel: Evidence from Europe

Growth, innovation and cheap finance induced a capital spending boom in steel during the 1960s and early 1970s. There was wholesale plant construction across Europe as steel producers switched to basic oxygen steelmaking and continuous casting.[11] The process reached its limit with construction of Generation III hot strip mills at Taranto, Bremen and Fos-sur-Mer. These mills have an ultimate capacity some six times greater than Nucor's initial thin slab rolling unit at Crawfordsville.

There has been declining investment since, but with cyclical fluctuations. This picture is common to Japan, the USA and Europe although it is best documented in Europe.[12] Real investment at the heavy end of ironmaking and steelmaking is perhaps less than a third the level of twenty years ago in Europe (Figs 4 and 5). Spending on ore preparation has been virtually non-existent for 20 years, for example. Similarly, real investment in rolling mills is now half the level of twenty years ago in Europe.

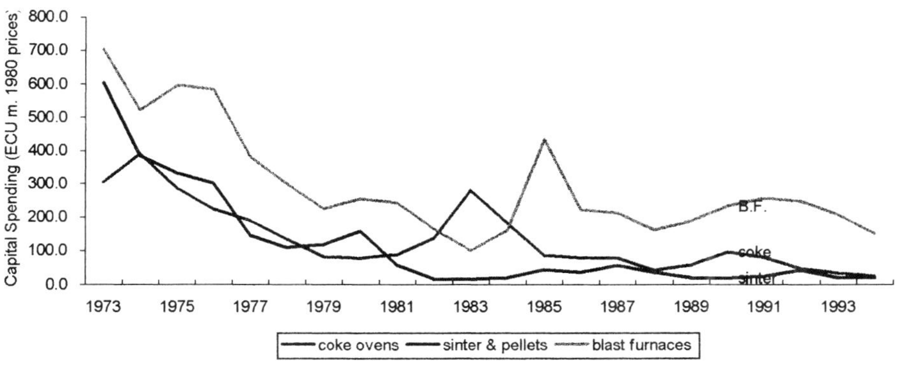

Fig. 4 European 'Ironmaking' investment EC9, 1973–1992 and forecast to 1994.
Source: European Commission.

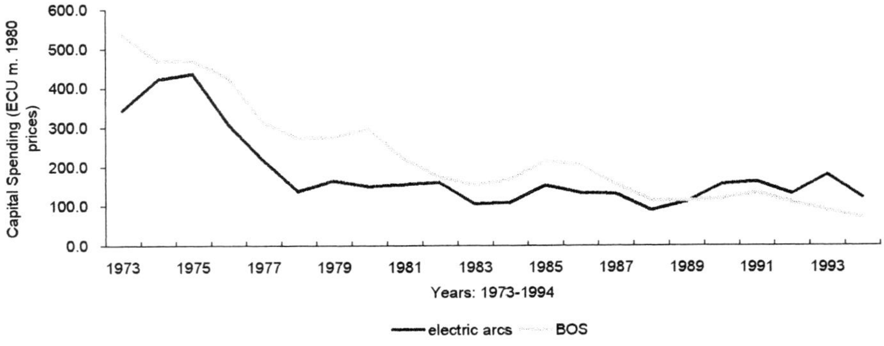

Fig. 5 European steelmaking investment EC9, 1973–1992 and forecast to 1994. Source: European Commission.

Rolling Mills: A Contrasting Picture

Europe's spending on rod, bar, sections was low during the 1980s but is now negligible, partly reflecting the maturity of the European construction sector. Plate mill spending has been almost non-existent for the past fifteen years, though a project is now under way in Sweden. This reflects a number of trends including the disappearance of the traditional shipbuilding market and the substitution of thicker hot strip for reversing mill plate. But, there has been buoyant spending for some processes, such as cold rolling mill rebuilds for continuous feeding and revamping of existing hot strip mills (Fig. 6). Another example is coating plant where there was a 'sunburst' of investment between 1988 and 1992, an issue to which we now turn.

JUST ANOTHER INDUSTRY

Strategic Moves by Leading Producers

It is misleading to focus on costs and capital equipment alone as a source of competitive advantage. Steel is becoming 'just another industry' where non-

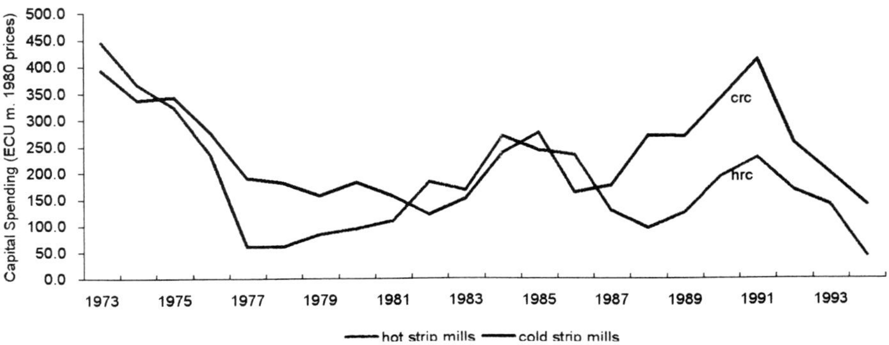

Fig. 6 European strip mills investment EC9, 1973–1992 and forecast to 1994. Source: European Commission.

price competition, marketing and distribution are of growing significance. In particular, the competitive process in steel is now strongly influenced by rapid product innovation and high levels of investment in process technology in certain specific areas. British Steel state 70 per cent of their products are now made to specifications that have been introduced within the past ten years. Product innovation is a natural response to the pattern of fierce price competition which has prevailed among the half-dozen leading steel producers in the USA, and within the European Community at those times when market forces have been allowed to operate.

Steel is a wide variety of products and leading firms introduce new and improved products to disadvantage their rivals in just the same fashion as any modern consumer goods industry. In this way, the strong producers 'up the stakes' on their weaker rivals. Product development has gone in two directions. Firstly, there have been moves towards raising the quality of existing products, *vertical product differentiation*. Examples include the costly adoption of continuous annealers for making strip, such as deep drawing grades for the car industry.[13] A modern, wide CAPL line for autobody sheet costs US$200 million at 1998 prices in a European location. Secondly, there has been a proliferation of new products with slightly different performance characteristics, namely *horizontal product differentiation*.

A major focus of innovation has been *coated steels*, especially for the vehicle and white goods industries. The output of coated strip in Europe more than doubled between 1982 and 1992 in tonnage terms. This growth echoes an earlier increase in coated steels production in Japan and subsequent growth in the USA (Fig. 7). Organically coated steels for construction and consumer durables have grown as fast as galvanised steel for the car market (Fig. 8).

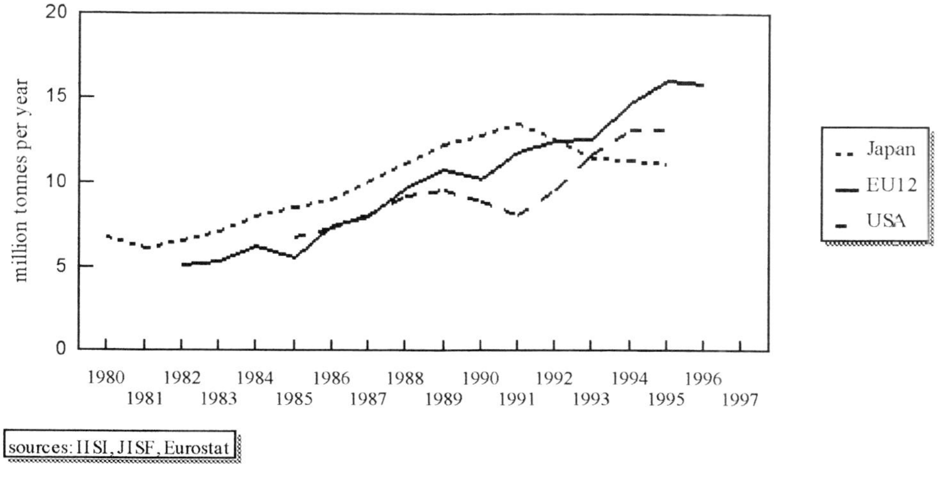

Fig. 7 Output of galvanised strip.

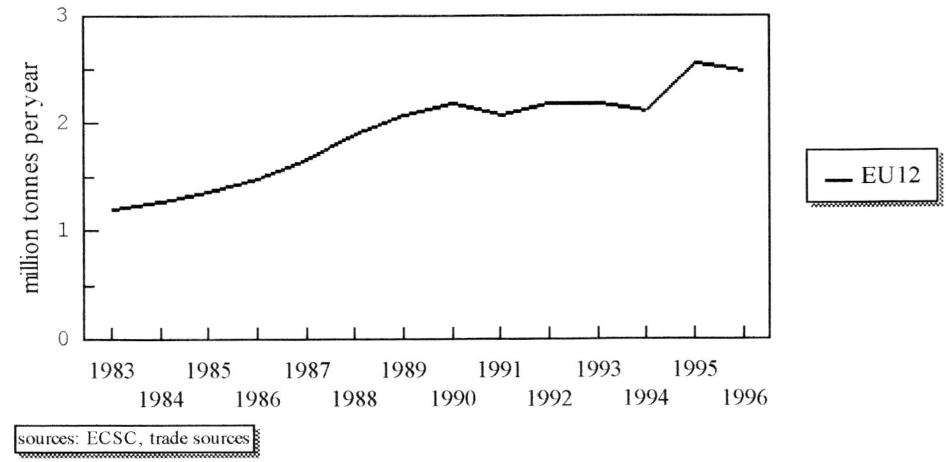

Fig. 8 Output of organically coated strip.

Measurement of growth in terms of weight severely understates the market significance of coated steels since the trend has been towards coating successively lighter products. During the 1980s the predominant coated product was 'hot and heavy' galvanised steel for the construction and agricultural markets. During the 1990s galvanneal products have developed for the car industry which put a thin corrosion resistant coating with a high surface finish on much thinner and lighter material for car bodies. The past decade has seen a progressive substitution of coated steels for uncoated steels in the manufacture of car bodies.[14, 15] The share of the European 'body in white' accounted for by conventional cold rolled strip has declined from 87 per cent in 1985 to a projected 37 per cent in 1997 (Fig. 9). Coated steels also encompass a wider and wider range of products. An inventory of products from a leading organically coated steels producer includes literally a few million combinations of colour, width, thickness, base coat, surface finish and metallurgy.

Other sectors of the industry have not been immune from product innovation: Rolling and marketing of close tolerance sections, continuously cast piling material, asymmetric beams and parallel flange beams are cases in point from structural steels. Here we concentrate on coated steels.

New products enable steelmakers to establish 'first mover advantages' which arise from supplying new products to new customers. One illustration is the supply of durasteel to Nissan's UK plant by Usinor Sacilor and British Steel. This is an electrogalvanised steel with a thin organic coating on one side to help resist corrosion, for instance from condensation on the interior of door panels. Supply of the product required investment in an electrogalvanising line, a simple organic coating line and purchase of a licence from an established Japanese supplier. However, the selection of Usinor-Sacilor and British Steel as suppliers to Nissan (plus Sidmed to Nissan Iberica) meant these two firms were

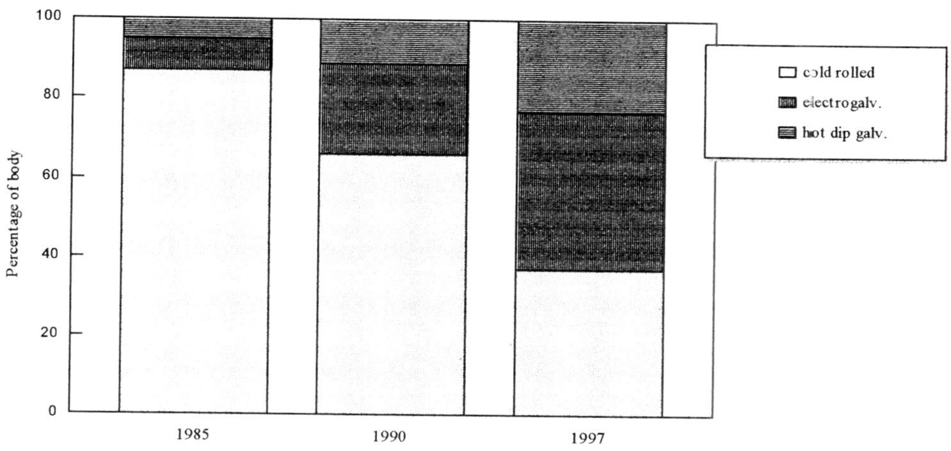

Source: Charbonnier (1996)

Fig. 9 Coated steels as a per cent of European body in white.

assured of the growing market at Nissan while their rivals were excluded. Similar examples include the supply of galvanneal product to Toyota at Derby. Product innovation of this sort involving an initial capital commitment by suppliers develops loyalty relationships with customers. Loyalty relationships are fostered by the growing dominance of distribution channels as steelmakers are called upon to provide added-value services such as just-in-time delivery, blanking, laser welding, profiling and slotting.

Development of coated products has reinforced the importance of economies of scope and scale. Producers need to be large, both to afford process lines and to provide a wide range of products to an individual customer. The scope of supply might include electrogalvanised steel for floor pans; various types of high strength low alloy and galvanneal steels for the body panels; heavy galvanised steel for wheel arches; conventional cold rolled materials for car roofs; lead coated materials for petrol tanks and organically coated products for interior trim of commercial vans. The trend towards economies of scale has been reinforced by a move towards wider sheet, heavier coil weights and lower cost line operation as volumes increase. This is more of an issue in hot-dip galvanising than electro-galvanising where the economies of scale run out much sooner.

The largest steel producers, Usinor Sacilor/Sollac; British Steel; Riva/ILVA and Thyssen have invested in a complete range of coated products (Fig. 10). Cockerill Sambre was an early mover in Europe and they maintain a full range of coated products, partly through co-operative ventures. To illustrate the commitment involved, a galvanneal line for autobody strip would cost US$200 million at current prices. Joint ventures are one way to circumvent such a

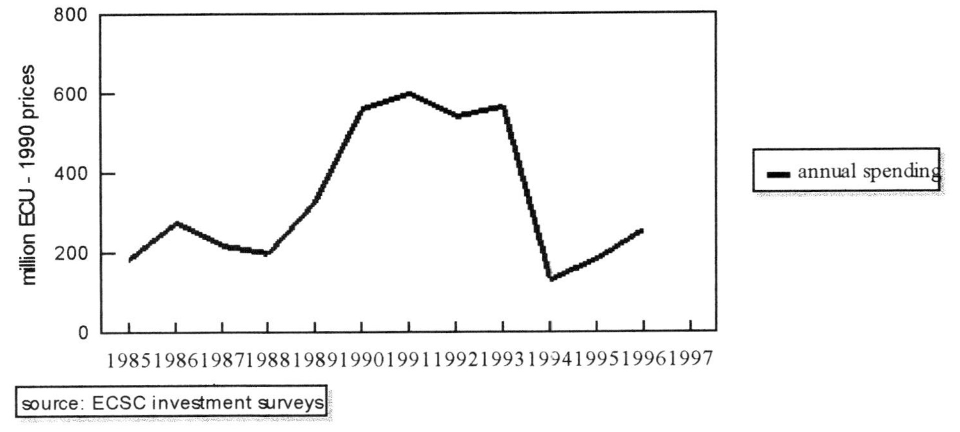

Fig. 10 Capital spending on coating lines, Eu12 – 1990 prices.

formidable barrier to entry. Segal of Belgium was set up by Hoogovens, Sidmar and Cockerill Sambre to make galvanised strip for the automotive market. This echoes other scale seeking joint ventures in coatings such as Baycoat formed by Dofasco and Stelco in Canada. Product variety reinforces the position of the biggest steelmakers. Some steelmakers foster a situation where individual carmakers each want a particular tailored product, sometimes to the extent of requiring a dedicated line. This gives a degree of countervailing power against the purchasing muscle of major car groups, while disadvantaging rivals reluctant to invest for such a narrow market.

The current position favours big producers and small niche players. Arguably, very large firms can survive since they couple economies of scale and a wide scope of products with heavy R&D and marketing effort and the ability to invest across a wide front. Niche producers can sell high value products to offset their high costs in markets which are difficult for large firms to enter in the face of specialist reputation. The strong market position of SSAB of Sweden in quench-and-temper plate for the yellow goods market is a case in point. Global sales support continuing investment in a new heavy plate mill at Oxelösund to foster an established reputation. Another example is the strength of both SSAB and Rautaruukki Oy in organically coated strip, fostered by widespread use of coated products in the Scandinavian construction sector. This enables them to compete with the biggest producers in organic products, notably British Steel which dominates organically coated strip for white goods and brown goods.

Rapid product innovation poses acute problems for medium size steel firms in Europe. Middle range integrated producers are too small to fund coating lines for the full range of products and have been unwilling or unable to match the rate of product innovation of leading firms. Medium sized firms are too big to rely on niche markets, too small to compete on costs and product range. Firms such as Klöckner, Sidmar (both now part of Arbed) and Hoogovens cannot provide the scope of products necessary to maintain comprehensive supply to the car industry, for example. As they lose an order for a new type of coated steel, they simultaneously lose the upstream loading on their plant and can only compete by offering lower margin commodity strip products. Klöckner was a very late mover in European coated steels. Two joint ventures, with Sidmar (Sikel) and with Rautaruukki (Bregal) came too late to save them from being one of the first casualties of this type of product differentiation. Other firms, such as Hoesch, failed to invest to exploit their early mover advantage, especially in organic coated products. Arbed only has a tiny organically coated sheet operation – Decosteel. Significantly, two firms – Sidmar and Hoesch – which lost out in coated strip backed continuous annealing as the best way to maintain their market share in conventional cold-rolled autobody sheet. Scarce investment funds were spent reinforcing the existing product range, rather than making a new family of coated products.

The new coatings technologies are not open to the thin slab based producers of America. The electric arc-thin slab route has severe metallurgical shortcomings which prevent production of the relevant base grades for the car industry.[9] In addition, the capital costs of a full range of coating lines would daunt even the most capable entrepreneur. Product proliferation in strip is a formidable barrier to entry.

An Efficient Industry Structure

At the moment, successful firms are prevented from buying up those enervated by the process of competitive investment. In truth, mergers are often euthanasia rather than marriage. The US brewing industry suggests failing firms avoid bankruptcy by selling out to successful firms.[16] Mergers are a civilised alternative to liquidation. But, in the current European steel industry the temptation remains for steel firms to lobby government rather than face up to exit.

We foresee fewer steel firms and fewer steel plants. Logic suggests scale economies and the need to make a wide range of products for powerful customers will result in greater concentration of production. Steelmakers have to pursue scale and scope – both size and product range, or find a niche. The outlook for European steel is closure of less efficient sites, mergers, fewer firms in each product line and concentration of production on a few, well-located coastal sites. Successful restructuring ought to attract new entry to European steel, in the same fashion as the US in the 1990s. A profitable steel industry is likely to be attractive to foreign firms using new casting technology to make sections or low quality strip for the European market via the electric arc route.

SUMMING UP: PROSPECTS AND OPPORTUNITIES

European producers face a stagnant market. Their options are growth from exports, takeovers of other companies, or overseas direct investment outside

Europe. Rapid product development has proved a successful strategy for the largest steelmakers and some specialist niche producers. At the same time, European steel also faces a problem of ageing capital stock. This poses the dilemma, how do you replace at home, adopt new technology and expand abroad?

Conventional integrated steelmakers in the US are staging a phased withdrawal, extracting profits from depreciated assets they cannot justify renewing. The emerging mini-mill sector in the USA has enjoyed growth by supplanting imports and existing domestic producers in a buoyant economy. Ultimately, they too will have to become multinational in their search for rising profitability.

The blast furnace/BOS route has extremely low costs in some European and East Asian locations. But conventional integrated works are also capital hungry, require skills to maintain and represent a huge sunk cost commitment in risky locations.

Management skill at cutting costs, strategic moves into new markets and access to capital rather than political lobbying power are likely to determine success and failure in steel in the next millennium.

REFERENCES

1. D. F. Barnett and R. W. Crandall: *Up from the Ashes: the rise of the Steel Minimill in the United States*, Washington D.C., Brookings Institution, 1986.
2. R. Preston: *American Steel: Hot Metal Men and the Resurrection of the Rust Belt*, New York: Prentice-Hall Press, 1991.
3. P. Ghemawat: *Games Businesses Play: cases and models*, Cambridge, Mass.: MIT Press, 1997.
4. European Communities: 'Commission Decision of 16 February 1994 relating to a proceeding pursuant to Article 65 of the ECSC Treaty concerning agreements and concerted practices engaged in by European producers of beams', *Official Journal of the European Communities*, 6 May 1994, **37**, L116.
5. J. Aylen: 'Privatisation of the British Steel Corporation', *Fiscal Studies*, August 1988, **9**(3), 1–25.
6. J. Aylen: 'British Steel: a retrospective', *Steel Times International*, May 1991, **15**, 28–30.
7. Select Committee on the European Communities of the House of Lords: *Restructuring the EC Steel Industry*, Session 1992–93, London: HMSO, 1993, 24th Report.
8. J. Aylen: 'Choice of process route in steelmaking', *Ironmaking and Steelmaking*, 1990, **17**(2), 110–117.
9. A. Ritt: 'How good is Nucor's thin-slab steel?', *New Steel*, May 1994, 22–25.

10. T. Gill: 'British Steel – a sterling job', *Metal Bulletin Monthly*, January 1998, 14–21.
11. J. Aylen: 'Innovation in the British Steel industry', *Technical Innovation and British Economic Performance*, ed. K. Pavitt, Macmillan, London, 1980, chapter 12.
12. J. Aylen: 'Investment outlook for the world's steel industry', *Steel Technology International*, 1989, 17–21.
13. J. Aylen: 'Curious case of the costly CAPL's', *Steel Times International*, December 1988, **12**, 27 & 58.
14. B. Attwood: 'Development of coated flat rolled products for automotive end use', *Transactions, IISI-28*, Brussels: International Iron and Steel Institute, 1994.
15. J. C. Charbonnier, D. Quantin and J. P. Birat: 'Technological innovations: finishing and coating', *The Steel Industry in the New Millennium*, Terni, July 1996 (unpublished).
16. J. Sutton: *Sunk Costs and Market Structure*, Cambridge, Mass.: MIT Press, 1991.

DATA SOURCES

Eurostat, *Iron and Steel*, Luxembourg: Eurostat, Theme 4, Series B, monthly including quarterly supplements.

European Commission, *Investment in the Community Coalmining and Iron and Steel Industries, Report on the 199– Survey*, Luxembourg: Office for Official Publications, annual.

International Iron and Steel Institute, *Steel Statistical Yearbook 1996 (and 1997 supplement)*, Brussels: IISI, 1997 and 1998.

Japan Iron and Steel Federation, *Monthly Report of the Iron and Steel Statistics*, Tokyo: JISF, monthly.

P. Marcus: 'Ironies in Steel: the Steel Industry in the New Millennium: innovation, strategy and markets', *The Steel Industry in the New Millennium* Vol 1, IOM Communications Ltd, London, 1998, 7–32.

CO-OPERATION BETWEEN STEEL PRODUCERS AND STEEL USERS: A MAJOR DETERMINANT OF NATIONAL COMPETITIVE ADVANTAGE

M. KIPPING

Department of Economics, The University of Reading, Whiteknights, P.O. Box 218, Reading, R66 6AA

INTRODUCTION: MAJOR CHANGES IN WORLD STEEL PRODUCTION SINCE 1945

The world steel industry has undergone considerable changes since the end of the Second World War. This fifty-year period has seen the relative decline of the United States and, to a lesser extent, Western Europe and the emergence of East Asia as a major steel producing region. The growth of Japanese steel production between 1950 and 1973 has been particularly impressive, when output soared from less than 5 to about 120 million metric tons. The development of the South Korean steel industry since the early 1970s was almost as remarkable. In a period, marked by declining production in most other countries, the Pohang Iron & Steel Company (POSCO), which started operations only in 1973, managed to become the world's second largest producer after Nippon Steel. The situation in Western Europe also saw considerable change. While Germany retained its position as the region's dominant producer, the Italian steel industry greatly expanded its production capacity after 1945 and, in the 1970s, surpassed the 'earlier movers' Britain and France.[1]

The explanations given in the relevant literature for the emergence of East Asia as a major steel producing region are manifold, but usually centre around two main aspects. First of all, the successful nations were 'latecomers' and, as a result, were neither constrained by the location of raw materials nor locked into existing technologies.[2] Instead they could achieve lower costs by building modern, large-scale integrated plants in coastal locations. Constant technological upgrading and the exploitation of scale economies allowed the Japanese and later on the Korean steel industry to achieve low production costs.[3] In addition, they benefited from the development of bulk shipping, which lowered the transportation cost of the necessary raw materials. As a result, Japan and South Korea could import high-grade iron ore from Australia whereas many other countries, such as France and the United States, continued to use domestic deposits, usually of a lower quality and more costly to extract and transform.[4]

Secondly, government intervention was seen as an important factor for the success of the East Asian producers. Concerning the Japanese steel industry for example, most authors so far have highlighted the role of the Ministry of International Trade and Industry (MITI) in promoting investments and in protecting the domestic producers against foreign competition.[5] Many American academics – somewhat accusingly – have attributed the Japanese success to the cartel-like co-operation between the steel firms, which would have been prohibited under US legislation.[6] Paul Tiffany went even further and suggested that the US government had encouraged and subsidised the growth of foreign producers, while curtailing the domestic industry through price controls and anti-trust rules.[7]

This wide-spread view regarding the apparently crucial role of government intervention has, however, not remained unchallenged. For Japan, Seiichiro Yonekura has suggested that individual firms and entrepreneurs played a more important role in the development of the steel industry than public policy. The direct influence of MITI was largely confined to the immediate post-war years and, even then, its recommendations were not always followed by the producers.[8] In more general terms, Michael Porter has highlighted the importance of what he calls 'domestic rivalry', i.e. competition in the home market, for the success of companies world-wide.[9] 'Nowhere, according to Porter, is the role of fierce rivalry more apparent than in Japan, where there are 112 companies competing in machine tools, 34 in semiconductors, 24 in audio equipment, 15 in cameras'.[10] There are also five major integrated Japanese steel producers: Nippon Steel, NKK, Kawasaki Steel, Sumitomo Metals and Kobe Seiko.[11]

In the South Korean case however, POSCO is the only fully integrated domestic competitor and therefore does not confirm Porter's hypothesis about the crucial role of competition in the home market. Porter himself contends that the company 'though the exception, proves the rule', but resorts to factors outside his explanatory framework, i.e. 'a sense of national purpose and priority' in order to explain POSCO's success in 'providing low-cost steel to Korean industry as well as contributing to Korean exports'.[12]

Against this background, the objectives of the following article are twofold. First of all, it attempts to demonstrate that, in addition to factor cost advantages, the co-operation between the steel producing and the steel using industries played a major role in determining the developments of world steel production after 1945. Secondly, with respect to government involvement, it will show that in cases where steel producers and users failed to co-operate closely, public intervention could, at least temporarily, achieve the necessary co-ordination. But before illustrating these arguments by a detailed analysis of the post-war developments in Japan, Germany, South Korea, France and Britain, the article will provide a framework which helps to understand the crucial importance of a close producer-user relationship.

COMPETITIVE ADVANTAGE IN STEEL PRODUCTION: A TRANSACTION COST FRAMEWORK

So far most of the relevant research on competitive advantage in steel production has focused on raw material inputs and steel-making itself. Differences and changes in the relative position of companies and countries were largely explained as a result of (1) the availability and the price of the necessary inputs, mainly iron ore, coking coal and labour, and (2) productivity improvements due to the exploitation of scale economies and investment in new technologies. The government was seen as an independent variable which could influence the outcome either in a positive or in a negative way. This article goes beyond this framework by drawing attention to the finishing stage in the so-called 'value chain'. It will argue that the relationship between the steel producers and the steel users at a national level is an important determinant of international competitive advantage. The following model (Fig. 1) shows both the different stages of transformation from raw materials to finished products and the economic environment in which firms operate.[13]

Along the value chain the inputs at each stage determine the competitiveness of producers at the following stage. In the case of steel, the price and the characteristics of the iron ore, the availability and price of scrap metal etc. constitute a vital part of the production costs and also influence the choice of technology.[14] For example, one of the main reasons why the Japanese steel industry adopted the basic oxygen furnace more rapidly than its foreign competitors was the lack of scrap required for the open-hearth technology.[15] In a similar way, the availability, the price and the quality of steel products play an important role for the transforming industries, such as shipbuilding and automobile production.

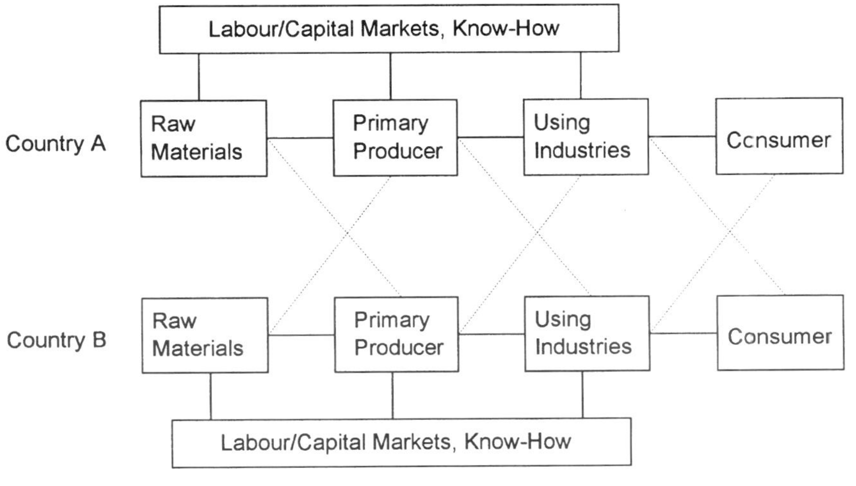

Fig. 1 A systems view of the economy.

Other inputs are indicated horizontally across the different stages, because they constitute a pool from which all firms in the economy can draw, even though some of them tend to be more sector-specific than others. Capital for example is usually mobile, whereas labour might be limited to certain industries. Steel production is capital-intensive, requires skilled labour and, while not a hi-tech industry, is still relatively technology driven. The lack of these resources could limit its development, but can be overcome, as the South Korean case has shown. The country had a large pool of skilled labour, but lacked both the necessary capital and technological know-how. After the refusal of the World Bank to provide funding for POSCO, the government used reparation payments from Japan. Technical assistance was first obtained from the Japanese steel industry and then from Western Europe, but local technological know-how increased steadily.[16]

The government itself is not explicitly incorporated in the above model of the economy, because it can influence all of the different variables. It might intervene directly in the activities of firms at each stage of production, mainly via subsidies and partial or total ownership. It also regulates the relationship between the firms, both within and along the different stages, through a variety of means, including for example anti-trust policy and price controls. The education, R&D or general economic policy influence the availability and the quality of labour, capital and know-how. Last but not least, the government can also protect the markets of each firm from foreign competition (in Fig. 1 this is indicated by the dotted lines). In the case of the European Coal and Steel Community (ECSC) established in 1952, member countries have lost the latter power, but are still able to exercise a significant influence on their national industries via all the other mechanisms described above.[17]

Concerning the transactions between the independent firms along the different stages of the production process, neo-classical economic theory assumes that these are costless, because all actors have unlimited information and act rationally. This means in our case that exchanges between raw material suppliers and steel producers or between steel producers and users would take place in the 'market' with price as the only purchase criterion assuming perfectly homogeneous goods. Prices will differ as a result of production cost differentials and transportation costs.

However, other economists have challenged this model and its underlying assumptions. In addition to production and transportation costs, they have introduced the concept of 'transaction and information costs'.[18] It refers to all the efforts required to ensure the supply with the necessary inputs, including for example the drafting of a contract, the 'haggling' over its terms and the legal enforcement in case of one party's failure to meet the agreed conditions. The more each side has to invest in the relationship, e.g. by acquiring specific assets, the higher the costs resulting from a disruption and an eventual break-down. In these cases, exchanges will not take place in the market, but will be cemented by 'explicit' or 'relational' contracting (with the latter referring to a trust-based relationship). If transaction costs are extremely high, it might be more advantageous for firms to make rather than buy its inputs, and vertical integration will be the preferred solution.

The steel industry resembles the transaction cost framework much more than the ideal neo-classical world. First of all, the inputs for the production and transformation of steel are far from being homogeneous. Even the suitability and the quality of iron ore and coking coal can differ significantly and exercise considerable influence over the production process and the steel output. The same is true for steel products, with some variation according to the level of transformation. Thus, while merchant bars could be considered a commodity, many of the higher value-added products, such as thin sheet or special steel, are not. In their case, the price is complemented and possibly supplanted by other purchase criteria, viz. the quality, both in terms of specifications and regularity, and delivery reliability.[19] A good example is automobile production, where the just-in-time delivery of high-quality body sheets at reasonable prices is an important element of competitive advantage.

As a consequence, firms at the receiving end have to make sure that these criteria are met. However, the predominant form of transaction does not necessarily remain stable, but varies over time as a result of changing circumstances. In the case of raw material inputs for example, considerations of assured supply, especially in the case of a war, and high transportation costs have historically favoured countries with domestic iron ore and/or coking coal deposits, such as the United States, Britain, France and Germany. In order to secure and control at least part of their vital supplies, steel producers preferred full or partial ownership over market transactions. The French steel industry for example, which already owned most of the iron ore deposits in Lorraine, since the late 19th century also acquired stakes in German coal mines.[20] Today, following the discovery of rich deposits in other parts of the world, the development of bulk shipping and the liberalisation of trade, long-term supply contracts have become the predominant form of transaction.

The using industries also have to ensure the regularity and the quality of their supply with steel products. This is made particularly difficult by the economic characteristics of steel production and the differences in 'market power' and political influence between steel producers and users. As seen above, steel production is characterised by significant economies of scale and high fixed costs due to the necessary capital investments. Demand fluctuates with the business cycle, but is relatively 'price inelastic', i.e. shows only a limited response to changes in price. As a result, steel prices and consequently profit margins of the producers are extremely volatile.[21] In order to keep prices at a sufficient level even in a downturn, producers have tended to establish cartels, usually at a domestic, but sometimes also at an international level. This has obviously hampered the competitiveness of using firms. On the other hand, because of the high fixed costs, it is crucial for the steel producers to operate production units at close to full capacity. In case of a shortfall in domestic demand, producers therefore have an incentive to export, even 'dump' their products in foreign markets, in order to achieve a high capacity utilisation rate. Once again, this can be potentially harmful for domestic users, because they have to compete with foreign firms which obtained the same steel products at a lower price.

The collusion between steel producers, through price leadership or the formation of cartels, has been facilitated by the high level of ownership and, very often, geographical concentration in most countries. In addition, during most of this century governments have tended to favour domestic steel production, mainly for military, and laterly for political reasons, i.e. in order to avoid large-scale unemployment in certain regions and reduce the risk of social unrest.[22] Due to the concentrated nature of the industry and the government's receptiveness, the trade associations of the steel producers have usually exercised considerable political influence, and often obtained protection against foreign competition, which in turn deprived the using industries of access to alternative sources of supply. Moreover, the fact that most of the steel users, with the exception of shipbuilding and automobile production, are small and medium-sized enterprises has reduced both their political influence and their buyer, or in economic terminology, monopsony power, vis-à-vis the steel industry.

All of the above constitutes a potential – and very often actual – source of problems between steel producers and users all over the world. But the ways in which these problems are avoided or solved differ considerably across countries. The following section provides a detailed analysis of the developments in most of the major steel producing countries in what used to be called the 'free world'.[23] It will identify two different ways to achieve the necessary co-ordination between steel producing and steel using industries.

AN INTERNATIONAL COMPARISON OF THE STEEL PRODUCER-USER RELATIONSHIP

A comparison of the relationships prevalent at different times in different countries reveals that purely market-based exchanges are the exception in transactions between steel producers and users. If at all, such a 'model' might have existed in the United States during the first half of the century. In this period, US Steel dominated the industry and acted as a price leader, but refrained from obvious collusive behaviour and allowed other firms to increase their market share through efficiency improvements and product innovations. During the 1920s for example, Armco Steel developed a continuous hot strip mill which improved the quality and lowered the price of flat rolled products significantly and thus contributed to the rapid growth in the production of various using industries, including automobiles.[24]

In all other cases, transactions have been co-ordinated explicitly either by the firms and their trade associations or, alternatively, by the government. For most of the post-war period, Japan probably represents the best example for a close producer-user relationship, without government interference. Research on inter-firm relations in Japan in general has highlighted their long-term and stable nature. Japanese firms at different stages in the value chain are known to co-operate closely, viz. with respect to joint product development, quality control and just-in-time delivery.[25] The same is true for the relationship between the steel industry and its major customers.

One reason for the phenomenal growth of the Japanese steel industry between 1950 and the mid-1970s was a 'virtuous circle' of continuous investment in steel production which lowered prices and made the Japanese using industries more competitive, thus leading to a further increase in demand for steel and additional investments.[26] But it appears that the direct co-ordination of activities between steel producers and users has also played a significant role in this development. The co-operation between the Japanese steel firms and the car manufacturers seems particularly close. Their relationship, which has only recently been studied in some detail by Françoise Guelle, shows a number of characteristics.[27]

Prices and quantities are negotiated each year. This protects the steel industry from the volatility of the market, while providing the car producers with advance information about production costs. However, the co-operation extends for beyond this relatively short time frame. Usually, one steel producer co-operates closely, but not exclusively, with one car producer over many decades. Thus for example, Nippon Steel is the main supplier of Toyota. Both jointly develop new products, for instance anticorrosive steel sheets which contributed to the high quality – and good reputation – of Japanese cars in the 1980s. In recent years, joint development efforts have focused on the need to reduce the weight of automobiles without compromising their safety.

The car manufacturer usually negotiates on behalf of the whole enterprise group, i.e. also for its suppliers. Thus only 40% of the steel products sold to Toyota go directly to the assembly plants. The rest is used in the production of car parts by its network of suppliers. On the one hand, this system of 'concentrated purchasing' increases the market power of the car manufacturers vis-à-vis the steel industry and provides the parts suppliers with cheaper steel inputs. But on the other hand, it also benefits the steel producers by reducing the number of direct contacts and separate negotiations. Costs are reduced further through the intervention of trading companies who co-ordinate the delivery and the financial aspects of the transactions between steel producers and users. In addition, so-called 'service centres', located close to the using industries, in many cases 'customise' the steel products for the users.

Because of the long-term, stable nature of the relationship, the Japanese steel industry does not hesitate to invest in specific assets for the automobile industry. The best example is certainly their investment in the United States since the 1980s, leading to a 'New Iron Age' in America.[28] This was partially due to the fear of protectionism, but mainly a result of the Japanese car transplants. The Japanese automobile producers found the quality of US steel products insufficient and therefore asked their suppliers to follow them. Subsequently, all of the five major steel producers either acquired an American steel company or established joint ventures for the modernisation of existing or the construction of new facilities geared towards the needs of the Japanese automobile producers.

Martin Kenney and Richard Florida have highlighted the close co-operation between the automotive transplants and the Japanese steel producers, especially regarding the supply of high-quality galvanised steel. They stress that, unlike their American counterparts, the Japanese steel makers did not offer standardised products, but supplied each of the car manufacturers with products geared to their specific needs. According to an American steel executive, whom they interviewed, the attitude of US car manufacturers resembles a 'club', i.e. a stick, whereas the Japanese transplants offer 'more of a helping hand. If your raw materials have escalated or something has changed materially in your cost structure, they have an interest in keeping you profitable'.[29]

In recent years, the close relationship between Japanese steel and car producers has come under some pressure because of the relatively high cost of domestic steel production. A number of automobile manufacturers have begun to import some low-value added steel products, viz. from the South Korean POSCO. But the Japanese steel industry still supplies all of the high-value added products where quality and delivery reliability are crucial for the competitiveness of the car manufacturers. In addition, the steel producers have made efforts to reduce production costs.[30] It seems therefore likely that the mutually beneficial relationship between both industries will continue in the foreseeable future.

In Western Europe, Germany comes closest to the same 'ideal type' as Japan, because co-ordination takes place directly between steel producers and users, largely without government interference. Throughout most of this century, German steel producers and users followed what could be called a 'value chain logic', i.e. their actions took the whole transformation process from raw materials to finished products into account. They maximised their returns by regarding not only their direct output, but also the output of the following stage in the value chain, which in turn increased their own sales. The producer–user relationship was not without problems, but these were usually solved between them in a mutually beneficial way.

The producer–user co-operation in Germany has a long tradition. It can be traced back at least to the so-called AVI agreement which lasted from 1925 to 1941.[31] A domestic cartel and protective tariffs had led to high steel prices in Germany, but the Working Party of the Iron Transforming Industries (Arbeitsgemeinschaft der Eisen verarbeitenden Industrien = AVI) obtained an agreement under which steel producers would pay rebates for the steel used in exported goods, thus effectively bringing its price down to world market levels. In this way, German steel users remained internationally competitive, while steel producers benefited indirectly from the exports of finished goods. The agreement even survived the adverse effects of the Great Depression, when rebates paid by the steel industry increased significantly, because world prices came down and, at the same time, German steel users stepped up export sales to compensate for the shortfall in domestic demand. It was only formally terminated by the Nazi government in 1941 when exports had become insignificant as a result of the war.

Similar patterns can be observed for the post-WWII period. In 1948, when the steel industry's Trade Association, the Wirtschaftsvereinigung der Eisen- und Stahlindustrie (WV Stahl), re-established an export commission, its member firms highlighted the importance of a close co-operation with the using industries in order to favour 'indirect' exports, i.e. exports of goods which incorporated steel.[32] On the other hand, in the early 1950s, the steel users contributed financially to the reconstruction and modernisation of the German steel industry, by accepting to pay a temporary levy on sales, when efforts to obtain funding from the US had proved largely unsuccessful.[33] Such a contribution to investments in modern facilities was obviously in their own interest, since it provided them with the quantity and the quality of steel necessary for their production.

Another telling example comes from the same period. The outbreak of the Korean War in 1950 resulted in a rearmament boom, leading to a sharp rise in steel production which in turn increased the demand for coking coal. Since the domestic coal supply was insufficient, German steel producers had to import additional quantities from the United States to be paid in US dollars. Initially, they therefore increased their exports to the dollar area. But the export of steel products exacerbated the shortages of steel in Germany itself, where the using industries, especially in machine tools, also benefited from the rearmament boom. The steel producers finally agreed to stop their exports to the United States and to supply the quantity gained to domestic users. These could in turn increase their export sales and then pay the additional US coal needed from the dollar proceeds. Since their high value-added products achieved higher prices than steel, more coal could be imported which benefited both the steel producing and using industries.[34]

All of the above examples, especially the last one, demonstrate how German steel producers and users acted in concert based on an understanding of the transformation process from raw materials to high-value added finished products. Over time, they established a mutually beneficial relationship and managed to do so without outside, i.e. government help. During the post-war decades, the German steel industry was not as successful as the Japanese. However, it managed to maintain its leading position among the Western European steel producing countries and, unlike most of its counterparts in the ECSC, also restructured largely without government interference and public subsidies.[35]

Compared to Japan and Germany, South Korea most closely resembles the opposite 'ideal type' where the government assumes most of the co-ordination between steel producers and users. In a recent dissertation about South Korean industrial policy, Ha-Joon Chang has highlighted potential benefits in terms of transaction costs resulting from the co-ordinating role played by the government. He concludes that the 'active intervention by a strong developmental state in industrial development' can minimise the risk of co-ordination failures and ensure productivity growth at the same time.[36] Chang does not provide any evidence from steel production, but accounts of the industry's evolution suggest that the co-ordination of steel producing and steel using activities by the South Korean government was an important determinant for the success of both industries.[37]

Even before trying to develop an integrated, large-scale steel industry, the Korean government had taken measures to promote domestic steel production, including a 25 per cent tariff on imported steel products. However, in order not to harm the international competitiveness of the steel using industries, the tariff was refunded for exported products. This measure highlights the difference between the German and the South Korean pattern very clearly. As seen above, in the former case similar results were brought about by producer–user co-ordination, i.e. the AVI agreement. From 1962 to 1972, the four private domestic steel producers in South Korea indeed raised their output, and self-sufficiency climbed from less than 4 to 23 per cent of demand. But in the late 1960s, it became obvious that these efforts were insufficient to support the intended development of the domestic heavy industry, including shipbuilding, as part of the so-called 'Big Push'. From 1967 until 1982, the percentage increase in steel consumption indeed exceeded general economic growth by far.[38]

While 'import substitution' was as a major motive the reasons behind the decision to establish a new, state-owned enterprise, the Pohang Iron & Steel Company (POSCO) in 1968 are unclear.[39] Instead, the government could have supported the modernisation and expansion of one or several of the existing producers. But its intention seems to have been to achieve a clear break with the past by constructing a modern, large-scale integrated coastal steel work. Given their outmoded facilities and small size, the existing producers had neither the technological nor the organisational capabilities required. With respect to the latter, it is quite telling that a retired army general, Tae Joon Park, was appointed as POSCO's first Chairman. Production technology was initially provided by the Japanese steel industry, and later on obtained from Western Europe.

Alternatively, the Korean government could have asked one of the large family-owned and run business groups, known as 'chaebol', to develop steel-making activities. There are other cases, like shipbuilding, where the government is known to have wielded considerable pressure and its control of the credit market to direct the chaebols' business development.[40] But since steel was a vital input for many using industries, such a decision would have given one of the groups quite a significant degree of control over some of the others. This did not seem like a wise move in light of the extremely fierce rivalry among the chaebol which, as Michael Porter notes, prompted the government repeatedly to intervene in order 'to prevent "destructive" competition'.[41] To spread the activity between the different groups was also impossible, at least initially, because the necessary scale economies required the construction of one large facility.[42]

Whatever the exact reasons behind it, the government's strategy proved highly successful. POSCO initially focused on the production of low-cost steel for domestic users. By the end of the 1980s, Korea's self-sufficiency exceeded 80 per cent, a remarkable achievement given the two-digit growth of steel consumption during the whole period.[43] In addition, based on its cost advantage, the company exported on average just under 30 per cent of its output. However, domestic steel users always came first, which is apparent by the fact that the government kept the domestic below the export price. POSCO has also been attentive to its clients' need and adapted to changes in demand.[44] When a customer survey, conducted by the Mitsubishi Research Institute on its behalf in 1986, had found the quality of its products wanting, POSCO made major efforts to improve it.[45] The company invested heavily in new technology for the production of high-quality steel for the fast growing Korean automobile industry. It has recently announced plans to expand capacity in order to become the world's largest steel producer. Based on the positive experience in the past, co-operation between steel producers and users in South Korea is likely to continue in the future, even though the government has recently been relaxing its direct control over POSCO.[46]

In Western Europe, France comes closest to the South Korean model. However, the French government only gradually after 1945 assumed a co-ordinating role between steel producers and steel users. During the first half of the 20th century, the producer–user relationship was marked by confrontations rather than co-operation.[47] The frequent conflicts between the French steel industry and the car producer Renault provide a very good illustration for the acrimony of these confrontations and for the efforts made by a major steel user to secure the supply with one of his most important input materials.

Already during the First World War, Louis Renault, the automobile company's founder and owner, accused the 'steel barons' of limiting the necessary steel imports to keep prices at a high level. In 1918/19, he managed to persuade a number of other users to create their own steel company, in order to secure at least a partial independence of supply from the steel industry. The Union des Consommateurs des Produits Métallurgiques et Industriels (UCPMI), in which Renault held a 20 per cent stake, acquired large steel works in Lorraine, which had been sequestered from Thyssen after the war. But these efforts appeared insufficient and, in 1931, Renault established his own steel firm, the Société des Aciers Fins de l'Est (SAFE).[48]

The developments in the immediate post-WWII period demonstrate quite clearly that these confrontations were not driven by Renault's personality, but had much deeper roots. The engineer Pierre Lefaucheux, who became the first

President and CEO of the nationalised car producer, almost immediately followed in Renault's footsteps by claiming a continuous hot strip mill for SAFE in order to ensure the company's supply with high quality and low price body sheets. This attempt to save transaction costs by integrating backwards were unsuccessful, because both the French and the American governments supported the installation of such a rolling mill by a joint venture of all steel producers in Lorraine. But Lefaucheux obtained a widening of the strip mill from 66 to 80 inches with the argument that these specifications corresponded better with the requirements of the fast growing, modernising and exporting automobile industry.[49]

This decision already revealed a shift in the attitude of the French government, or at least parts of it, towards the interests of the steel using industries. But the steel users continued to complain about the high prices and the poor and irregular quality of French steel products which they attributed to a large extent to the cartelised structure of the country's steel industry. In Renault's 1949 annual report for example, Lefaucheux blamed the company's difficulty to compete internationally on the high domestic steel prices. He also criticised the steel industry's policy to export low-value added steel products instead of finished goods such as automobiles. At another occasion, he claimed that 'blast furnaces are pissing gold', implying that the steel producers were highly profitable at the expense of the using industries.

These criticisms proved successful, because in the early 1950s the government restricted the activities of the domestic steel cartel quite severely and opened the French steel markets to foreign competition in the framework of the European Coal and Steel Community (ECSC) which originated from a proposal drafted by Jean Monnet and presented by Foreign Minister Robert Schuman on 9 May 1950. To no surprise Lefaucheux and other steel users welcomed the Schuman Plan and played an important role in the ratification of the ECSC treaty against the resistance of the powerful steel lobby.[50]

But the confrontations between steel producers and users continued in the following decades. Due to the fact that for much of the post-war period demand exceeded supply, most steel producers remained relatively unresponsive to the claims of the using industries who therefore continued to complain to the French government. And the government indeed intervened occasionally in their favour, viz. through the control of domestic steel prices until the 1970s. This was an obvious contradiction to both the spirit and the treaty of the ECSC and the steel industry blamed the restrictions for their inability to finance investment programmes through retained earnings.[51]

The sharp drop in sales and production since the oil crises of the 1970s exposed the inherent weaknesses of the French steel industry, viz. their failure to rationalise and modernise production.[52] It also made the steel producers more attentive to the needs of their clients. But by then it was too late. At the end of the decade, the French government rescued the steel industry from bankruptcy. After the election victory of François Mitterand in 1981, the new socialist and communist government made the *de facto* control of the industry official by converting its debt into share capital. Under public ownership, the French steel producing firms were rationalised and finally, in 1987, combined into one entity, Usinor Sacilor.[53] Following further modernisation, the company has become quite successful and was privatised in July 1995.

There are many reasons for the relative decline of the French steel industry. But the strategy to produce low-value added steel products for export rather than domestic consumption, pursued by many companies in Lorraine, and, more importantly, the lack of co-operation between the steel producers and users certainly occupy prominent positions among them.[54] Usinor Sacilor, now the third largest steel producer in the world after Nippon Steel and POSCO, seems to have learnt the right lessons from the past. It has decided to focus on high-value added steel and discontinued the production of beams or merchant bars where price is the only purchase criterion. A company representative also declared recently that another major objective was to develop a close partnership with clients, viz. producers of automobiles, packaging and domestic appliances.[55]

Britain is another case where the government contributed significantly to an improvement in the relationship between steel producers and steel users. In the post-war period, the development of the British steel industry was marked by selective government intervention, partially driven by ideological conflicts between the two main parties and usually resented by the industry itself. Labour had nationalised the industry in 1950, a decision which the following Conservative government reversed shortly thereafter. In 1967, the remaining steel companies were taken once again into public ownership and combined to form the British Steel Corporation (BSC).[56]

Hearings conducted by Parliamentary Committees during the 1970s revealed that many users were highly dissatisfied with the price and the quality of steel supplied by BSC.[57] A customer survey, commissioned by the company itself in 1976, came to similar conclusions. Most automobile producers therefore obtained an increasing proportion of their inputs from foreign suppliers. In the early 1980s, the government, in the person of Patrick Jenkin, Secretary of State for Industry, had to intervene directly to settle a long-standing dispute about steel prices between BSC and the car manufacturers.[58] During the 1980s, the company improved its record, and subsequently its reputation, with respect to price and quality, but found it difficult to regain the market share lost earlier.

When it was privatised again in 1988, British Steel had achieved a turnaround which Jonathan Aylen has characterised as 'one of the most remarkable in business history'.[59] But while the company had become more competitive, it was also significantly smaller in size. This was partially due to the contraction of its domestic market, caused by the decline of manufacturing industries in Britain, a long-term trend which was only reversed in recent years with the arrival of the Japanese car transplants. But, as seen above, the company has to take part of the blame, because it neglected the demands and needs of the steel users, viz. the automobile producers, who therefore turned to foreign suppliers in order to obtain steel products of superior quality.

The situation in the United States during the 1960s and 1970s showed a number of characteristics similar to the British case.[60] Most of the American steel producers exhibited a 'take it or leave it' attitude towards their customers. Repeated complaints from the using industries about the poor quality of domestic steel products were not taken seriously. Like their British counterparts, the American steel users therefore imported low price and high-quality steel in increasing proportions. During most of the post-war period, the US government intervened occasionally in favour of the steel users, by limiting price increases and anti-competitive behaviour. But, unlike the French and British governments, it shied away from assuming more direct responsibility, largely for political and ideological reasons.

However, when the situation of the US steel industry became increasingly difficult, since the late 1970s successive administrations gave in to demands from the producers for import restrictions. The result, like in many other industries, was a downward spiral of protectionism, loss of competitiveness of the domestic producers, a further increase of imports, leading to yet more restrictions etc.[61] Most steel producers used the 'breathing space' to diversify into other areas, though not always very successfully. The renewed dynamism of steel production in the United States in recent years, is actually not driven by the established producers but, as seen above, by the investment of the Japanese steel companies on the one hand, and by a number of new entrants, such as Nucor, on the other. The latter rely on low-cost mini-mill technology and therefore mainly compete on price with commodity-like products. These are exchanged in market-type transactions and do not require a close co-operation with the using industries.

CONCLUSION

The preceding overview has highlighted the crucial importance of the relationship between steel producers and steel users as an explanatory variable for the changes in national competitive advantage during the post-war period. It appears that both industries can benefit from a joint development and close co-operation to maximise returns along the process of transformation from raw materials to finished goods. Such a value-added strategy however, requires a close co-ordination and co-operation between the firms in the different stages of the production and transformation of steel, in order to limit transaction costs.

Ideally, this co-operation should take place between the producers and users themselves, without outside interference or arbitration. Japan and Germany come closest to this ideal type model, since steel producers and users have limited their potential conflicts and instead acted in concert according to a 'value chain logic'. In cases where the companies and industries concerned are unable to establish such a mutually beneficial and stable relationship, the government might be able, at least temporarily, to assume a co-ordinating role. South Korea has quite clearly demonstrated that government-induced co-ordination can successfully minimise transaction costs.

Selective intervention on the other hand, as practised in France and the United Kingdom during much of the post-war period, appears less likely to succeed and might have created additional problems instead. However, in both countries public ownership seems to have played a rather positive role in the 1980s, when the decline in steel production was reversed. Since then, Usinor Sacilor and British Steel have been developing a more co-operative relationship with the steel users. However, given the rather strained relations prevalent for several decades, both companies found it hard to regain the confidence of its customers and to recapture the market share lost.

In general, by identifying and comparing the relationship between the steel producers and users in a number of countries, the article has demonstrated the importance of history in understanding national patterns of conflict and co-operation in the process of steel production and transformation. It seems that in most cases, the existing relationships were the result of a long-term evolution. As a consequence, it appeared difficult, but not entirely impossible, to change the relationship itself and, even more so, the perceptions which had been formed over a long time period.

Turning from the past to the future, what, if any, are the lessons to be learnt from the analysis of the relationship between steel producers and users? First of all, it suggests that downstream activities, in addition to raw material costs and technology, are among the most important factors to be considered in order to obtain, sustain or regain competitive advantage in steel production. This refers not only to the existence of domestic using industries, but also to the co-ordination between the production and transformation stages. Secondly, government involvement is not necessarily harmful. Under certain conditions it can, at least temporarily, play a co-ordinating role between the steel producing and the steel using industries. It seems, however, crucial that public authorities actually try to (re-)establish a positive relationship between producers and users which continues to function well once the companies are returned to private ownership.

NOTES

The author would like to thank the conference organiser, Ruggero Ranieri, the conference participants as well as his colleagues Mark Casson and Geoffrey Jones for a number of helpful suggestions. The usual disclaimer applies.

1. Eurostat, *Iron and Steel 1952–1982* (Luxembourg, 1983), p. 19, Seiichiro Yonekura, *The Japanese Iron and Steel Industry, 1850–1990. Continuity and Discontinuity* (Basingstoke, 1994), pp. 1–3, for the development of one of the Japanese steel producers Etsuo Abe, 'Anatomy of a Japanese Steel Firm:

NKK – Its Strategy and Performance Against the Tide, 1951–1990', in Takeshi Yuzawa (ed.), *Japanese Business Success. The Evolution of a Strategy* (London, 1994), pp. 97–118, and for South Korea Sung-il Juhn, 'Challenge of a Latecomer: The Case of the Korean Steel Industry with Specific Reference to POSCO', in Etsuo Abe and Yoshitaka Suzuki (eds.), *Changing Patterns of International Rivalry. Some Lessons from the Steel Industry* (Tokyo, 1991), pp. 269–93.

2. See esp. Yoshitaka Suzuki, 'The Rise and Decline of Steel Industries: A Business Historical Introduction', in Abe and Suzuki (eds.), *Changing Patterns*, pp. 1–17.
3. Estimates of the economies of scale in steel production vary. However, they increased significantly in the post-war period and are now situated between 5 and 10 million metric tons for integrated steel works; a detailed overview of the relevant research is given by Anthony Cockerill, 'The Steel Industry. International Comparisons of Industrial Structure and Performance', University of Cambridge, Department of Applied Economics, Occasional Paper 42 (1974), pp. 67–92.
4. Ibid., pp. 93–107. See also Donald F. Barnett and Louis Schorsch, *Steel: Upheaval in a Basic Industry* (Cambridge, MA, 1983), passim.
5. Patricia A. O'Brien, 'Industry Structure as a Competitive Advantage: The History of Japan's Post-war Steel Industry', *Business History* 34, No. 1 (January 1992), pp. 128–59 and, very similar, James E. Vestal, *Planning for Change. Industrial Policy and Japanese Economic Development, 1945–1990* (Oxford, 1993), chapter 5.
6. Thomas K. McCraw and Patricia A. O'Brien, 'Production and Distribution: Competition Policy and Industry Structure', in Thomas K. McCraw (ed.), *America versus Japan* (Boston, 1986), pp. 77–116, here pp. 84–100.
7. Paul A. Tiffany, *The Decline of American Steel* (New York, 1988), esp. chapter 4. In addition, he blamed organised labour and its inflexible high-wage policy for the decline of US steel production. See also his contribution 'The American Steel Industry in the Postwar Era: Dominance and Decline', in Abe and Suzuki (eds.), *Changing Patterns*, pp. 245–65.
8. Seiichiro Yonekura, 'The Postwar Japanese Iron and Steel Industry: Continuity and Discontinuity', in Abe and Suzuki (eds.), *Changing Patterns*, pp. 193–241; in more detail his book *The Japanese Iron and Steel Industry*.
9. Michael E. Porter, *The Competitive Advantage of Nations* (London, 1990), esp. chapters 3 and 4. According to Porter, firm strategy, structure and rivalry is the most important of four inter-related determinants of competitive advantage which he calls the 'diamond'. The others are: factor conditions, demand conditions, related and supporting industries.
10. Quoted from his article 'The Competitive Advantage of Nations', *Harvard Business Review* (March–April 1990), pp. 73–93, here p. 82.
11. See for details Yonekura, *The Japanese Iron and Steel Industry*.
12. Porter, *Competitive Advantage*, pp. 473–4.

13. The graph is inspired by Mark Casson, *Entrepreneurship and Business Culture* (Aldershot, 1995), figure 8.1.
14. Cockerill, *The Steel Industry*, pp. 21–36.
15. Yonekura, *The Postwar Japanese Iron and Steel Industry*, pp. 198 and 223–6.
16. Juhn, 'Challenge of a Latecomer' and Alice H. Amsden, *Asia's Next Giant. South Korea and Late Industrialization* (New York, 1989), chapter 12.
17. The ECSC Treaty also prohibits subsidies, but the relevant provisions have not been enforced very strictly until now, see Damien J. Neven and John Vickers, 'Public Policy Towards Industrial Restructuring: Some Issues Raised by the Internal Market Programme', in Karel Cool et al. (eds.), *European Industrial Restructuring in the 1990s* (Basingstoke, 1992), pp. 169–79.
18. The idea goes back to an article written by Ronald H. Coase, 'The Nature of the Firm', *Economica*, new series, Vol. 4 (November 1937), pp. 386–405. It has recently been developed in more detail by others, viz. Oliver Williamson, see esp. his book *The Economic Institutions of Capitalism: Firms, Markets, Relational Contracting* (New York, 1985).
19. A survey of 100 steel users, conducted in the United States in 1980, identified the following purchase criteria: price, quality, supply protection, sales attitudes and various marketing services; mentioned in Robert E. Ankli and Eva Sommer, 'The Role of Management in the Decline of the American Steel Industry', *Business and Economic History* 25, No. 1 (Fall 1996), pp. 217–31.
20. Jean-Marie Moine, *Les barons du fer. Les maîtres de forges en Lorraine du milieu du 19e siècle aux années trente*, (Metz/Nancy, 1989).
21. See for the economic logic behind this phenomenon Frances Ferguson Esposito and Louis Esposito, 'Excess Capacity and Market Structure', *Review of Economics and Statistics*, Vol. 56, No. 2 (May 1974), pp. 188–94.
22. In his memoirs, the former French steel industrialist Roger Martin has criticised this almost 'fetishist' pre-occupation of successive governments with steel production and highlighted its negative consequences, *Patron de droit divin ...* (Paris, 1984), pp. 226–7. In the 1970s, shortly after becoming CEO of Pont-à-Mousson, he divested the company of all its steel-making activities.
23. This article explicitly focuses on steel production in capitalist countries where decision-makers are accountable either to shareholders or to voters, in case of private or public ownership respectively. Until very recently, the steel production in the former Soviet Union and in mainland China evolved under very different conditions and therefore has to be analysed separately. The depth of analysis for the chosen countries is largely driven by the availability of the necessary information. For more detail on the French and the German cases see Matthias Kipping, 'Inter-Firm Relations and Indus-

trial Policy: The French and German Steel Producers and Users in the Twentieth Century', *Business History*, Vol. 38, No. 1 (January 1996), pp. 1–25.

24. Tiffany, 'The American Steel Industry', pp. 248–50 and United Nations, Economic Commission for Europe, *The European Steel Industry and the Wide-Strip Mill. A Study of Production and Consumption Trends in Flat Products* (Geneva, 1953).
25. E.g. W. Mark Fruin, *The Japanese Enterprise System. Competitive Strategies and Cooperative Structures* (Oxford, 1992) and Kazuo Wada, 'The Development of Tiered Inter-firm Relationships in the Automobile Industry: A Case Study of Toyota Motor Corporation', *Japanese Yearbook on Business History* 8 (1991), pp. 23–47.
26. Vestal, *Planning for Change*, pp. 140–1 stresses, however, that this mechanism also resulted in significant over-capacity when demand dropped following the oil crises. See also Yonekura, 'The Postwar Japanese Iron and Steel Industry', p. 227, who highlights the fact that the investments in steel production generated around 20–25% of the demand for steel products and Abe, 'Anatomy of a Japanese Steel Firm', pp. 99–100 with respect to 'indirect steel exports'.
27. A first summary of her ongoing work is contained in 'Les liens entre l'automobile et la sidérurgie japonaises', contribution to the 4th International GERPISA Colloquium: 'L'industrie automobile mondiale: entre homogénéisation et hiérarchisation', Paris, 19–21 June 1996. The author would like to thank Ms. Guelle for providing him with a copy of this paper. See also Jonathan Aylen, 'Japanese Steel: A Matter of Relationships', *Steel Times International*, June 1985, p. 39.
28. Martin Kenney and Richard Florida, *Beyond Mass Production. The Japanese System and Its Transfer to the U.S.* (New York, 1993), chapter 6.
29. *Ibid.*, pp. 164–5.
30. Guelle, 'Les liens'; see also Seiichiro Yonekura and Françoise Guelle, 'The Japanese Steel Industry in the 1990s. Rationalization, Diversification, and Globalization', *The Steel Industry in the New Millennium*, Vol. 2: Institutions, Privatisation and Social Dimensions, pp. 151–70.
31. It has been described and analysed extensively in two dissertations, Ulrich Nocken, 'Inter-Industrial Conflicts and Alliances in the Weimar Republic: Experiments in Societal Corporatism' (unpublished Ph.D. thesis, University of California, Berkeley, 1979) and Daniel P. Barbezat, 'International Cooperation and Domestic Cartel Control. The International Steel Cartel, 1926–1938' (unpublished Ph.D. thesis, University of Illinois, Urbana, 1988).
32. Minutes of the Trade Association's board meeting on 10 June 1948, Archives of the WV Stahl, Düsseldorf.
33. Matthias Kipping, 'Competing for Dollars and Technology: The United States and the Modernization of the French and German Steel Industries

after World War II', *Business and Economic History* 23, No. 1 (Fall 1994), pp. 229–40, here pp. 237–8.

34. Franz Lammert, 'Das Verhältnis zwischen der Eisen schaffenden und der Eisen verarbeitenden Industrie seit dem ersten Weltkrieg' (PhD. dissertation, University of Cologne, 1960), pp. 193–4.
35. Neven and Vickers, 'Public Policy', esp. table 7.7. Only German reunification and the threat of large-scale unemployment and subsequent social unrest have led to massive government subsidies for the ailing steel plants in the former GDR, see Werner Plumpe, 'Das Ende der Schwarzmetallurgie – Die Stahlindustrie in der früheren DDR im Übergang', *The Steel Industry in the New Millennium*, Vol. 2: Institutions, Privatisation and Social Dimensions, pp. 83–94.
36. Ha-Joon Chang, *The Political Economy of Industrial Policy* (Basingstoke, 1994), quote pp. 134–5.
37. The following is based mainly on Amsden, *Asia's Next Giant*, chapter 12 and Juhn, 'Challenge of a Latecomer'.
38. 28.1 vs. 9.6% from 1967 to 1972, 16.7 vs. 9.7% from 1972 to 1977 and 13.5 vs. 5.8% from 1977 to 1982, ibid., p. 270.
39. See on this 'mystery' also Amsden, *Asia's Next Giant*, pp. 293–4.
40. Chang, *Political Economy*, p. 108; on shipbuilding Amsden, *Asia's Next Giant*, chapter 11 and on the business groups in general Richard M. Steers et al., *The Chaebol: Korea's New Industrial Might* (Cambridge, MA, 1989).
41. Porter, *Competitive Advantage of Nations*, p. 474.
42. With approx. 1 million metric tons per year the production capacity of POSCO's first stage, completed in 1973, was lower than the minimum efficient size of an integrated steel work, but this initial handicap was largely offset by the lower labour costs in South Korea.
43. See note 38 above. Between 1987 and 1992 steel consumption still grew at over 10 per cent.
44. This is mainly true for its domestic clients, since the exports are not handled by the company directly, but by the trading arms of the chaebol.
45. Amsden, *Asia's Next Giant*, pp. 309–14.
46. The company was floated on the stock exchange in 1988. A majority of the shares is owned by private investors, but the government controls an overall blocking majority of 35%, Juhn, 'Challenge of a Latecomer', p. 277.
47. See in more detail Matthias Kipping, 'A Confrontational Model of Interfirm Relations? The Case of Steel Producers and Users in France', in Mila Davids et al. (eds.), *Proceedings of the Conference on Business History, October 1994* (Rotterdam, 1995), pp. 248–68.
48. See Patrick Fridenson, *Histoire des usines Renault,* Vol. I: *Naissance de la grande entreprise 1898–1939* (Paris, 1972), pp. 134–7, 220–6 and 329–30.
49. Details about Lefaucheux's position and actions can be found in Matthias

Kipping, 'Les tôles avant les casseroles. La compétitivité de l'industrie française et les origines de la construction européenne', *Entreprises et Histoire*, No. 5 (June 1994), pp. 73–93.

50. See for the role of the steel users at the origins of the ECSC Matthias Kipping, *Zwischen Kartellen und Konkurrenz. Der Schuman-Plan und die Ursprünge der europäischen Einigung 1944–1952* (Berlin, 1996).
51. Jean Sallot, 'Le contrôle des prix et la sidérurgie française 1937–1974' (thèse pour le doctorat, University of Paris I, 1993).
52. See Michel Freyssenet and Catherine Omnès, *La crise de la sidérurgie française* (Paris, 1984) and René Gendarme, 'Le casse-tête de la sidérurgie. Analyse d'une crise à travers le cas lorrain', *Revue française de gestion,* No. 39 (January–February 1983), pp. 66–84.
53. Eric Godelier, 'De la stratégie locale à la stratégie globale: la formation d'une identité de groupe chez Usinor (1948–1986)' (thèse de doctorat, EHESS, Paris, 1995), pp. 330–54 and 370–2.
54. For the former in more detail Gendarme, 'Le casse-tête de la sidérurgie', pp. 71–2.
55. Presentation of Guy Dollé, member of Usinor Sacilor's strategy and planning department at the Terni Conference, 4–5 July 1996.
56. For details Keith Ovenden, *The Politics of Steel* (London, 1978).
57. See for this and the following Geoffrey F. Dudley and Jeremy John Richardson, *Politics and Steel in Britain, 1967–1988. The Life and Times of the British Steel Corporation* (Aldershot, 1990), esp. chapter 10.
58. Dudley and Richardson characterise the government in this instance as 'cross-sectoral mediator' and welcome its efforts to achieve what they call 'co-ordinated rationality', *ibid.*, pp. 253 and 290–5.
59. Jonathan Aylen, 'Privatisation of the British Steel Corporation', *Fiscal Studies* 9, No. 3 (August 1988), pp. 1–25; see also Naresh R. Pandit, 'British Steel Corporation: Probably the Biggest Turnaround Story in UK Industrial History', Manchester Business School, Working Paper No. 324 (March 1996).
60. Ankli and Sommer, 'The Role of Management'.
61. Robert B. Reich, *The Next American Frontier* (New York, 1983).

The Markets

THE STEEL MARKET IN ITALY THE TREND OF STEEL CONSUMPTION: MACROECONOMIC ASPECTS

F. AUDINO
Federacciai, Viale Sarce 336, 20126 Milano, Italy

STEEL IN THE ECONOMY: ITALY AND SOME OTHER INDUSTRIALISED COUNTRIES

In the century that is now coming to its end, the relationship between steel consumption and Gross Domestic Product (*steel intensity*) in various economies shows a certain variety in timing and trend (see Figs 1–6).

- in the USA a roughly stationary peak was reached during a relatively long period between the 20s and the 40s, showing a decline thereafter;
- in Germany the ceiling was hit in the late 50s at about the same level as the USA;
- France and the UK attained maximum level on the average in the 50s at a lower level than in the two previous countries;
- Japan reached its peak levels in the early 70s showing a very steep increase after the end of the Second World War;
- Italy also had the fastest development of steel intensity after the Second World War but hit a significantly lower ceiling than the other countries and the following down-fall was less steep and, with the exception of Japan, is now higher than in the other countries.

Steel consumption per capita

Steel consumption per capita in the current century also showed a variety of development patterns.

- in the USA, Germany, France and the UK the top was reached in the late 60s, early 70s at higher level in the two former countries than in the latter two (Figs 7–10);
- Japan and Italy showed again a rapid increase after the Second World War and have not moved along a declining path (Figs 11 and 12).

THE STEEL MARKET DEVELOPMENT IN ITALY

In 1995:

- production was about 28 million metric tonnes of crude steel: the second

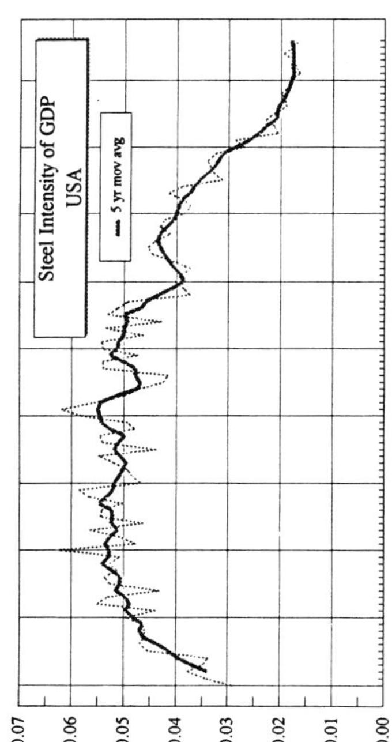

Fig. 1

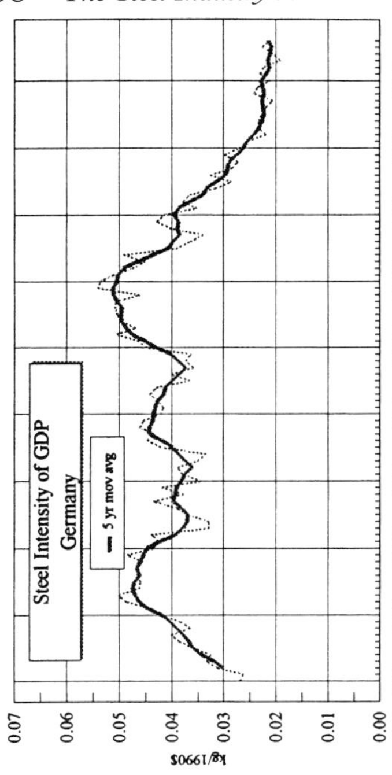

Fig. 2

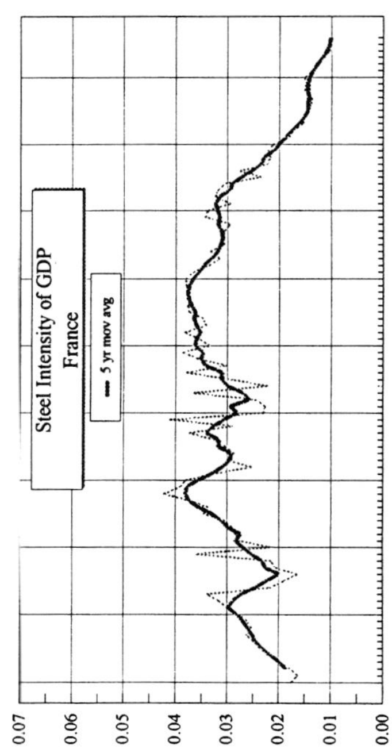

Fig. 3

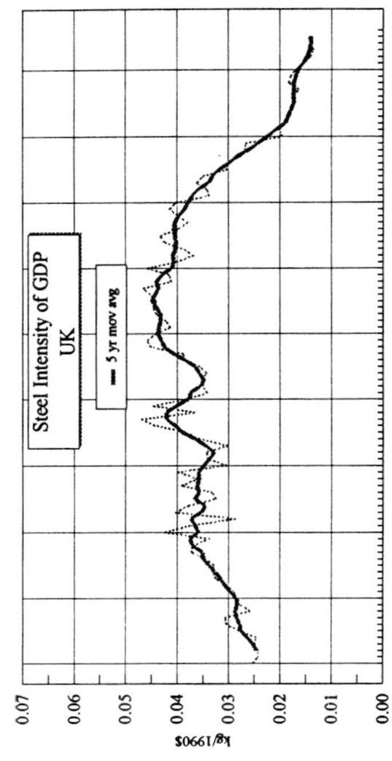

Fig. 4

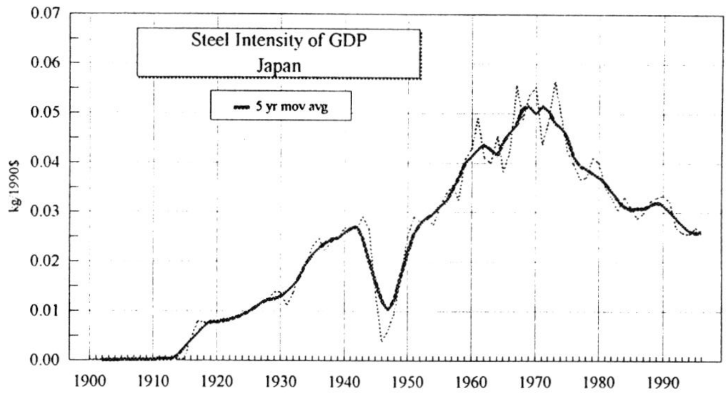

Fig. 5

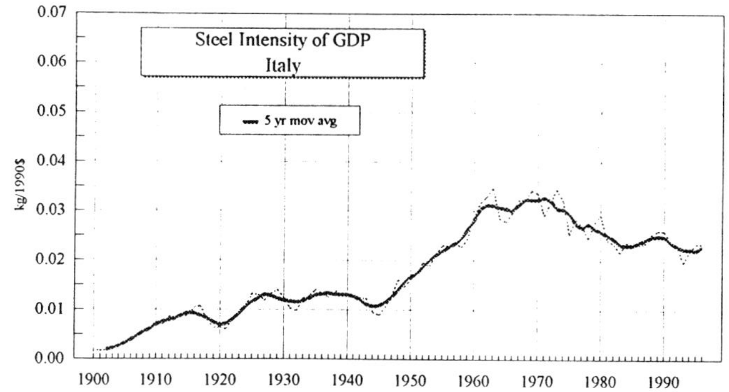

Fig. 6

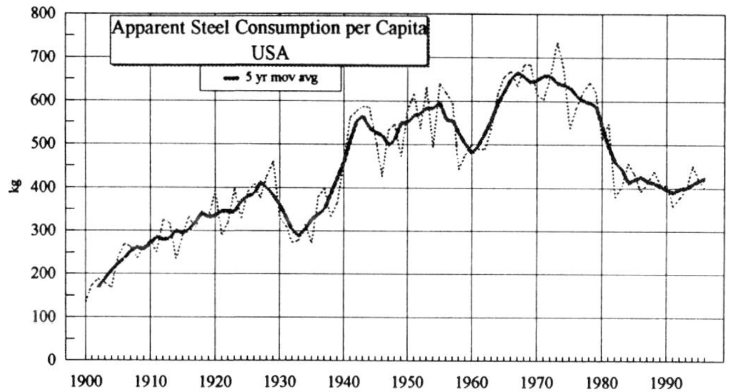

Fig. 7

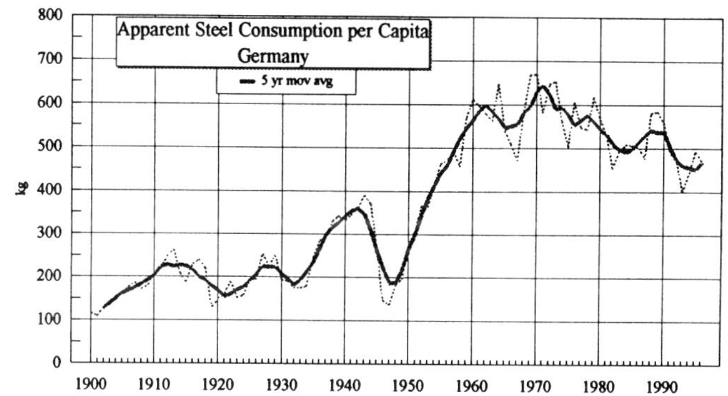

Fig. 8

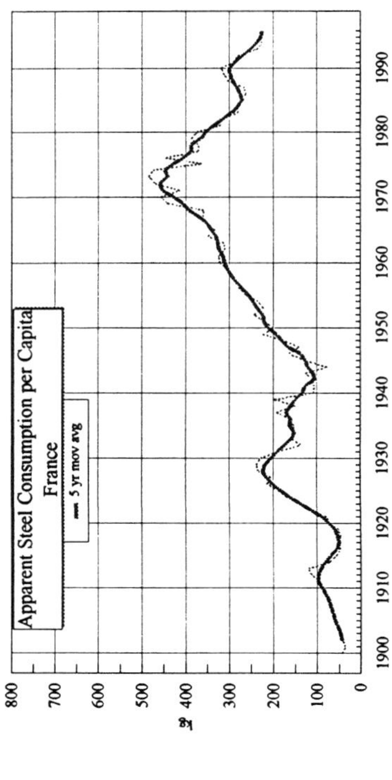

Fig. 9

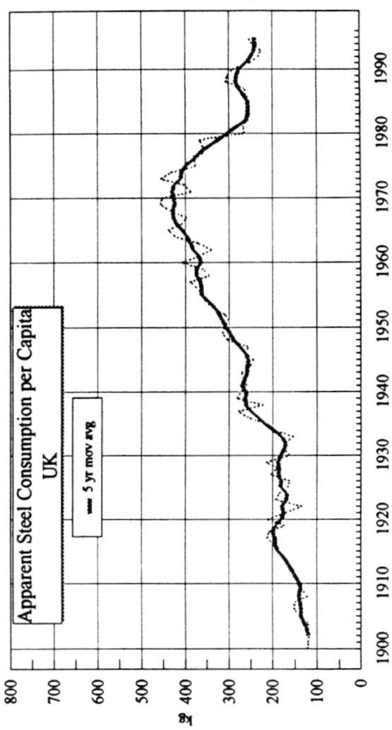

Fig. 10

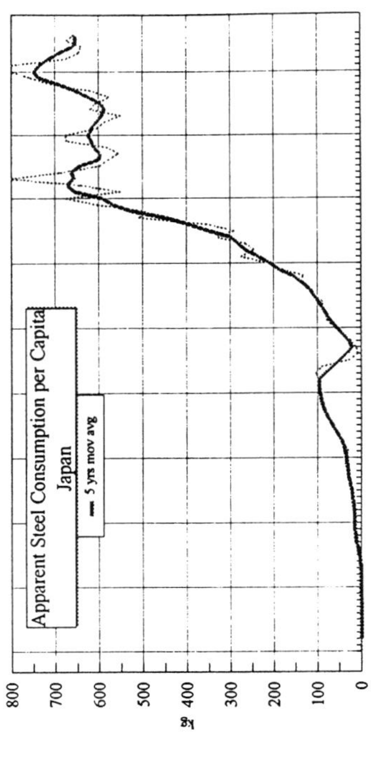

Fig. 11

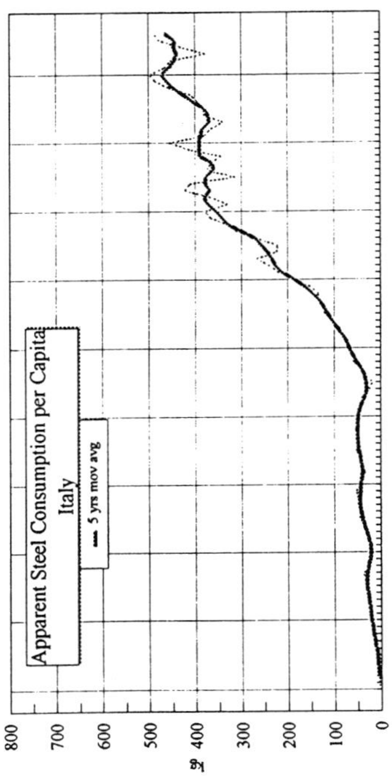

Fig. 12

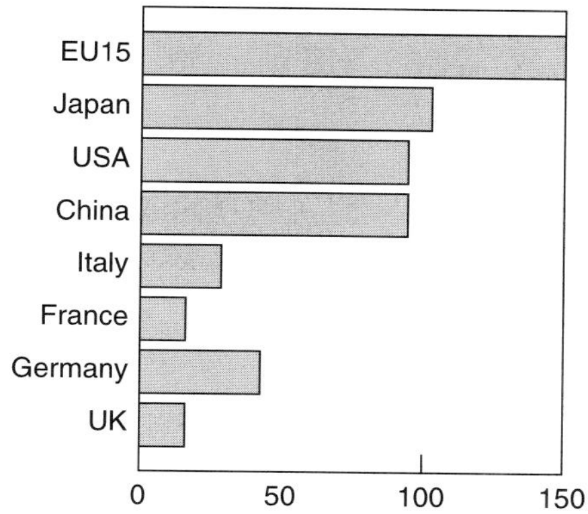

Fig. 13 Output of crude steel in selected countries in 1995 (Mt).

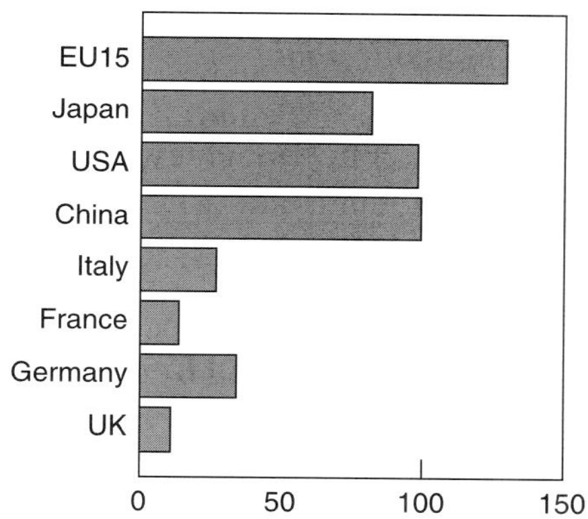

Fig. 14 Apparent steel consumption in selected countries in 1995 (Mt).

highest production in Europe; this compares to a production of 42 Mt in Germany, 18 in France and the UK (Fig. 13);

- apparent consumption was over 29 million tons of crude steel equivalent, again the second biggest steel consumer in Europe with Germany, the biggest economy, consuming about 37 Mt, France over 16 and the UK 14 (Fig. 14).

STEEL DEMAND

In 1950:

- apparent consumption of steel in Italy was about 3 Mt;

In 1995

- it was almost ten times as much (Fig. 15).

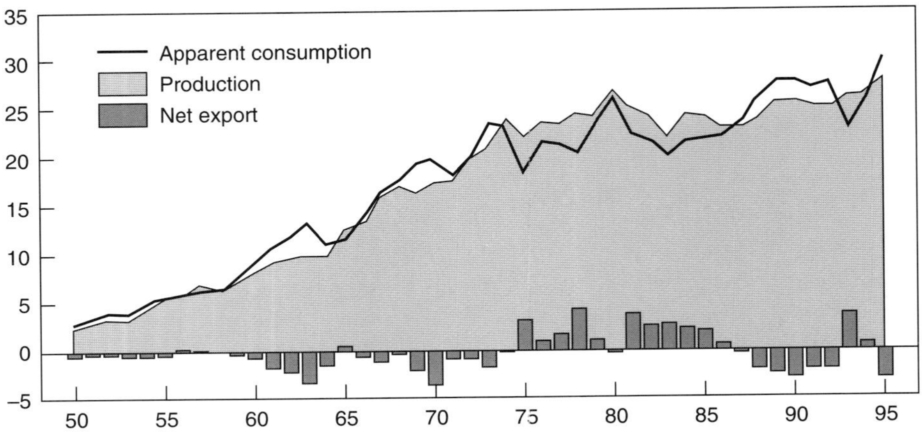

Fig. 15 Steel consumption in Italy.

This development is explained by the fast pace of the post-war reconstruction: the growth was steepest in the 50s and 60s and started to moderate thereafter.

A higher stage in the development of the economy and possibly also the adjustments induced by the oil shocks slowed down the trend and in the early 1980s it was currently assumed that it was going to flatten out (Fig. 16).

Reviewing the actual development that occurred afterwards it appears that the expected flattening of the trend has not materialised.

Net foreign trade of steel containing manufactured goods has always given a

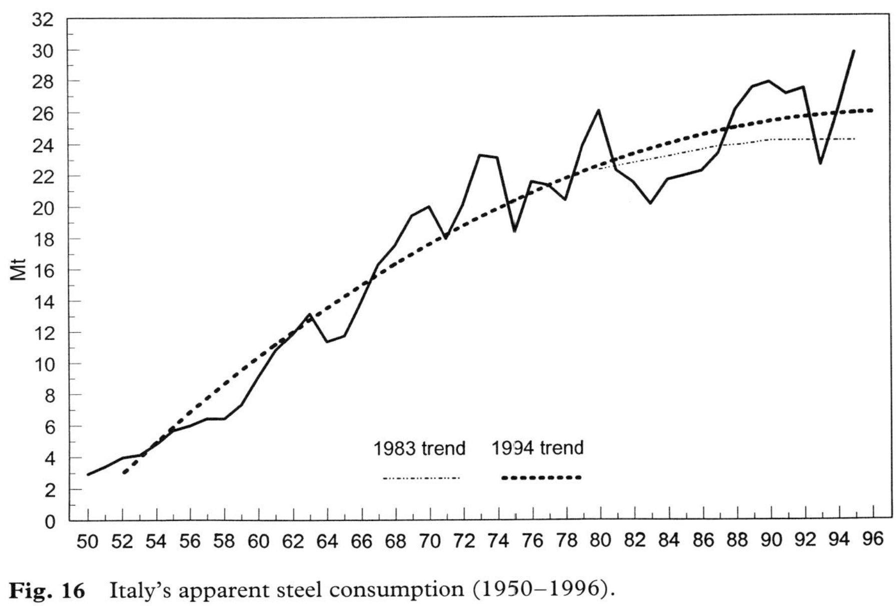

Fig. 16 Italy's apparent steel consumption (1950–1996).

substantial contribution to the development of apparent consumption: apparent final consumption on the average has been around 4 million tonnes lower (Fig. 17)

In 1993 and 1994, following a surge in the country's exports the contribution of net foreign trade increased to about 7 and 8 million tonnes respectively.

The relationship between steel consumption and macro-economic variables in Italy has shown a different pattern from other EU countries.

- in the EU as a whole, after the first oil shock, apparent steel consumption became substantially flat, despite continued economic growth;
- in Italy, despite the different magnitude of fluctuations, apparent consumption followed closely economic and industrial development.

However, both in Italy and to a lesser extent in the EU, changes in apparent consumption showed some correlation with GDP as well as with industrial production (Figs 18–24).

The cycle of apparent steel consumption has shown wide fluctuations, particularly pronounced at the time of past oil shocks.

In the 80s and the beginning of the 90s, fluctuations seemed to become more subdued but a renewed strength of the cycle appeared in the economic recession of 1993 and the subsequent upswing.

The cycle of real steel consumption (the steel products actually employed in the production of steel using manufactured goods) is of course less unstable than the cycle of apparent steel consumption: steel users' and traders' inventories' changes, often unpredictable, and speculative in nature, amplify the cycle of real steel consumption (Figs 25–32).

Steel Prices

Steel prices in Europe show a roughly common trend. Nominal DM denominated prices for hot rolled coil in the last 20 years appear to have a more or less stationary trend with wide fluctuations (Fig. 33).

The latest upswing of the past two years have abruptly come to an end in the last part of 1995 and in the second quarter of 1996 prices were getting closer to the lowest levels reached in the past.

Quite different is the picture given by the development of 'real' price:

- looking at the terms of trade between the average European steel price and inputs into the steel making process one can see a steeply declining trend (Fig 34).

The fast increase of imports from third countries has contributed to a great extent to undermine price stability.

Apparent final consumption = apparent consumption before net foreign trade; i.e. steel products actually used in the country excluding steel products incorporated in exported manufactured goods, net of imported manufactured goods.

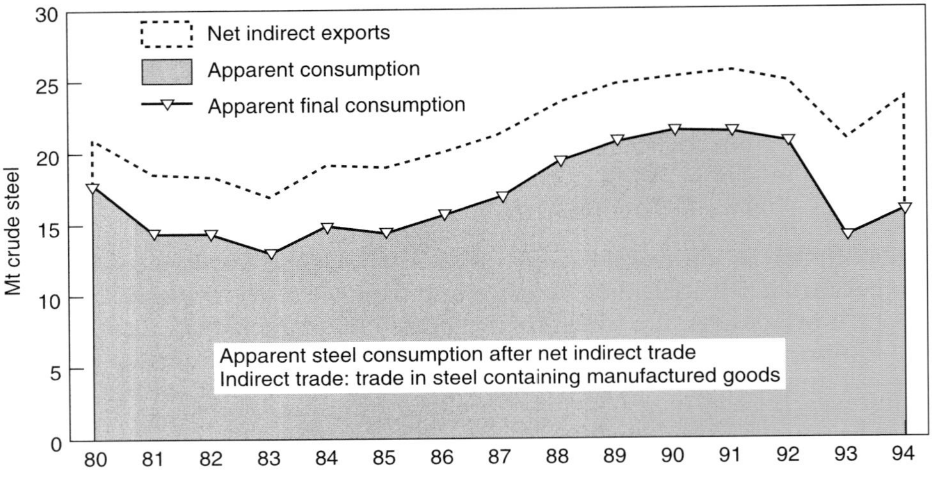

Fig. 17 Apparent steel consumption in Italy after net indirect trade (1980–1994).

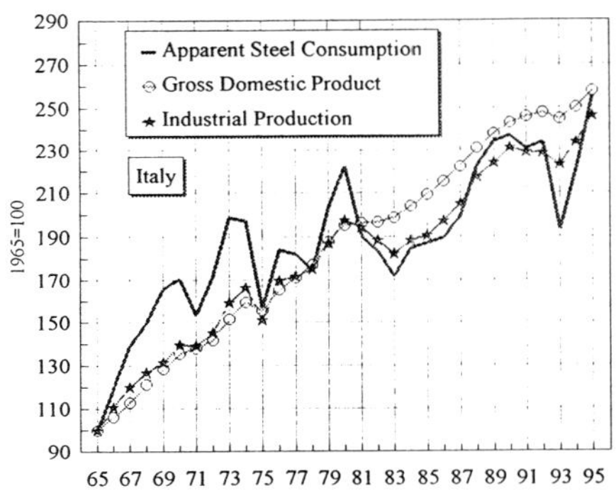

Fig. 18

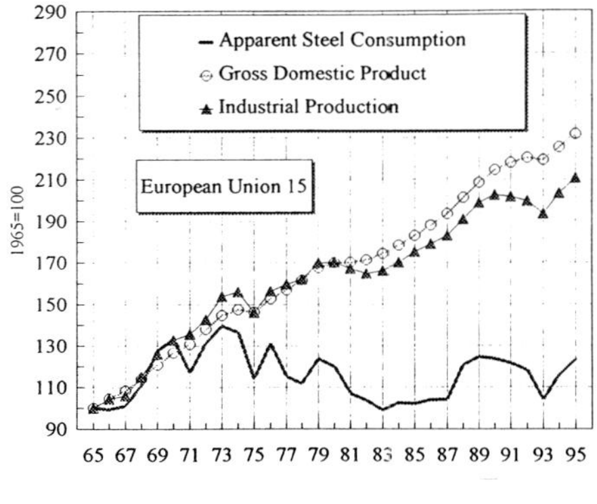

Fig. 19

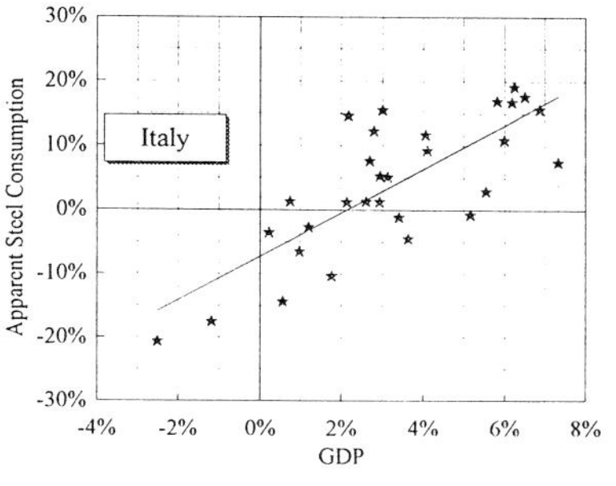

Fig. 20

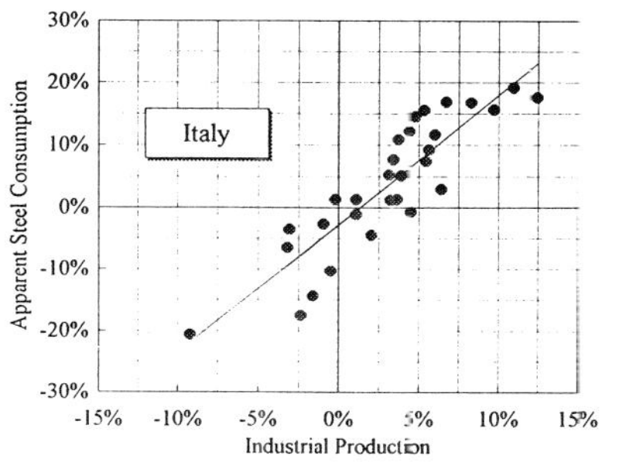

Fig. 21

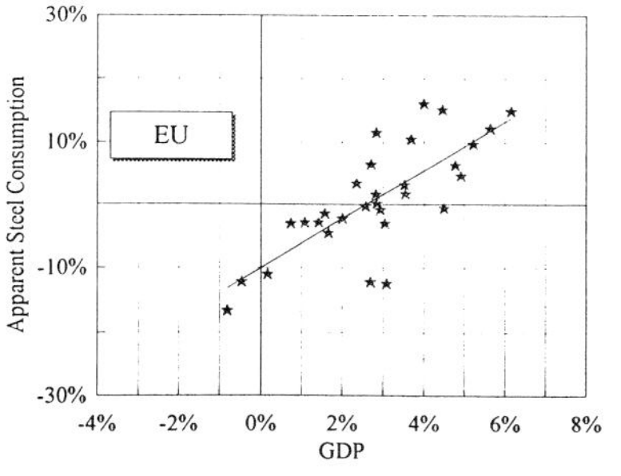

Fig. 22

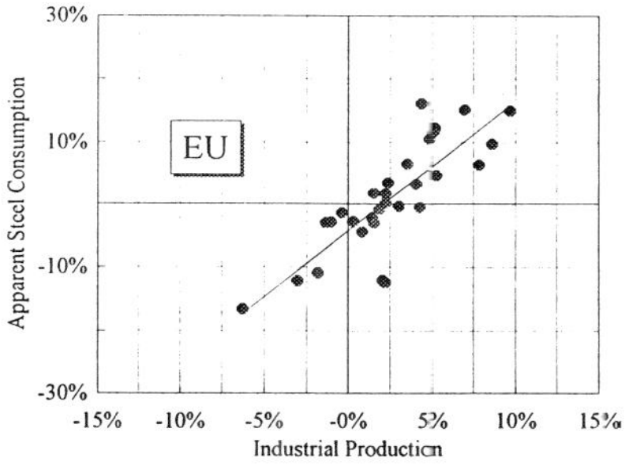

Fig. 23

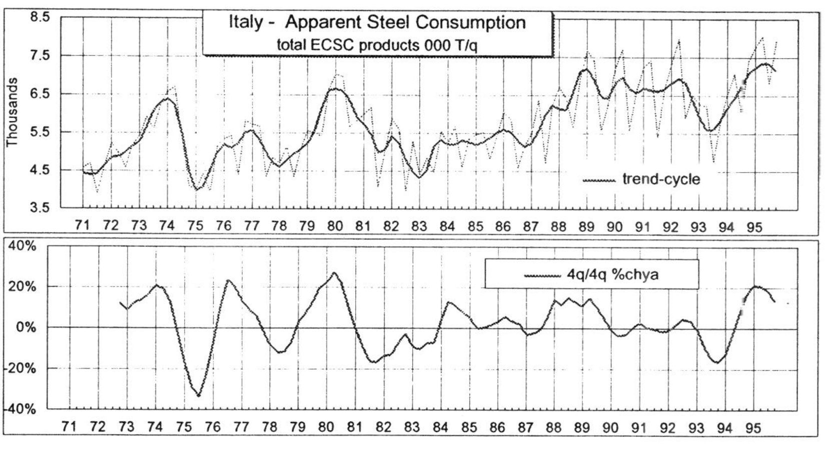

Fig. 24

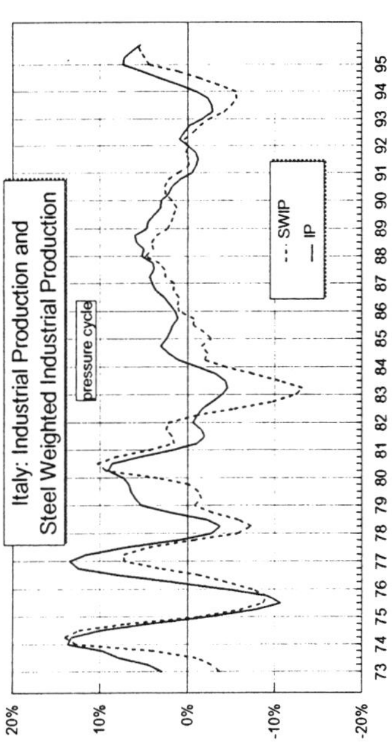

Fig. 25

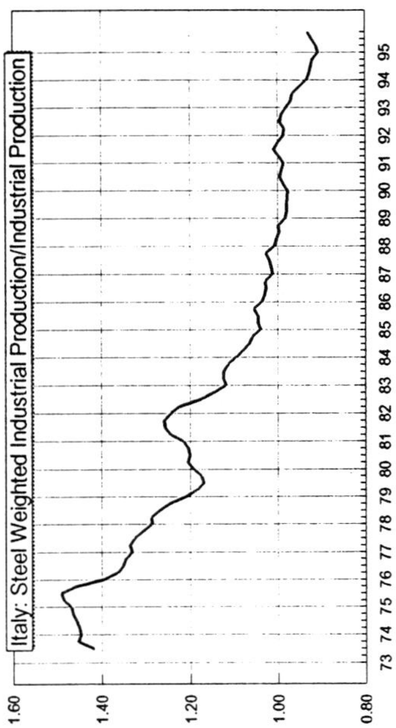

Fig. 26

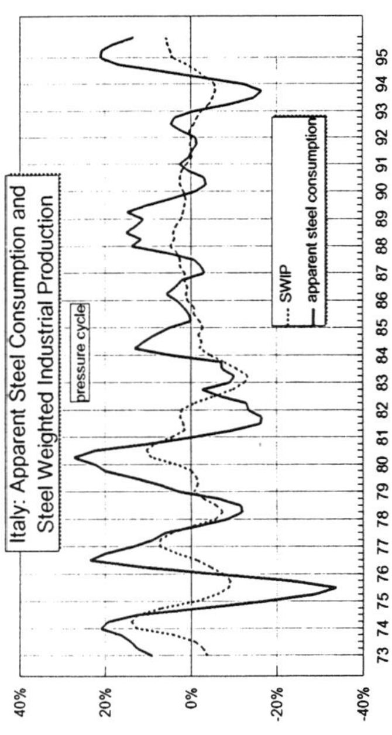

Fig. 27

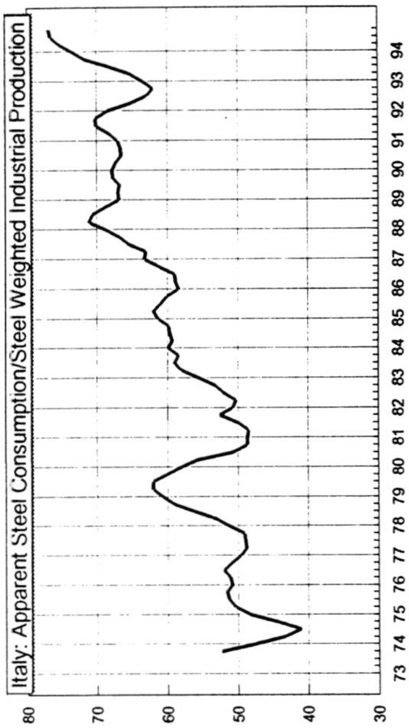

Fig. 28

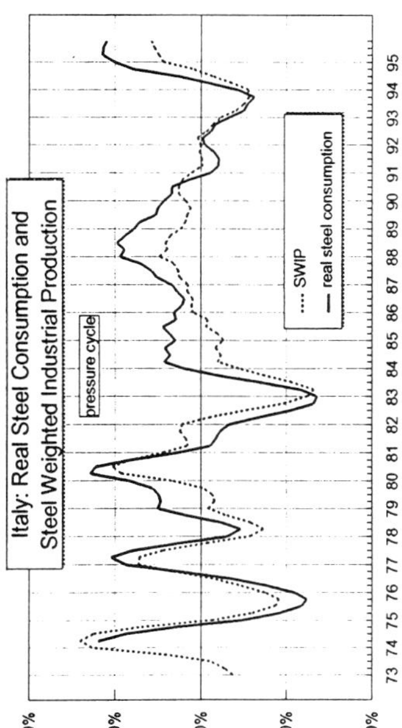

Fig. 29

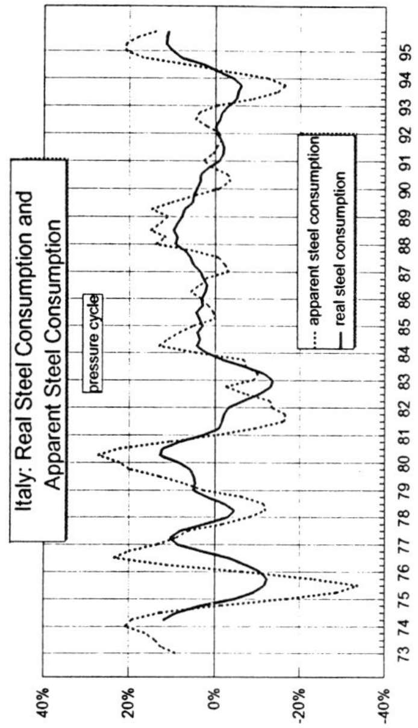

Fig. 30

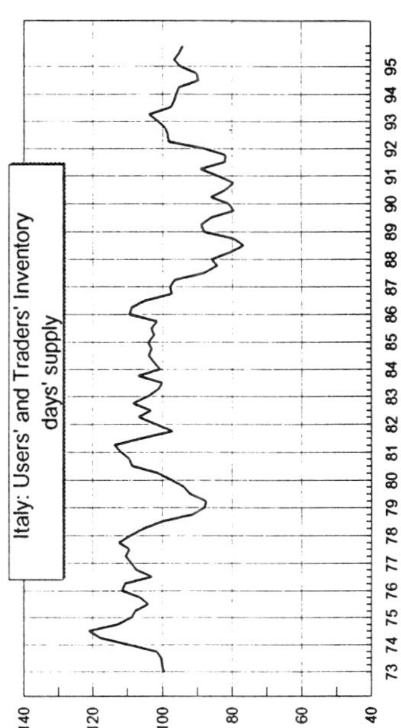

Fig. 31

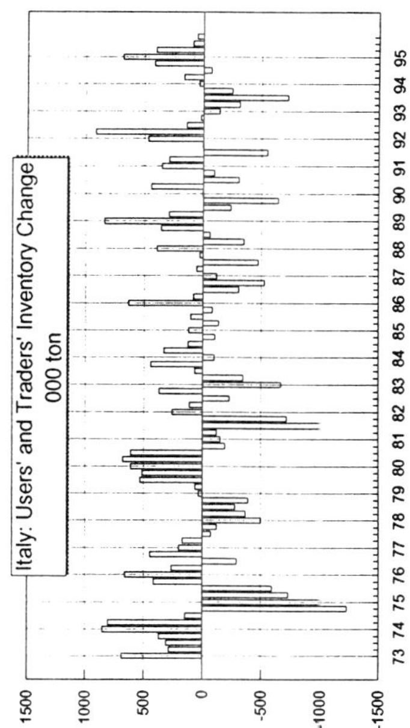

Fig. 32

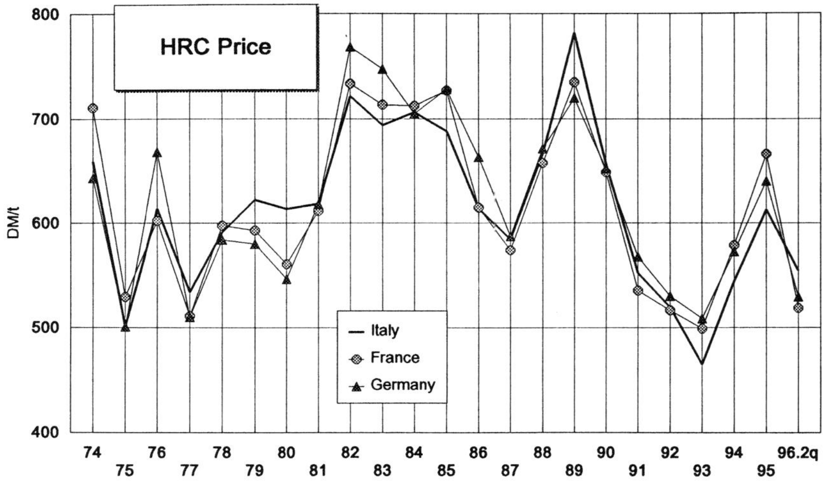

Fig. 33

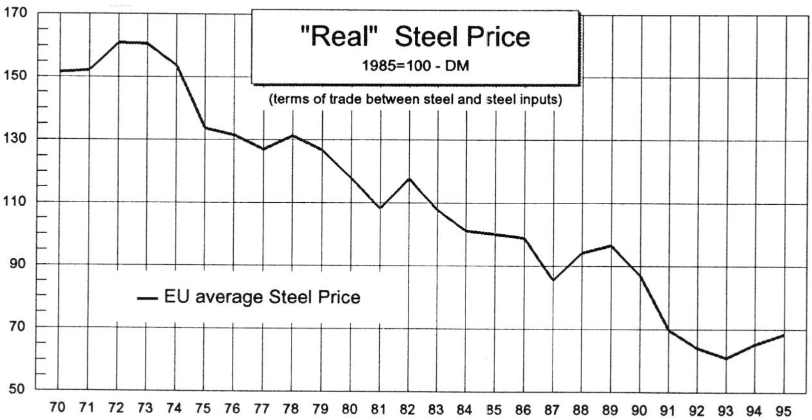

Fig. 34 'Real' steel price index in the EU (1970–1995).

When demand for steel products becomes tight and prices start to grow, a speculative spiral is magnified by imports from third countries which is reflected in all EU countries (Fig. 35):

(a) imports into the EU market tend to accelerate more rapidly than consumption and domestic deliveries of EU producers.

(b) this contributes to an excessive oversupply that, together with intensified stock movements, exacerbates price volatility bringing about an inversion of the price cycle.

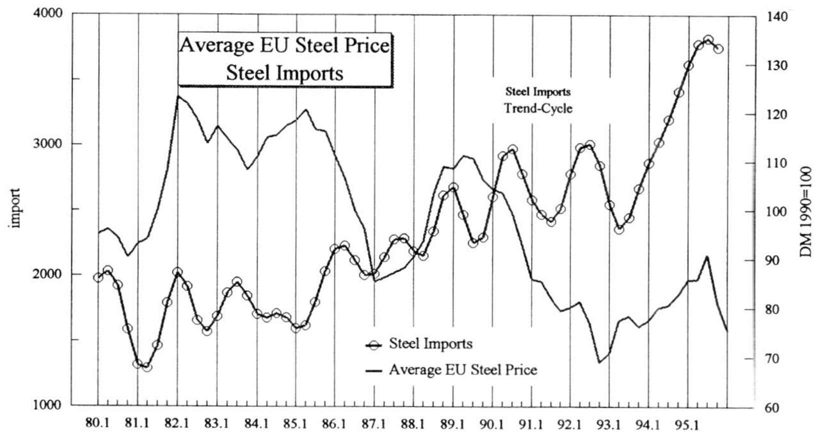

Fig. 35 Average EU steel price against the price of steel imports (1980–1995).

(c) imports at heavily discounted price lead to unintended stockpiling of steel products that contribute to aggravate market instability to the detriment of both consumers and producers.
(d) delays in the delivery of imports to buyers may cause additional pressure even when the demand cycle has already flattened or is even declining.

Statistical analysis seems to support evidence of a stronger impact of imports on prices than domestic deliveries.

Domestic deliveries of EU producers have echoed the development of apparent consumption avoiding to induce an over supply of the market as well as speculative build up of inventories which are always bound to determine price

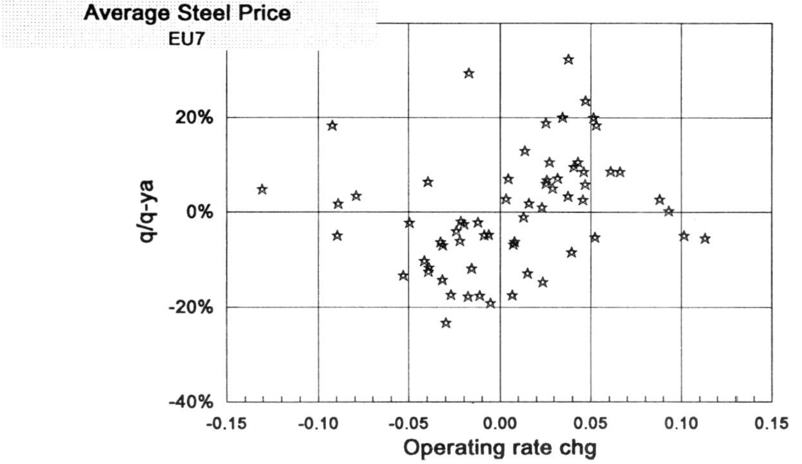

Fig. 36 Correctation between average steel prices and operating rates in the EU.

instability and bring about an inversion of the price cycle. Moreover the operating rate of EU steel producers does not seem negatively correlated with average price changes (Fig. 36).

Due to their often artificially favourable cost position, foreign producers can sustain their penetration in the EU market and consolidate their market share even when prices decline to levels that are not compatible with profitability thresholds.

In the long run, economically unjustified imports would displace more efficient EU steel producers thus reducing competition and market efficiency, eventually at the ultimate disadvantage of consumers.

The gains in productivity in the European steel industry appear clearly from the falling trend in employment and from the stable, and in some cases rising, trend in steel production (Figs 37–42).

Technological change as illustrated in Figs 43–46, showing evolution in steelmaking processes in Italy and in the main industrialised nations, has supported the rise in productivity.

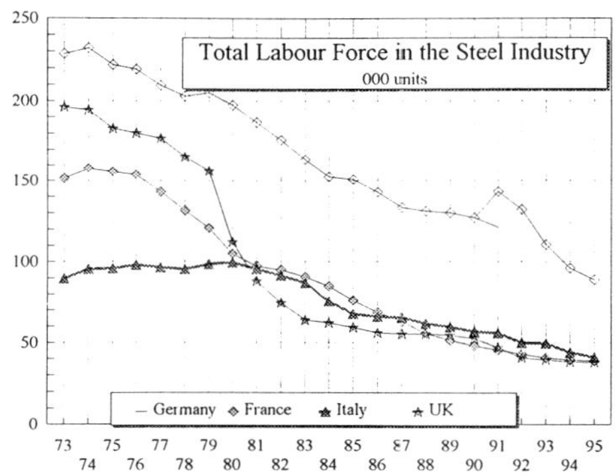

Fig. 37

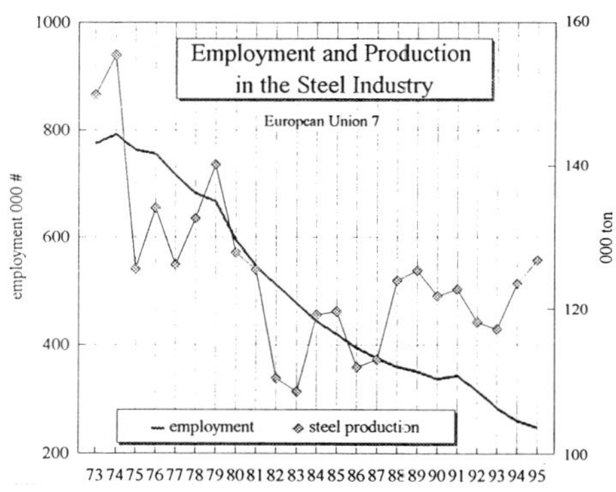

Fig. 38

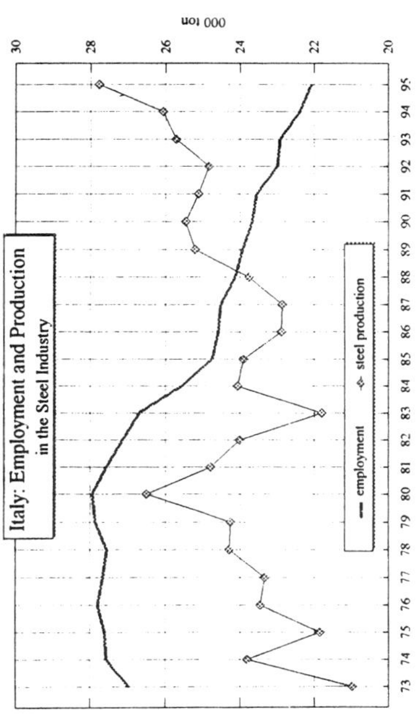

Fig. 39

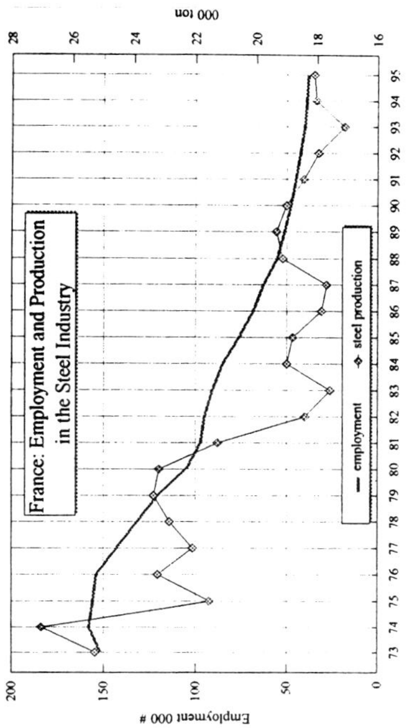

Fig. 40

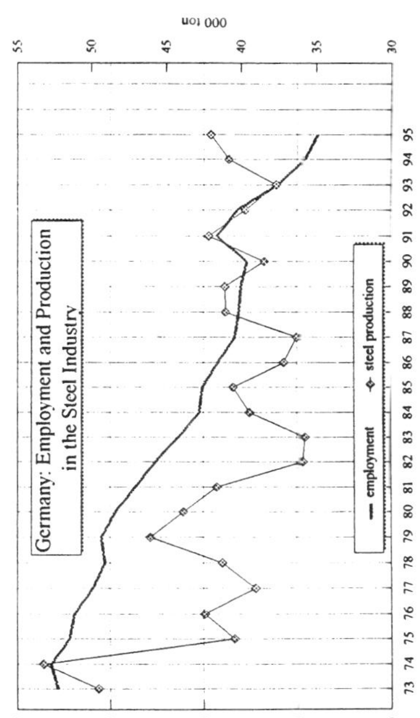

Fig. 41

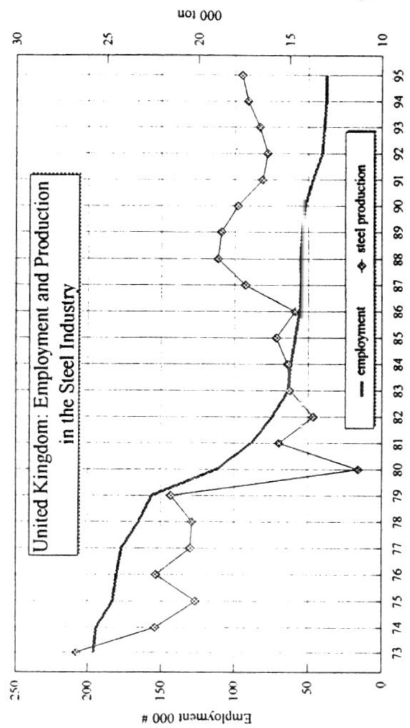

Fig. 42

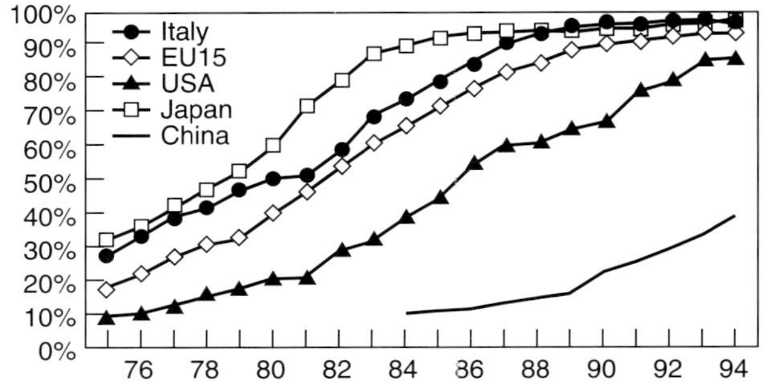

Fig. 43 Production of continuously cast steel % of total crude steel.

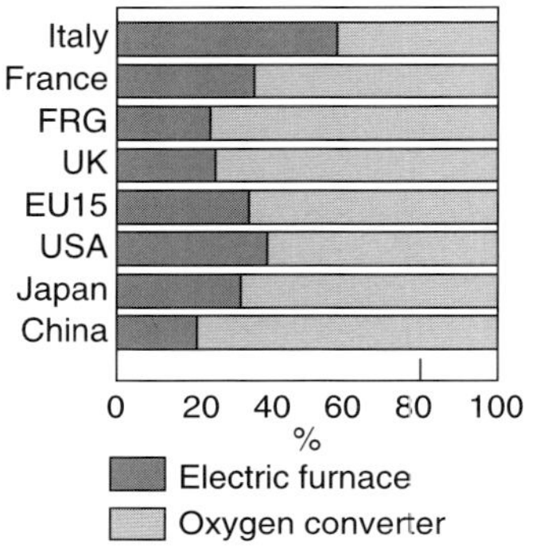

Fig. 44 Crude steel production by process in selected countries in 1995.

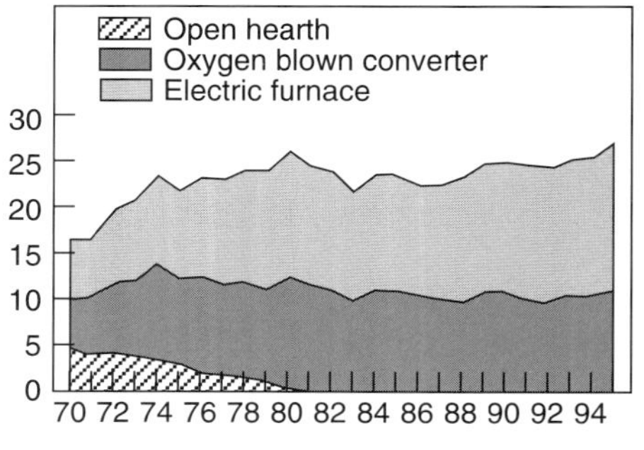

Fig. 45 Italy: steel production by process (000 t).

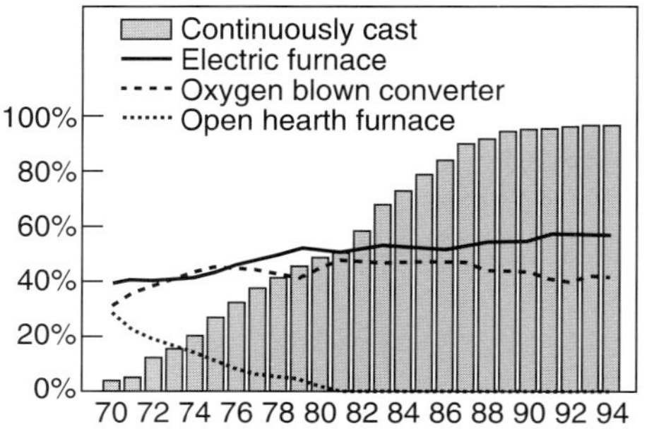

Fig. 46 Italy: steel production by process share.

SPECIAL STEEL MARKETS: A PERSPECTIVE

P. SOULIGNAC*

Acciai Speciali Terni SpA, Viale B. Brin, 218, I–05100 Terni, Italy

WHAT IS SO SPECIAL ABOUT SPECIAL STEELS?

When I was asked to present a contribution to this conference, I was told that I was free to speak about anything concerning the specialty steel markets. Such a wide framework is particularly adapted to a class of steels which is extremely clear to people who are working in the field (you are in the special steel area when you feel different from 'big steel') but which are not easily categorised, if you consider that special steels include (Fig. 1):

- Heat resistant steels
- Stainless steels
- Mould and Tool steels
- High speed steels
- Cryogenic steels
- PH and Maraging steels
- Electrical steels with improved magnetic properties
- or non magnetic steels and many others

A general definition could be the following:

Special steels are designed, produced, distributed and processed to offer to

	World (Mt)	W. Europe (Mt)
All steels	660	135
Special steels	71	18
of which:		
Engineering steels	45	12
Stainless and heat resisting	12.5	3.5
(Cr = 25%, CrNi = 75%,		
Electrical steels	6	1.5
(NO = 50%, GO = 20%, SP = 30%,		
Spring steels	2.5	0.5
Bearing steels	3.5	0.7
Tool, mould steels	1.5	0.45
High speed steels	0.15	0.03
High nickel alloys	0.2	0.05

Fig. 1 Consumption of Special Steels (1995).

* Now with CLI, BP56–F71202 Le Creusot Cedex France.

the whole user's chain, from processors to end user, unique combinations of properties which are not normally found in conventional steels or materials.

Obviously, it is also extemely difficult to give market trends applicable to such diverse and changing applications, which in some cases correspond to niche-type markets and in other cases to well established markets where cost-down and scale effects are essential to success.

This introduction to special steels could be illustrated by an example: the transport of natural gas from producing countries to consumers. (Fig. 2)

Even if you know a lot about the future trends in world gas trade, this example shows how difficult it can be to deduce from this a forecast of the consumption of the special steels listed in this slide, if you do not include in your analysis a huge amount of other variables of a different nature (technical, industrial, and even political . . .).

This leads us to our first set of conclusions:

1. Special steel producers are not able to control the decisions which finally determine the future of their business. At best, they help potential customers (in this case plant suppliers and shipbuilders, etc. . . .) by helping them achieve products *more adaptable to the final use*, more *available*, *easier to process* and more *competitive*.
2. The second conclusion could be reported as 'you have a lot of work to do, but the result could be nothing'. In a more optimistic perspective, I will say: Special steels have a *huge potential for growth*, substantially higher than that of carbon steels, but you cannot rely upon 'incremental' progress or

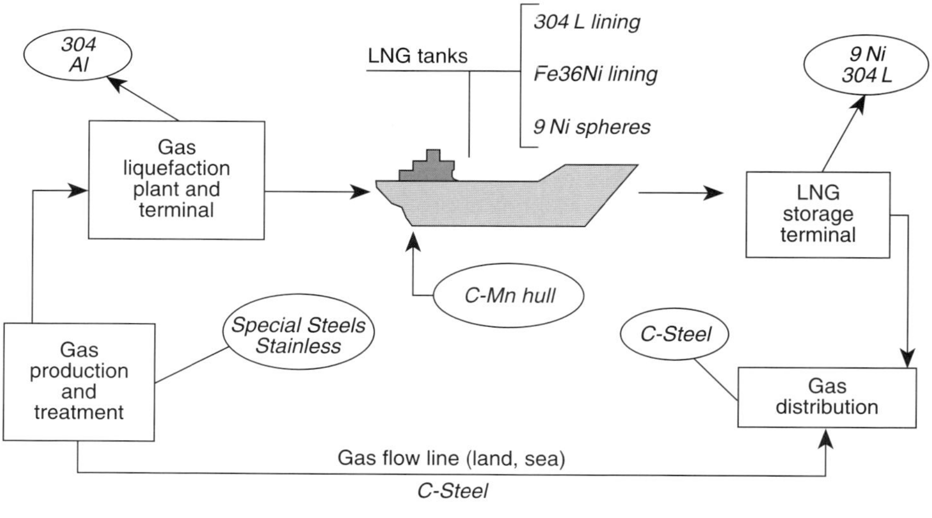

Fig. 2 Special Steels for Natural Gas Distribution.

	Consumption	
	World (Mt)	W. Europe (Mt)
Cement	1000	200
Plastic	120	30
Glass	100	24
Aluminium	25	7
Copper	15	4
Industrial roundwood	2	0.3
FRP/GRP	2.5	0.5

Fig. 3 Non-Steel Competition to Special Steels.

'end use market trends'. You will have to use your brains and work out solutions, in partnership with companies sharing the same objectives (in this case engineering companies, LNG carriers, shipyards using tank or membrane solutions, etc.).

3. The third conclusion is that for special steel producers, competition is *everywhere* (including sometimes inside your own company or group as different types of steel compete against each other for some applications). Competition is usually provided by *common steel* but only because carbon steel is used to a great extent in plant and machinery construction. It may also be represented by cement, plastics, wood, aluminium or other metals. (Fig. 3).

Obviously the word 'special steel' implies some idea of higher performance, probably also of higher price and reduced volumes.

Figure 4 shows how each family of materials evolves to cover more specialised needs from the high volume applications (such as building) down to high performance markets (such as professional sports or aerospace).

The special steel family, in this context, may be viewed as the adaptation or the reaction of the steel industry to changing end uses. Stainless steels, in particular, are probably with specialty plastics, technical ceramics and light metals one of the most successful examples of Darwinian evolution in the world of materials (Fig. 5): improve or die! As all marketing people know, when a product or a process cannot be improved, it is ready for substitution by another solution.

STAINLESS STEELS: THE GROWTH FACTORS

To introduce the stainless steel family, I have selected a very simple example which corresponds to one of the basic needs of man: drinking! (Fig. 6). On Italian hot summer days, individual cans or bottles are extremely popular. They are made of glass, polyethylene, cardboard, aluminium or steel and correspond to huge volumes of materials (and refuse to recycle). But if you go into an (air conditioned) bar, the beer or soda you drink will probably have been transported in a stainless keg or container, and produced in stainless equipment.

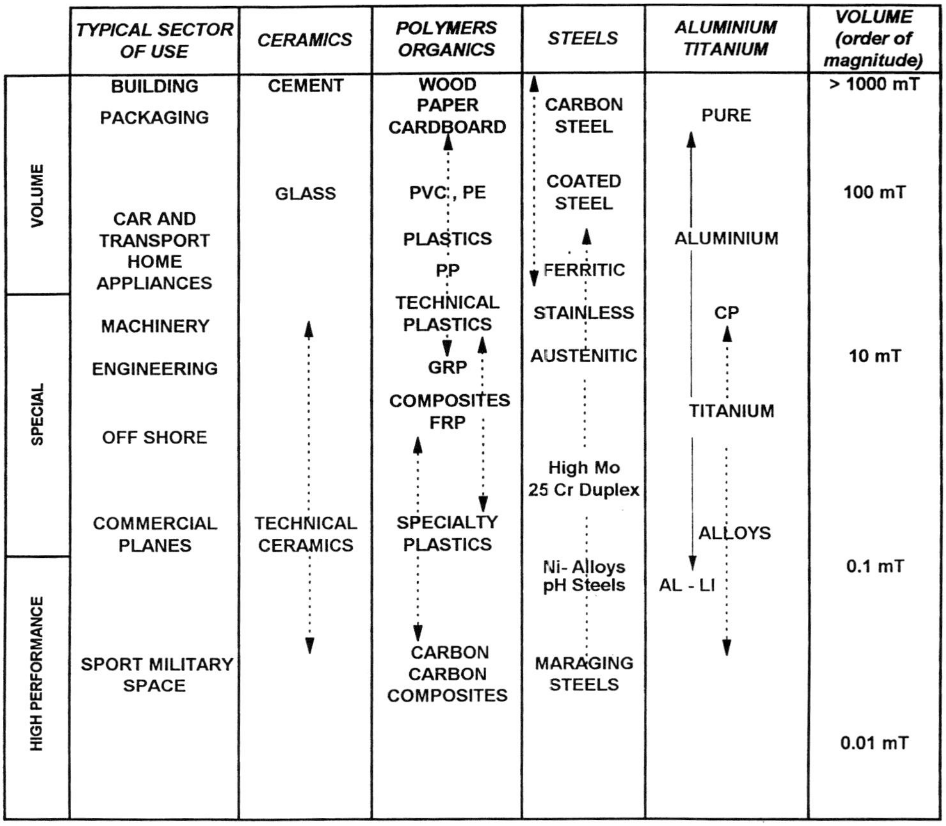

Fig. 4 Materials Families.

The use of stainless steels grows when the requirements of:

- *Hygienic* properties (inert, clean surface);
- *Easy maintenance* (easy to clean, no corrosion);
- *Durability* (10 Years to 100 years);
- (no downtime, no trouble) and *safety*.

are considered more important than the 'usual' design properties such as a strength, ease of fabrication (which they also possess) and low unit costs (which is their apparent handicap).

This explains the fact that capital goods account for 70% of flat stainless steels, compared to 30% accounted for by consumer *durables* such as electric appliances, car exhausts, sinks, etc. . . . (Fig. 7).

At this point, rather than producing growth rates or market forecasts I would like to list some of the factors driving the development of stainless steel.

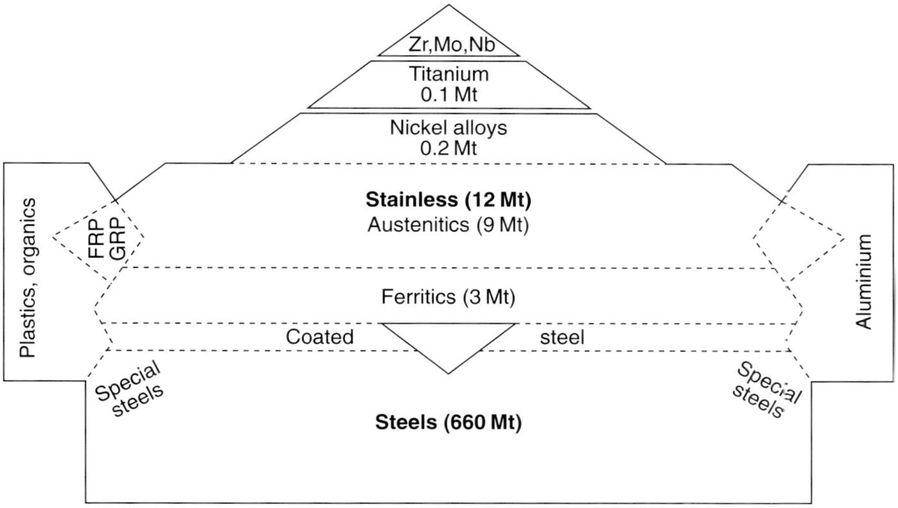

FRP = Fibre Reinforced Plastics
GRP = Glass Reinforced Plastics

Fig. 5 Stainless steel position in international competition.

Can/bottle Individual	Bottle/pack Family	Keg Commercial, community	Containers/vessels Production, distribution	Sector
Once	10 times	10 years	20 years	Use
30%	30%	>90%	100%	Recycle
200 billion	500 billion	4 million	10000	Yearly production
0.25–0.351	1–21	10–501	10.0001	Volume
		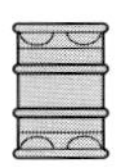		Application
← Cardboard → ← Glass → Coated steel Aluminium ← Pet, plastics →		Stainless steel (50%) Aluminium Wood	Stainless steel (80%)	Materials

Big volumes low cost, fashion → Design, life cycle cost and usage properties

Fig. 6 Use of materials for beverage distribution.

Better Product Design

As the number of parts and the weight of products and equipment are reduced

- each part become essential to global efficiency;
- and each component fills several functions.

	% of total consumption	Stainless steel penetration	Examples
Consumer goods			
Household appliances and kitchen equipment	17%	High (washing) Medium (cooking)	Washing, cooking, sinks, kitchen hoods, water heating, air conditioning, small electric appliances
Tableware	7%	High	Cutlery, pots, pans, trays
Car and automotive	5%	High (car exhaust) Low (other)	Car exhaust, including welded tubes
Capital goods			
Community and catering	10%	High	Industrial kitchens, vending machines, bars and shop equipment, hospital furniture and equipment
Chemical and process industries	26%	Medium to High	Food and beverage industries, chemical/petrochemical, paper and textile plants, energy production
Transport	4%	High (food and chem.) Low (others)	Containers, frames and panels, chemical tankers, road and railway tanks, containers, leisure boats, bus structures
Welded tubes, fittings	24%	Medium	Decorative/structural, food services, process industries, heat exchangers
Building and construction	3%	Low	Roofing and wall cladding, rainwater channels, chimneys, hand rails, water treatment, swimming pools
Other	5%	–	Mechanical parts, flat springs, coinage

Fig. 7 End Use Sectors for Stainless Flat Products in Italy.

This normally favours materials combining attractive service and design properties, and stainless steels or special steels in general.

Design requirements will also tend to include the cost and extent of maintenance programs and help the end user to make better choices of materials instead of continuing to reproduce choices and practices of the past (Fig. 8).

Production Without Waste, Closed Cycle Processes

The 1960 pulp and paper plant was described as *dirty*, *rusty*, and *stinking*. Nowadays, such parts incorporate a fume and water treatment plant, as many

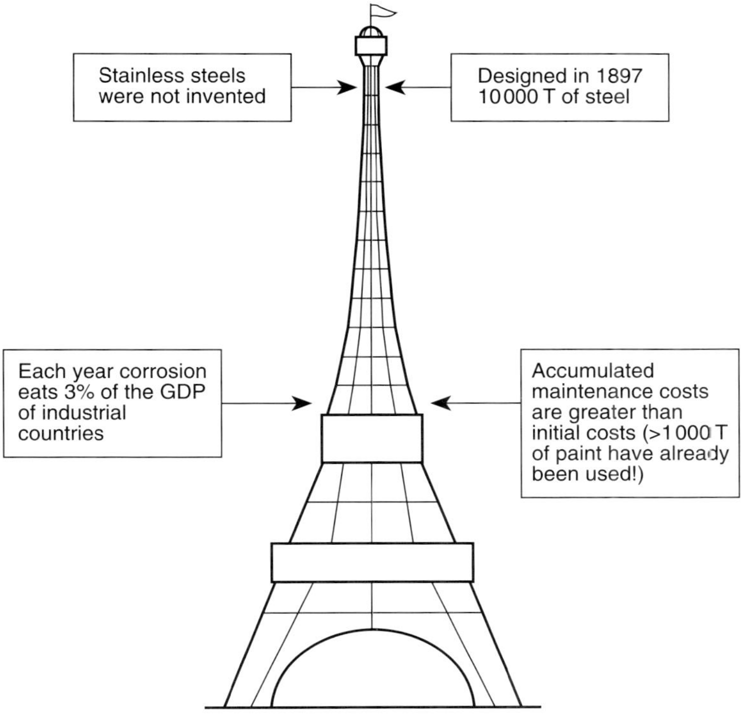

Fig. 8 Is stainless a better choice?

closed loop processes as possible and recycle old paper. Carbon steel is replaced or lined with stainless steel and plants are able to work at nameplate capacity, with limited downtime and minimal maintenance.

But production without waste also means more concentrated solutions, higher temperatures and increased corrosion!

Less Taxes and Better Services

Cities, communities and nations will have to consider not only the initial costs of equipment (buses, commuter trains, street furniture, buildings . . .) but also their global *life cycle cost* including maintenance costs (refurbish, repaint, clean and repair) and the accompanying losses of production or service efficiency. Stainless buses cost only 5% more to buy and save a lot of money after a few years of service: a well managed city should prefer them!

The same evaluations have been made for municipal water treatment plants, with the same type of conclusions.

Consumer Power

All customers, from chemical plants, car manufacturers and car owners require extensive guarantees from their suppliers. Failure analysis and warranty-induced costs will modify the initial cost-based material selection rules in favour of more sophisticated approaches.

Extreme Conditions of Service

Fifty years ago, when a sour gas field was discovered, the usual reaction was to plug the hole and look for a better place to dig. In 1995 few countries can afford to neglect the contribution of new resources, however difficult to exploit and we need materials resistant to sour gas, sea water and Alaskan blizzards and, perhaps in the future, to hydrogen fusion conditions or efficient solar collectors.

'Green Power'

From Bhopal to Chernobyl, as in everyday life, the impact of man's activity upon the environment has become part of the rules of the game. Stainless steel is 100% recyclable in first choice products. Each heat of stainless material in an EAF yields 50% scrap. Stainless steel is also used in catalytic car exhaust systems (over 10 kg/car in Europe and more than 20 kg/car in North America), water treatment plants, and many other pollution control systems.

NEW CHALLENGES FOR SPECIAL AND STAINLESS STEELS

As you can see from this incomplete list of potential opportunities, stainless steels will help in solving new problems as they helped solving the old ones (Fig. 9).

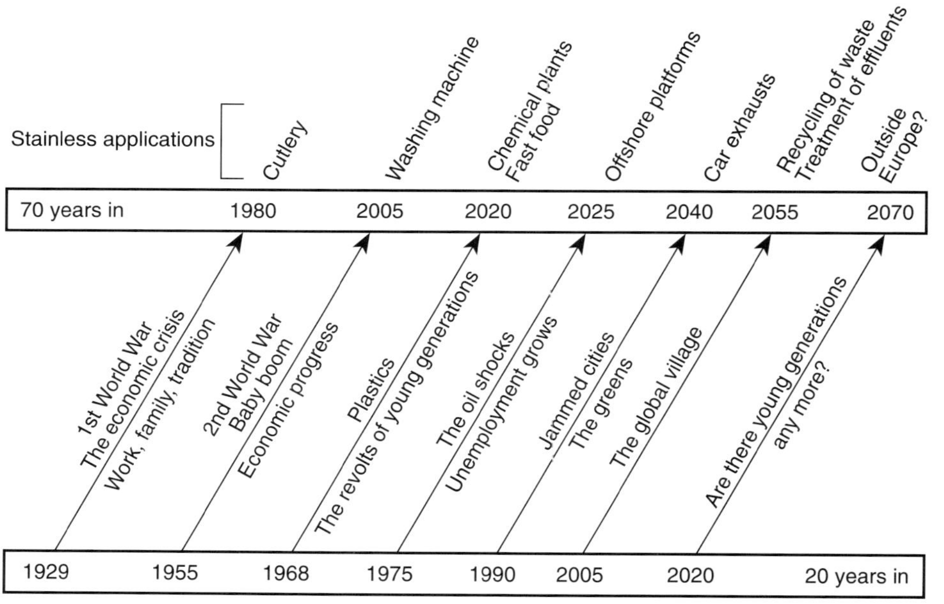

Fig. 9 Changes in Social Needs in Europe.

Maybe the incognita are more than the problems we will have to solve in the future. The first oil shocks have produced huge effects upon steel consumption (from a $\cong$ 4% rate of growth in tonnage to near zero growth) and only limited effects upon other materials such as stainless steels, titanium, aluminium.

Which materials will be shock(s) sensitive and why? What will be the next shock(s) in Europe and the rest of the world? Waste management, water shortage, growing pollution, greenhouse effect . . .?

I will not answer these questions myself, since your own answer is probably as good as mine. But I hope to have convinced you that stainless steels and special steels will continue to offer innovative and efficient solutions even if, as suggested in Fig. 9, they will probably have to find new growth opportunities, new horizons, new frontiers, away from their European birth places.

INTERMATERIAL COMPETITION BEYOND TECHNOLOGICAL AND ECONOMIC FACTORS

A. MOREAU

Usinor Sacilor, Immeuble Pacific, TSA 10001, 13 cours Valmy, La Défense 7, 92070 La Défense Cedex, France

ABSTRACT

When we address the subject of intermaterial competition, we usually assume that the factors determining it are technological and economic. In reality, however, competition among materials increasingly depends on other factors.

The choice between materials for an end use application is not primarily determined by technical properties nor by comparative prices. The substitution of one material by another has to be considered *the result of competition between alternative processing chains*. That is to say that in order to understand the substitution process, we need to look at the exact functional requirements at the different stages of the processing chain. These tend to be closely linked to the way the final product is manufactured and to the nature of the demand for it. Furthermore we need to take into account the fact that substitution is not just about other materials competing with steel products, but is also about competition between different qualities of steel. After all the processing methods needed to turn out flat steel products are not the same as those for alloyed steel and the two steel products can easily be in competition with each other.

The decision chain for substituting materials goes upwards (and not downwards) from final users to producers passing through different processing levels such as stockholders, intermediary processors, final processors, etc. The decision to use one material rather than another is not merely based on the technical features of the materials, but, as we will see below, on a more complex set of factors.

It is not necessary to consider the whole chain, since intermaterial competition hinges mainly on three factors: *changes in final demand and in environmental perceptions; structure and organisation of the processing chain and strategies of the different actors* such as final users, manufacturers of the finishing stages, intermediate processors and raw material producers. It should therefore be clear that we can not reduce the problem simply to one of comparative advantages in technology and prices. An oversimplified marketing view can in fact lead to overlooking the real issues behind inter and intra-material competition.

1 CHANGE IN FINAL DEMAND AND IN ENVIRONMENTAL PERCEPTIONS

It is well known that for the last two decades the final markets both for steel products as well as for competing materials have been saturated, at least in the more developed countries. As a result competition, differentiation, market segmentation have all increased.

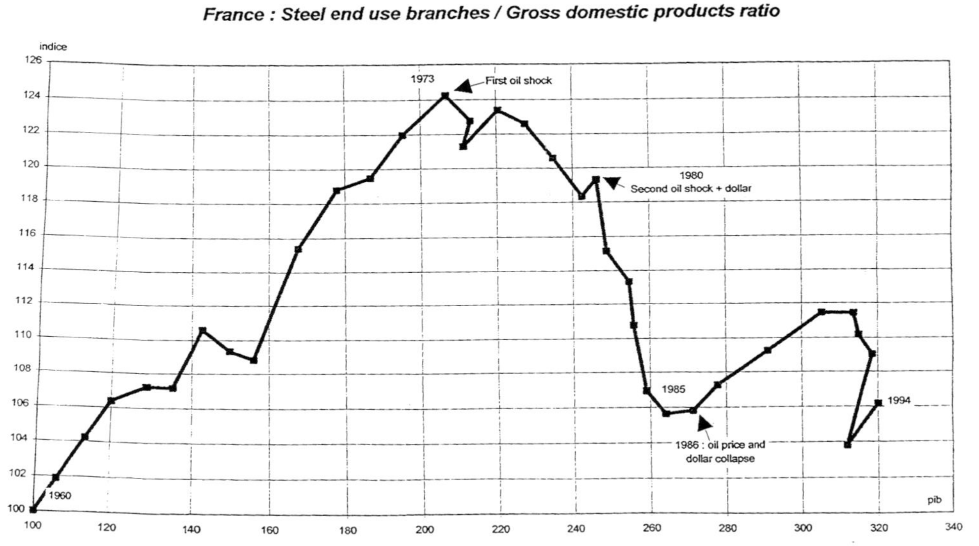

Fig. 1 France: Steel and use branches/Gross domestic products ratio.

Mature Final Markets

Among the main end use sectors for steel products are building construction, mechanical and electrical engineering, household equipment manufacturing, railway equipment manufacturing, shipbuilding and a few others. As shown by Fig. 1, since the first oil shock of 1976 steel using branches of the economy have generally shown a declining trend and have made little contribution to economic growth.

Increasing Competition, Differentiation and Market Segmentation

The French automotive market gives us a good idea of the magnitude of such changes. The number of car models sold in France has risen from 100 in 1970 to 161 in 1994. For Renault alone the number of models on the market has more than doubled between 1980 and 1993, rising from 47 to 111. The peak was in 1990 with 123 different models of Renault cars. Furthermore it is interesting to note that variations in the number of models have tended to take place in a relatively short time period and that they have been erratic.

Changes in the market such as the ones reported above are very important in shaping the choice among competing materials. Generally speaking, it is fair to say that models which are manufactured on a large scale tend to incorporate a larger amount of steel in all parts of the autobody. In the case of models produced in more limited numbers, even though flat steel products remain the major component, there is a tendency for other materials to increase their share, thus reaping the benefit of market segmentation. Thus, for example, in the case of Renault's Espace or the Audi 8, more plastics and aluminium are used for components such as wings and hoods as well as for the auto body itself.

On the other hand, the case of the packaging industry suggests quite the opposite conclusion. Whereas, in fact, the shape of the can has become an important factor in the differentiation policies of the soft drink and beer producers leading to the use of materials such as aluminium. Steel products, appear well placed to fulfil the increasing demand for personalised packaging.

Increasing Role of Environmental Awareness

The growing importance of environmental factors has become a major criterion in the competition among materials. In particular three factors stand out:

(a) awareness of pollution and effects on the environment;
(b) need to reduce oil consumption;
(c) recycling strategies.

Recycling in particular requires to be considered in the early stage of conceiving and designing an industrial good.

Increasing Dematerialisation of Industrial Goods in Modern Economies

The service sector accounts for an increasing share of GNP in the more developed countries. Furthermore services are increasingly providing a huge input into the manufacturing process itself.

An analysis of the input-output matrix, for example, shows that mechanical engineering and the automotive sector for the last fifteen years have been significantly increasing their use of immaterial inputs, namely service and electronic goods (containing little physical materials). In the French case, these two inputs now account for over 50% of intermediate consumption of the two sectors considered. This trend, therefore, goes a long way in explaining both the decrease in the use of all materials in the economy, as well as the variations of intensity of materials' use within single branches. In other words it appears that different materials are competing in order to have a larger share of a shrinking market: some of them have managed to increase the intensity with which they are used, whereas for others intensity of use is falling.

Furthermore the competition among materials itself has the effect of increasing the immaterial component of GNP. Distribution networks, marketing divisions, R&D, as well as a range of customer services are being increasingly developed within manufacturing industry in order to secure an advantage for the producers of a given material.

Example 1 – automobiles. Between 1980 and 1994 the share of services in the purchases of the automotive branch increased by more than 120%. This share now accounts for nearly 20% of direct purchases, and for probably more than 40% if we take account of services incorporated in other intermediary inputs (Table 1).

Example 2 – mechanical engineering. A similar evolution can be found in the mechanical engineering sector. The share of services directly consumed by the

Table 1. Changing shares of purchases of the automotive branch in France.

% of current Francs	1980	1985	1993	1993/1980
Metals	10.0	8.1	5.8	−42%
Metal working, iron	20.5	18.8	17.8	−13%
Plastic products	3.4	3.8	4.8	+41%
Glass	2.0	1.8	1.6	−20%
Pneumatics	7.1	6.7	6.0	−15%
Equipment	3.9	3.9	3.8	−2%
Traded services	8.7	13.9	19.1	+120%

Table 2. Share of intermediary inputs in mechanical engineering industries (%).

	1977	1980	1986	1990	1990/1977
Primary steel	13.0	12.8	10.3	9.4	−28%
Other materials	17.0	15.2	13.4	13.1	−23%
Industrial goods	39.1	37.8	39.3	38.7	−1%
Energy, etc.	8.0	8.8	8.5	7.9	−1%
Services	22.9	25.5	28.5	30.9	+35%

industry exceeds 30% of inputs and exceeds 50% taking account of indirect flows. The share of services increased by 35% from 1977 to 1990, whereas steel input's share decreased by 30% and that of other materials also decreased by more than 20% over the same period (see Table 2).

During this period steel consumption decreased by an average 2.3% a year. The decline of the specific weight factor, measured by the quantity/value ratio, is commonly called lightweighting. Lightweighting can be explained by the improvement in the steel used in this sector and by a rapid growth of other materials.

The trend is also explained by the increase of all inputs incorporating fewer or no materials. Services represent the most important share of this input. The conclusion is that the decrease in steel specific consumption in this sector corresponds to the increase in an invisible matter: the so called 'grey matter'.

Other examples feature the growing importance taken by 'intelligent goods' such as the intelligent house, the intelligent car and so on.

2 THE STRUCTURE AND ORGANISATION OF PROCESSING CHAINS FOR STEEL, ALUMINIUM AND PLASTIC

It is important, in the first place, to observe that we are talking about steels, rather than steel. This is because there are different categories of steel, and because for example, stainless steels are closer to aluminium than is the case for common steel. In the structure of processing chains, as we shall see, alloy steel products are quite different from flat common steel products. Moreover the comparison between the aluminium and plastic chains on one side and that of steel on the other is very instructive.

The Aluminium Processing Chain Compared to that for Steel

- a narrower range of final applications for aluminium semi-products (castings, wire, extruded and rolled shapes) has to be compared to the wider variety of uses for steel products (from pins to sky scrapers!);
- a narrower range of final products. There are in fact only three predominant final uses for aluminium: can making, building construction, transport; within each of them, moreover, there are relatively few final uses: for example in buildings the main use are window frames and curtain walls (whereas relatively small quantities of aluminium are used for roofing and cladding applications). In comparison steels are used in the building structure, in interior works, in fittings and installations and so on;
- consequently aluminium products are closely linked to their final uses: for example aluminium extruded shapes for window frames.

Concerning the Aluminium and Plastics Processing Chain
Aluminium and plastic semifinished products, such as foils, extruded profiles and moulded counterparts have to be compared to steel mill products or to the products of the first processing of steel. With the difference, however, that aluminium and plastic semifinished products, such as cylinder heads, aluminium automobile wheels or plastic dashboards are final objects, i.e. they do not need further processing.

In most cases, moreover, the plastic industry directly supplies a complete system to the automobile industry. This allows the plastics industry to incorporate more and more services in its goods. The competition, therefore, is now focusing on service provision. As an illustration I will quote the statement of the Chairman of the company Plastic Omnium:

> *Producing dashboards, bumpers, tanks has become the business of service providers. These products are technically very elaborate and incorporate more and more functions (tanks with pumps and gauges, bumpers with lighting and pre-wired devices, intelligent windscreens). A lot of grey matter is involved in these components; we are now selling less and less primary material, handworking incorporated in our products is also declining to the benefit of more and more R & D and logistics.*

About the Steel Processing Chain

- For *flat steel products*, the processing chain is nearly the same as with the other two materials: for example to become an autobody part, steel sheet has to be cut into shape and stamped. These two operations can be considered as two processing stages, which are increasingly located outside the automobile industry. But for body parts, the steelmakers and automakers are still in direct contact, even though the parts are externally sourced.
- For the processing chain of *alloy steel* the problem is quite different. An alloy steel bar has to be processed in various steps (drawing, forging) and then

incorporated in a system by an original equipment manufacturer and finally installed in the vehicle by the automaker. So for alloy steel there are often as many as three intermediate actors between the steelmaker and the automaker, whereas in the case of plastic, aluminium and flat steel products there are none.

3 STRATEGIES TO ADAPT TO CHANGES IN FINAL DEMAND

In the car market we currently notice a double movement: downstream integration for the materials producer and upstream de-integration for the carmakers. At the intersection of these two movements lies the original equipment manufacturer, whose business is 'system conception', and who is taking on a more important role in the production chain.

We should point out that some material suppliers to automakers are also their component suppliers. In the aluminium industry this is the case of:

- Alcoa supplying extruded shapes and die-castings to Audi;
- Hydro Aluminium supplying Aluminium and magnesium die-castings to American and European automakers;
- Montupet supplying aluminium wheels, cylinder heads and blocks in Europe and America.

In the plastics industry we have already talked about Plastic Omnium. In the steel industry we might cite:

- Cockerill-Sambre/Ymos supplying automobile doors;
- Thyssen/Budd who are currently supplying the twist beam axle for Renault Twingo, steel body panels in the USA and are planning to supply aluminium tailored blanks and stamped aluminium tailored blanks to the European automobile industry;
- Finally the Ascometal group has taken over the spring steel manufacturer Allevard. These strategies could not take place without a strong partnership between materials producers, processors and final goods producers.

We can now sketch a typology of *different kinds of competition and company strategies*.

The simplest one is the *competition* between two *materials over supplying a component or final application*, such as the hood which can be made either of steel or of aluminium. Competition focuses on the comparative advantage of each material. Thus a direct strategy, aiming to show the direct advantage of using a material can be deployed. But already, if we consider other goods, such as cans or wheels, we see that competition does not hinge merely on technical specifications or price, but involves other factors.

Competition over function supply. The competitive arena is no longer related primarily to supplying a component, such as a torque rod or a spring, but it involves the suspension function itself. Such an approach allows one to widen the scope of the problem and to propose alternative solutions to satisfying a particular requirement and only as a consequence to suggest the use of different products (torque rods as against springs). This is *an indirect strategy*.

Competition over system supply. It is a case that plastic has already replaced steel for some car parts, aluminium has attempted to do the same, although it is finding it more difficult. A new approach to the problem, therefore, is the total redefinition and design of the body frame of the vehicle; for example the well known 'aluminium space frame', which represents a radical new departure from the traditional steel 'monocoque'. Here is a more indirect strategy which aims at changing the nature of the final product.

Competition over business. More and more final manufacturers are not particularly concerned by priority for a specific material or another. For example, a very large building construction group, previously specialising in concrete buildings, has taken over a steel construction company in order to respond to market requirements. We should also point out that most of these very large concrete building groups are now taking over water supply companies. This illustrates how the leading companies are intent on becoming primarily service companies, given the increasing importance of services in the supply of final goods.

Strategies of anticipation and mapping alternative routes. Here we have a range of examples. The first is an old one: it concerns the quasi monopoly acquired by aluminium in can making in the USA. This is not due to any technological advantage as may be seen from the wide use of steel cans in other countries, such as Germany (75%). The predominance of aluminium in the USA is due to an alliance between Alcoa and Coca Cola – an alliance, in other words, between supplier and customer which purports to resolve the entire range of problems along the chain of transformation, utilisation and recycling. It is a good example of the best strategy, that is to say anticipation, in that it allows aluminium producers to 'win without beating' in the manner of Japanese samurais!

The second example is more recent. It concerns aluminium in cars. What is very interesting here, is that in the technical literature the only argument for an aluminium space frame has been lightness. However accurate investigations have shown that neither the Honda NSX, nor the Audi A8 are any lighter than the models in steel which they are replacing. (The Audi 8 in steel was in fact lighter than the new Audi in aluminium). Audi's message to the consumer, however, ignores this issue altogether and opens a much wider perspective with the slogan: 'With Audi you are leaving the iron age for the new age of aluminium'. Such a formulation is the result of a partnership between Alcoa and Audi aimed at developing a differentiation strategy for an up-market product. The technical argument (weight), as well as the environmental and energy-saving arguments are thus replaced purely by a marketing strategy.

Prospects for the Steel Using Sectors

MICRO-ECONOMIC ASPECTS – INTERMATERIALS COMPETITION OPPORTUNITIES

CARLO PAGLIUCCI
C.S.M. Viadi, Castel Romano 100/102, I–00129 Rome, Italy

Expert international market analysts and strategists took part in this session together with representatives from industrial sectors noted for the quantity of steel which they use. These included representatives from the construction and infrastructure, shipbuilding, motor vehicle, railway, petroleum and offshore industries.

Particular points representing common denominators can be highlighted from the basic reports and associated discussion and reflect the state and broad trends affecting steel consumption, as well as the specific user industries participating in the session. This can be summarised in a number of points.

1. Competition between the various different materials is both significantly changing and becoming more demanding in nature. Traditional markets in industrialised countries are becoming saturated. Goods and markets are diversifying and fragmenting much more quickly and to a far greater degree. There exists an almost constant need for restructuring and/or changes in infrastructure and buildings as well as there being a constant stream of new applications. De-materialisation of industrial goods involves new more stringent cost/quality requirements on the part of the customer and end users.

   This background of non-stop ongoing changes has had a contrasting effect upon steel. On the one hand, and for certain applications its use has been threatened and diminished. At the same time, however, these changes have also presented the possibility of forging ahead with other applications and have lead to the discovery of new opportunities and challenges in the light of rather more sophisticated potential applications (e.g. gas transportation, oil and gas extraction, ship or rail transportation and civil infrastructure).
2. Competition is no longer between materials, or rather is no longer based on technical and economic factors which are specific to individual materials.

   A new area of competition is becoming evident in the area of 'added value and manufacturing process alternative chains' in which the material, with its particular functional characteristics and price constitutes but one of the elements which are to be considered. Indeed, environmental friendliness and impact are increasingly coming into the picture, along with factors such as potential for recycling and recovery (at each stage of the process), the life cycle cost of the product, energy consumption and savings, maintenance, the quality and efficiency of the service (research, marketing, distribution, end user technical support) to the customer and the user, and legislative rules, regulations and obligations of a social nature (product safety and reliability both in terms of the environment and in relation to the user).

Process chains are being investigated and changed as a result of pressure from above (end users' requests and their environment) and not from below (materials manufacturer).

The profile and length of the process chain varies according to the type of material in question and must be carefully evaluated in order to introduce any innovation and to consolidate its competitive position. For example, while on the one hand in the case of aluminium and plastics the transformation chain is short and compact, with steel there are several transformation/intermediary stages which are more varied and complex in nature, from flat sheets (pressing) right through to long products which are to be put to mechanical uses (which might involve several stages of fabrication).

3. Market changes and the growing importance of process chains means that new methods and strategies have to be adopted in order to compete effectively.

   There are no universal or predetermined models as such to fall back on. The approach varies enormously depending upon the material in question and the use to which it is to be put as well as from country to country (even the policies adopted by a particular government may affect the approach).

   Increasingly important to the materials manufacturer are strategies involving vertical organisation (forward integration). Apart from entering the field of transformation and manufacture of finished products (e.g. car doors manufactured from steel), strong links with customers can have a significant impact (e.g. Alcoa with Coca Cola for packaging) as can the absorption of companies or industrial sectors or even services (note the tendency for cement/concrete manufacturers to take over control of water distribution companies).

   In a few sectors and applications a multi-material approach will prove to be of particular importance (e.g. motor cars and bodywork).

   In each case the prerequisite which is necessary in order to be able to successfully compete is the ability to establish an effective system of interaction and communication with the market, namely with the customers and with the end users.

4. Overall steel prospects are favourable for all sectors but the steel industry needs to readjust and strengthen its strategies towards the market and its customers' needs, in particular via greater innovation, more dialogue and collaboration, improved anticipation and suggestions, more services; in short via a greater level of 'grey matter'.

   Future prospects are bright for this age-old material, although somewhat more complex than in the past.

PROSPECTS FOR STEEL IN BUILDING CONSTRUCTION (with a review of French domestic demand)

JACQUES BROZZETTI

Centre Technique Industriel de la Construction Métallique (CTICM), Domaine de St-Paul, B.P.64, F-78470 St Rémy-lès-chevreuse, France

This paper reviews the pattern of steel consumption in the construction sector at large (that is to say general buildings construction and civil engineering works) in Europe, Japan and the USA. It also aims to present a breakdown of the shares of the steel market accounted for by different branches of the construction industry, distinguishing new works from repairs and restoration. The latter market is expected to grow faster in the years to come, thus presenting an opportunity for steelmakers.

The paper goes on to deal with current trends and prospects in relation to steel consumption within new construction works. Two fields will be highlighted in this context: the buildings market and the construction of bridges in France. Finally data concerning the repair and restoration market in France will be presented, highlighting its importance in terms of both actual and potential steel consumption.

The construction and automobile sectors are the major steel using industries. The share of total steel consumption accounted for by the construction sector at large (i.e. general buildings construction and civil engineering works) ranges, according to the different countries, between 20 and 55%. In France, the share in 1995 was around 33%, a respectable one although much lower than in the UK, Japan and the USA, where it was above 40%. The prediction in France is that in the years to come the building and civil engineering sectors will increase their market share, as opposed to other using industries such as automobiles, shipping, general machinery and equipment industries.

Table 1. Share of total steel production used in general buildings construction and public works in different countries.

Countries	Percentage
France	33%
Europe (average)	30%
Japan, UK, USA	>40%

Table 2. Pattern of distribution of steel consumption within the building sector in different countries (in %).

	Japan	USA	Europe
Housing	21	4	2
Industry	34	33	58
Tertiary	34	45	31
Other	11	18	9

Within building construction the major markets for steel are to be found mainly in the industrial sector and the tertiary sector (offices, commercial premises etc.).

The use of steel for private housing, on the other hand, is limited both in the USA and in Europe, whereas it is much more extensive in Japan, due to the greater vulnerability of other materials to earthquakes. This is an area where there is much room for improvement in Europe. A great marketing effort should be addressed at increasing steel consumption, especially in one-floor and low-rise (up to three-floor) housing developments.

Another useful indicator for comparative purposes is provided by per capita steel consumption in the construction sector. Figure 1 provides a breakdown of per capita consumption for three different steel products: reinforcing bar, rolled shapes and steel plate. It shows that demand in a number of key European countries (as well as in the US) is much lower than in Japan.

Table 3 presents figures for consumption of different products, both in common and special steel, in the construction sector in France. It shows a fairly even distribution, although there is still a larger demand for reinforcing bars and constructional steel.

Figure 2 shows that, except for agricultural buildings, the buildings construction market in France (both public and private) shrank markedly between 1991

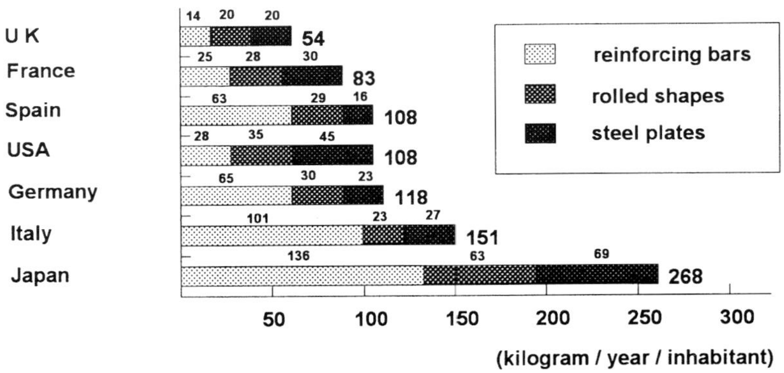

Fig. 1 Consumption per head of main steel products used in buildings construction (across total surface area) in different countries (kilograms/year per inhabitant).

Table 3. Steel consumption in French building construction in 1993 by main uses (in million tons).

	(in million tons)
Reinforcing bars	1,60
Constructional steel	0,90
HVAC equipment	0,40
Roofing, cladding, flooring	0,40
C. E. works (piling, sheet piling, . . .)	0,30
Metal work	0,45
Miscellaneous	0,30
Total	4,35

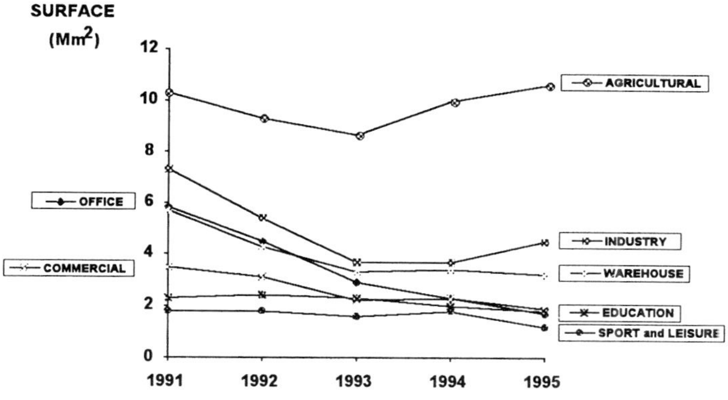

Fig. 2 Evolution of building construction in France by surface (in million m^2).

and 1995, due to a slackening of private sector investment in plant and equipment as well as repeated cuts in government spending on new housing. Private investment in housing also declined.

Despite the recession, however the use of steel is gradually increasing in France. Figure 2 measures the use of steel as a share of total materials used in different branches of building construction between 1991 and 1995. Whereas in industrial buildings, warehouses, agricultural and commercial buildings, the use of steel has been fairly high and, in some cases, rising over the period, the relative use of steel is fairly low in sports and leisure facilities, office and educational buildings. A key factor which affects the competition with concrete is steel's superiority in meeting anti-fire and anti-corrosion requirements.

To meet the emerging needs of consumers, a great variety of finished steel products are marketed in the building construction industry such as structural shapes, cladding, steel corrugated floor etc. In order to gain a better idea of the respective consumption patterns, it is interesting to break down the use of steel in the main areas of application according to three main destinations, that is steel structure, sheeting, and roofing. This is what is shown in Table 4, which

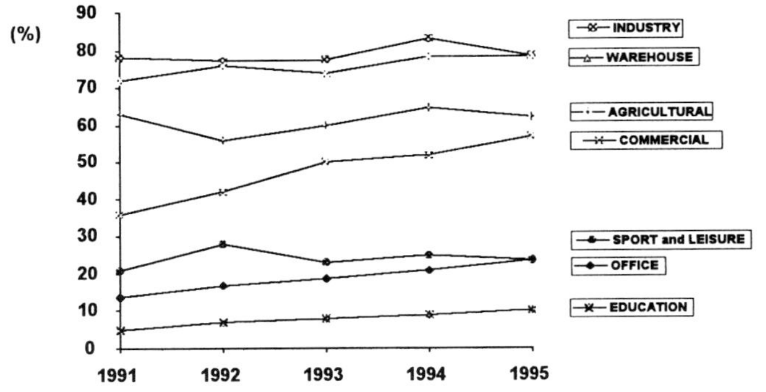

Fig. 3 Steel's market share of total surface area of building construction in France (1991–1995).

Table 4. Steel products' market share in main application fields within different branches of building construction in France in 1995, by surface.

	Area (× 1000 m²)	Market share (steel structure) (%/m²)	Market share (sheeting) (%/m²)	Market share (roofing) (%/m²)
Industry	7300	78	75	83
Warehouse	5700	72	73	76
Agricultural	10250	64	30	5
Commercial	3500	38	48	78
Sport & leisure	1800	21	36	60
Offices	5850	15	8	50
Educational	2300	7	5	30
Tertiary	4500	10	5	30
Housing (individual and collective	25000	0.2	0.1	1.5

gives the percentage share of the surface covered by steel (in relation to other materials) in the three main destination areas for each of the main branches of building activity in France. Structural steel accounts for a relatively small percentage of the surface structure of private housing, tertiary, offices and educational buildings. In relation to the latter three, however, the use of steel is much more extensive when it comes to walling and roofing, thanks to the use of a variety of steel shapes such as galvanised sheet and prepainted/precoated galvanised sheet.

Figure 4 concentrates on the use of steel for flooring purposes, in particular light-gauge floor systems. It shows that, except for industrial buildings, where the market share of steel has slightly increased, in all other branches there has been a decline, or at most, as in office buildings, a stationary condition. The composite slabs' system (corrugated steel sheet and concrete floors) must compete with cheaper concrete floor systems such as precast and pretensioned

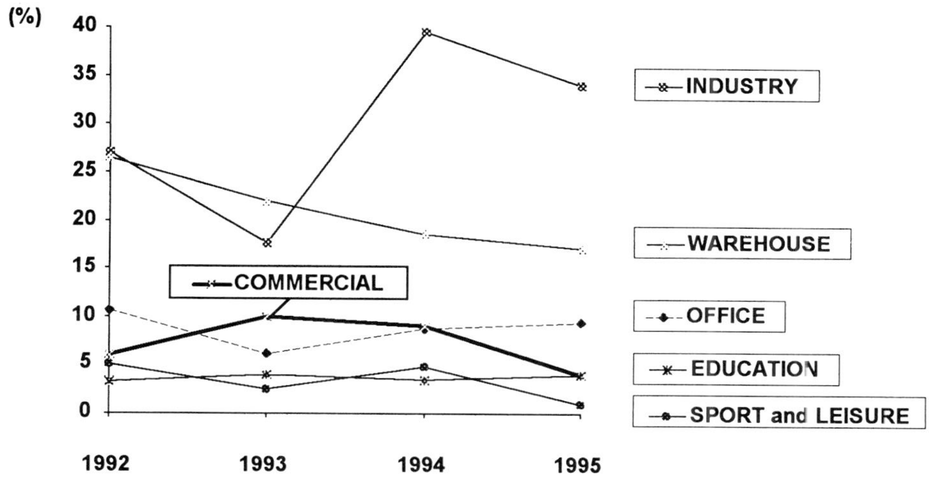

Fig. 4 Steel flooring as a share of total flooring surface in different branches of building construction in France.

ones. Despite their rather poor fire resisting performance, such systems tend to be accorded a preference in the building trade.

A bright market spot tends to be repair and restoration works, which have grown steadily over the last few years in France and which are expected to continue to grow faster than new building activity.

The evolution of markets for new work and for restoration work is charted in Fig. 5. The market for repair and restoration is constantly increasing. In 1995 the French market for new buildings was estimated at around 215 billion FF (without tax), whereas repairs and restorations were estimated at 261 billion.

What is needed is a closer investigation of the market for restoration, in order to assess the opportunities for structural steel products and steel products as a whole.

Figure 6 shows the distribution of all building activity in France for the year 1995 between new building and restoration work across both the residential and non residential sectors. The number of apartments built in France was around 300000. Forecasts for 1996 predict a fall of 8% in the collective apartment market, whereas individual housing may increase by 5 to 6% thanks to government incentives.

Figure 7 shows how the restoration market is distributed among the different sections of the building trade. It shows that the main shares are retained by collective and individual housing and by tertiary activity building, including both commercial and office building. Figure 8 on the other hand breaks down the repair and restoration costs into the different types of activities carried out.

About half the total of the restoration market is of potential interest to steel

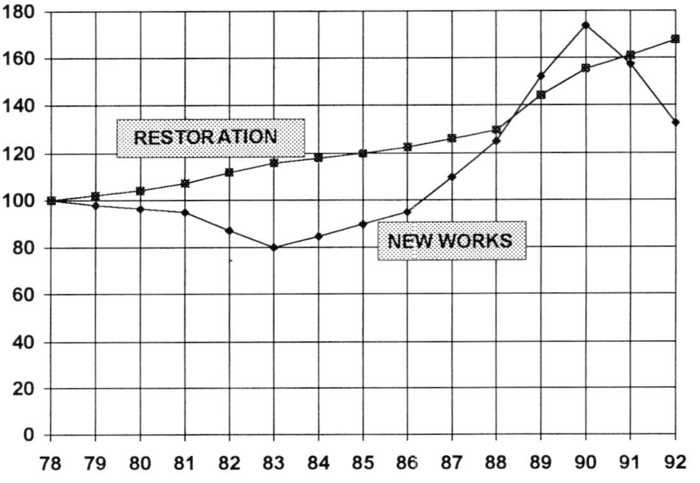

Fig. 5 Comparative trend of the markets for new buildings and restoration work in France (base index 1978=100).

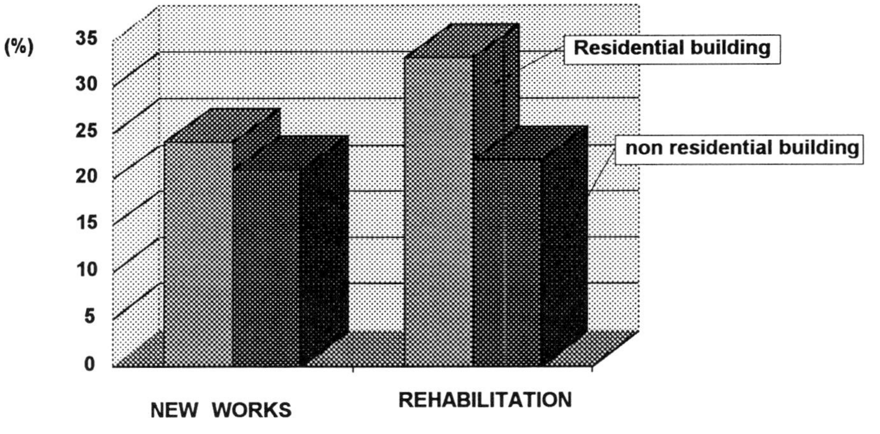

Fig. 6 Distribution of the French market for building construction in 1995 by type of building and type of works (%).

producers. However, because it is not a well structured market, in order to increase the use of steel products it is necessary to disseminate information and to promote a wider use of steel in building construction. This can be achieved by increased efforts on the part of steel makers and a greater exchange of information amongst designers, architects and structural engineers.

The next point to consider is the construction of bridges, which in France has been lately enhanced by a strong growth in rail and air infrastructure. The road

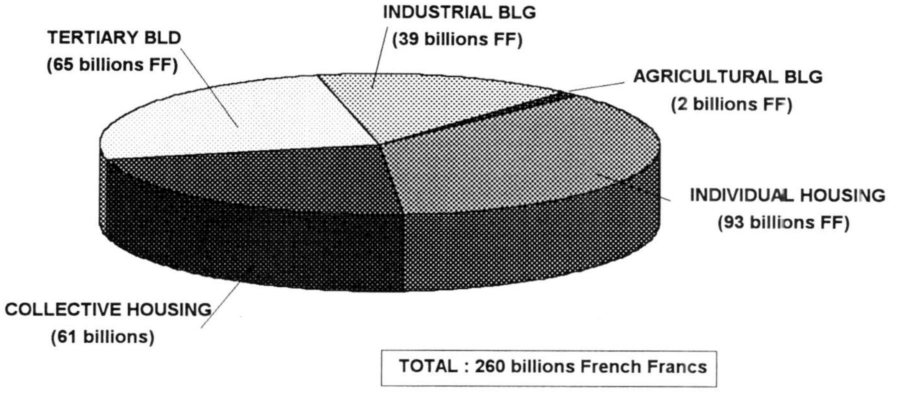

Fig. 7 Distribution of the French restoration market between different sectors in 1994, by value.

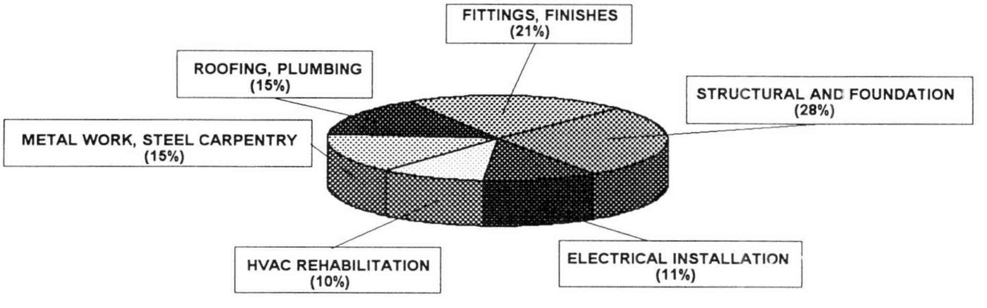

Fig. 8 Distribution of the French restoration market by type of works (in %).

building programme on the other hand is not expected to contribute significantly until after the year 2000.

In this section of the market, contractors are securing higher profit margins. At the moment, moreover, steel is beating off competition from other materials. Furthermore bridge projects tend to be large scale thus requiring large specialised contractors, whereas small steel fabricators are likely to be excluded from the bidding.

The first point to consider is the relative importance of bridge construction in terms of the overall steel market in different geographic areas. In 1990, according to IISI estimates, whereas a great number of bridges were built in Japan, North America and America, steel construction accounted for a relatively small number of them. Table 5 relates only to bridges of a certain size, leaving out short span bridges. Particularly in Western Europe, despite technological advances, steel bridge construction still commands a fairly low market share.

In France the construction of steel bridges has grown steadily over the last few

Table 5. Construction of steel bridges as a share of total bridges built in different geographic areas in 1990.

	Steel bridges number	Percentage of steel construction
Japan	760	8%
North America	500	9%
Western America	243	3.5%

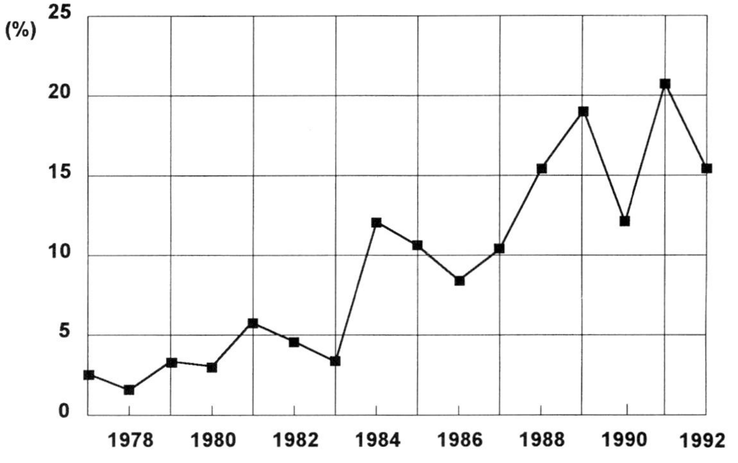

Fig. 9 Evolution of steel bridges as a share of stock of total bridge deck surface in France.

years (Fig. 9). This trend is mainly attributable to competitive gains made by composite bridges over reinforced and prestressed concrete ones.

In France, except for a few cases such as the Normandy bridge or the Millau bridge (currently being designed), there are no long-span bridges. In fact requests for bridges of a span length of more than 200 metres have lately been rather exceptional. The main market is for bridges with a main span length of less than 120 metres.

Analysis of the bridges built in France between 1990 and 1993 by various agencies (the State, private road companies, regional government authorities and the French national railway company, SNCF) shows that the most competitive span length range for steel bridges falls between 30 and 110 m with a very distinctive peak for span lengths of between 60 to 80 metres (Fig. 10). Within the latter range, 85% of the bridges being built are the composite type.

The analysis of unit deck costs (per square metre of surface) conducted in Fig. 11 reveals that the average price for a composite bridge of a span length in the range of 40–80 metres is less than that for an equivalent concrete bridge. Building a composite bridge within the 60 to 80 m range carries savings of 1500 FF per square metre over the equivalent prestressed concrete bridge. In other

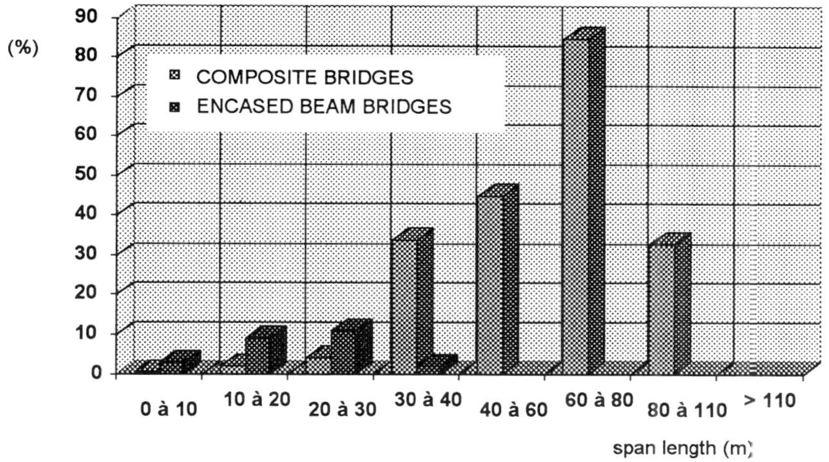

Fig. 10 Frequency distribution according to span length of composite bridges and encased beam bridges (by share of all bridges built in France 1990–1993).

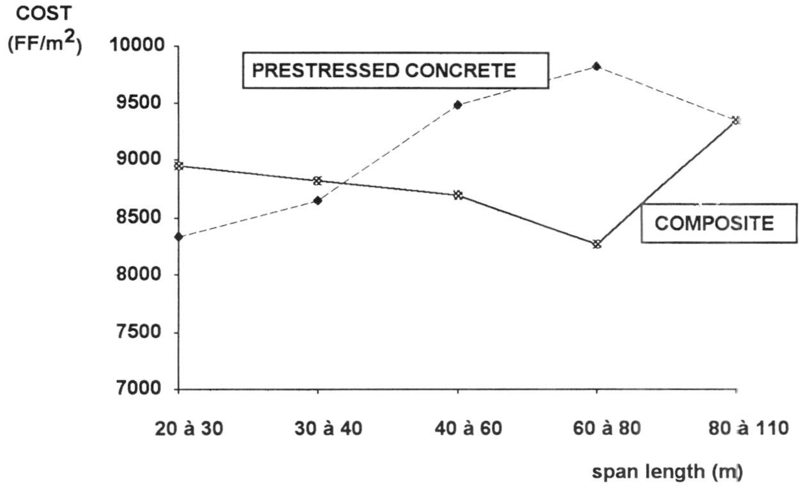

Fig. 11 Comparative cost curves for prestressed and composite bridges, by span length.

words these savings are equal to 18% of the bridge's final cost of 8200 FF per square metre (VAT excluded), which may be seen as a significant competitive gain.

Figure 12 highlights the distribution of existing bridges in France by type of material. There are more than 20500 bridges for a total deck surface of 7.3 million square metres, which corresponds to a bridge every 1.5 kilometre of

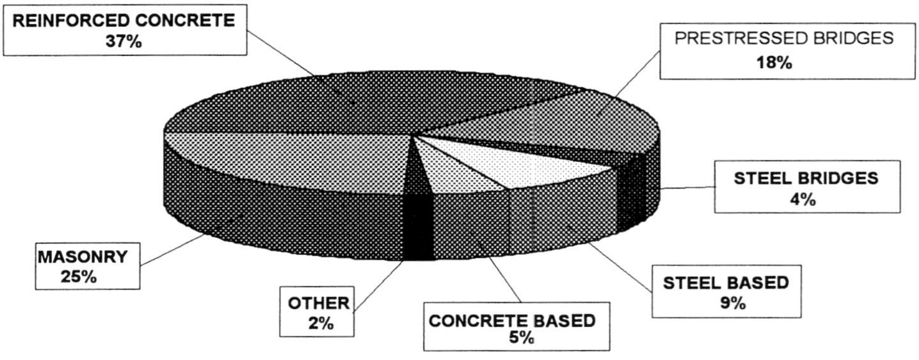

Fig. 12 Shares of French bridges by different types of materials, by number.

road. The whole stock of bridges is estimated as being worth roughly 70 billion FF. The masonry type accounts for 25% of all bridges, the reinforced concrete type for 37%, prestressed concrete type for 18%, whereas steel bridges make up only 4% of the total.

Figure 13 measures the same pattern of distribution by type of material, relating it, however, to the total bridge surface, rather than to the number of bridges. Still the reinforced and prestressed types account for the largest share, covering respectively 26 and 49% of the total surface, whereas steel bridges account for only 8%.

Figure 14 charts the results of a yearly survey of the Central Administration of Roads and Highways, covering the whole range of bridge types. Every year 800 to 1200 bridges are built in France for a total deck surface ranging between 300000 and 500000 square metres. Whereas before 1980 steel bridges accounted for just 3% of that surface, their share has gradually climbed, to reach 15–20% in recent years. This increase in market share is a consequence of the remarkable technical progress in steel bridge construction, consisting in better design (higher quality, improved aesthetics, simplicity), the use of innovative

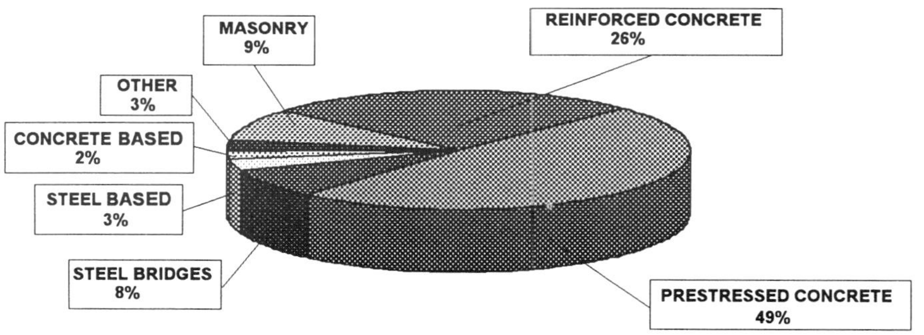

Fig. 13 Shares of French bridges by different types of materials, by bridge surface.

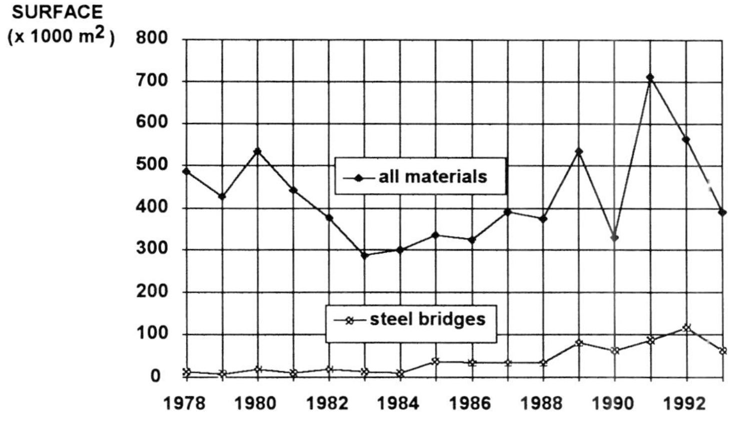

Fig. 14 Yearly bridge surface construction in France (1978–1993), by type of material.

materials (high-strength steels with a better weldability, steel plates of variable thickness), as well as improvements in launching techniques.

In 1993, bridge ownership followed the pattern shown in Fig. 15: the State owned 39% of bridges, the Departments 24%, Highway companies 22% and the SNCF 9%. There has recently been a notable increase in bridges privately built by Highway Concessionaires which are independent private limited companies. This should not however affect the competition between materials in any significant way.

Figure 16 shows the breakdown by type of material of the bridge surface and of the number of bridges built in France in 1993. 33% of bridge surfaces were made of reinforced concrete, 45% of prestressed concrete and 21% of steel. The

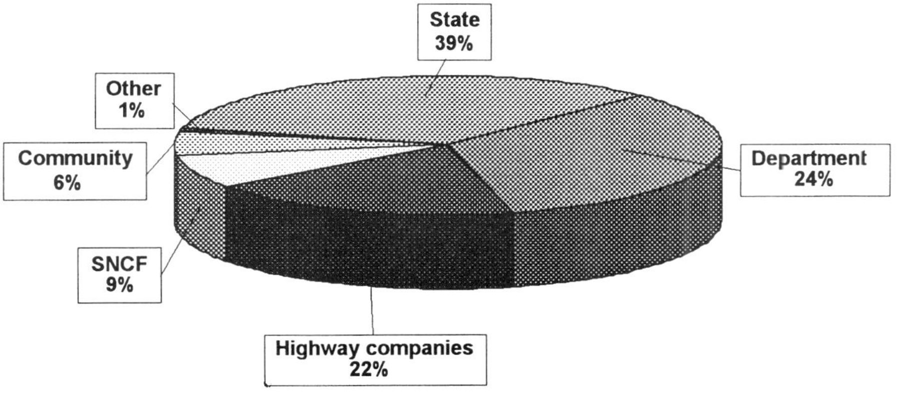

Fig. 15 Pattern of ownership distribution of French bridges in 1993, by number.

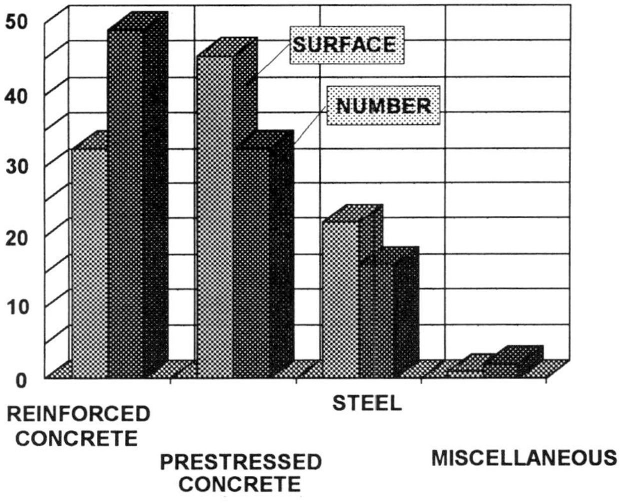

Fig. 16 Distribution of bridges built in 1993 by type of material, as a share of total bridge surface and of number of bridges.

majority of bridges, however, were short span: less than 10% had span lengths of more than 30 metres or surfaces greater than 1000 square metres. The majority of short span bridges are made of concrete, with the exception of railway bridges where steel beams encased in concrete are commonly used.

Taking the national road network under State ownership, it is noted (Table 6) that an average of 90 bridges are rebuilt every year (out of a total stock of 20500 bridges). This market represents about 15% of the total number of bridges being built every year. Table 6 indicates the age distribution of the bridges being rebuilt and the reasons for their reconstruction.

Table 6. Schedule of yearly bridge reconstruction on the French national road network.

Number:
- About 90 per year

Life period:
- < 50 years (25%)
- Between 50 years and 100 years (50%)
- > 100 years (25%)

Reasons for reconstruction
- Change in geometry (40%)
- Damaged (37%)
- Change in load carrying capacity (8%)

CONCLUSION

There a number of good reasons to believe that steel may improve its market share in buildings and bridge construction.

Housing activity, which consists of roughly between 250000 and 300000 new apartment units per year, has traditionally been dominated by reinforced concrete structures, owing to their comparative cheapness and good technological standard. Due consideration is now being paid to the market of low-rise apartment buildings and small individual housing units.

Research and development is now being carried out in France on types of individual housing made with cold formed steel gauge section and sheeting, following examples widely used in the US and Australia. One type which is now being marketed is the prefabricated light gauge steel frame home developed by Stylech, a subsidiary of Usinor. Its versatility has been proven by several individual housing units, both finished or under construction.

Another promising area is the development of low-rise apartment buildings of three to four floors. Due consideration had been paid to improve the ability of steel frame apartment buildings to comply with fire regulations and sound insulation requirements. In this respect the performance of these buildings is now very satisfactory and they have been certified to fulfil the specifications set by the Ministry of Construction.

This paper has shown that steel bridges possess many advantages over concrete bridges and that they are economically competitive. However they do have one major disadvantage: they have to be protected against corrosion. On the other hand bridge owners tend to demand longer repainting intervals. New European directives on the protection of the environment may thus have a great impact on the repainting cost and may greatly affect the global life cycle cost of a bridge.

GAS PIPELINES

M. PONTREMOLI

Centro Suiluppo Materiali SpA, Via di Castel, Romano 100, 00128 Rome, Itiay

1. BACKGROUND CONDITIONS

Figure 1 shows how a long-term increase in gas demand is expected for all market sectors, although the trend appears especially significant for electricity generators. On the other hand, as is shown by Fig. 2, the countries of the E.U. face a major difficulty, since their proven reserves are fairly small while the bulk of additional recoverable reserves is estimated to be located in areas remote from Europe such as central and S.E. Asia and the Pacific.

Figure 3 compares total demand of different areas in 1992, with the projected demand for 2030. In the case of Western Europe the share is supposed to remain roughly the same at 14% and it will only be possible to meet it by a large increase in imports. Among the options available to EU countries to cope with such a challenge are:

- improving the pipeline network for short–medium range trading;
- upgrading the LNG system;
- building an intercontinental network of large diameter pipelines for long-distance transportation of gas from remote areas;
- conducting 'deepwater' and arctic offshore exploitation (including 'marginal fields').

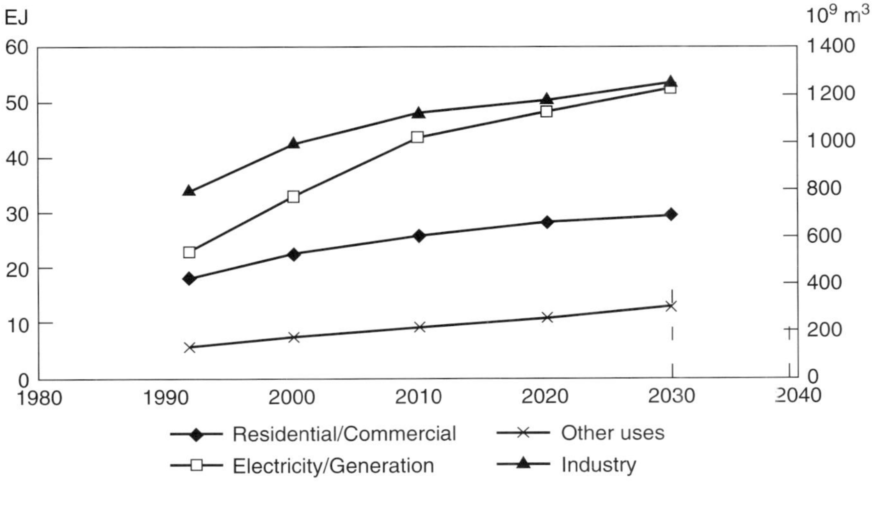

Fig. 1 Projections of world gas demand for different uses.

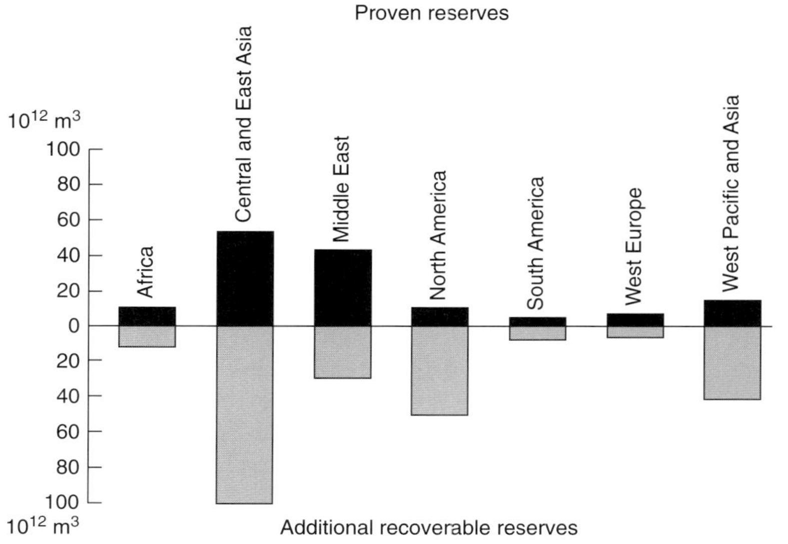

Fig. 2 Gas reserves by geographic area.

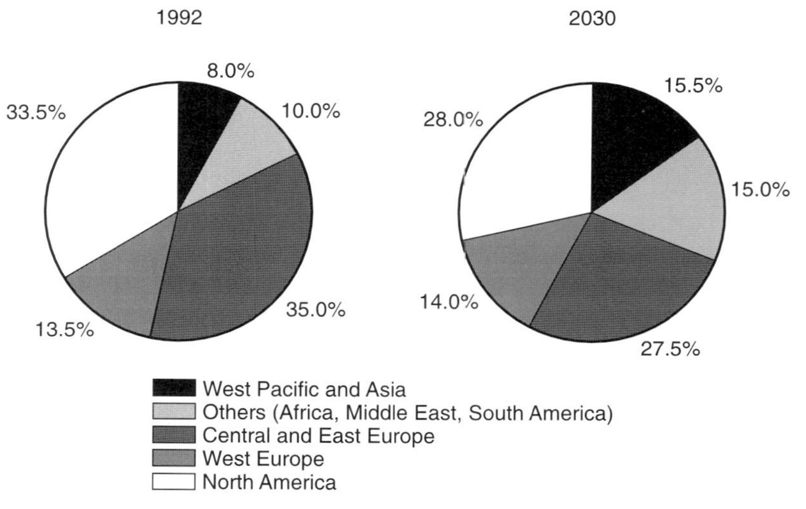

Fig. 3 Actual and projected shares of total gas demand by geographic area.

2. TECHNOLOGICAL OPTIONS FOR LONG DISTANCE GAS TRANSPORTATION

Long-distance transportation of gas offers a number of opportunities and challenges for steel producers.

At the moment the only realistic option appears to be the use of C-microalloyed pipes, for which there is no intermaterial competition. A fact to be taken into account, however, is that long-distance transport involves a very large quantity of linepipes, amounting to several hundred thousand tons for each project over the medium term, i.e. 4 to 6 years. Furthermore, given that it is very difficult to anticipate which technical options each gas company will take, it is mandatory that there be co-operation between producers and end-users at an early development stage. In addition it appears that pipes of new qualities and grades will be required, which means that suitable funds should be allocated for R&D&I and possibly for new investments in production plants.

Long distance transportation poses a considerable technological challenge since a drastic reduction in both capital and service costs is mandatory if projected developments are to be economically viable.

In our view the preferred option lies in an increase in the operating pressure (from 75 to 150–200 bars) and in the optimisation of the scale of diameters, taking into account how it affects the whole range of variable costs, such as trenching, transport, pipe-laying and maintenance.

3. NEW CHALLENGES FOR STEEL

I will now examine the implication of the trends described above for manufacturers of final steel products. The two alternatives open for the production of high pressure (150 bar) large diameter pipelines are:

- high grade steel (API X70–X80) for very thick pipes (25–32 mm);
- ultra high grade steel (API X100–X120) for pipes of conventional thickness (15–20 mm).

Although the second option is not yet available commercially, most pipe producers are trying to develop X100 steel so that it is likely to become available within three to four years.

In principle, the economic advantages of API X100 are remarkable. They include:

- lower costs for unit length of pipeline which more than offset higher costs per ton of steel;
- reduced costs for pipe transportation;
- lower on-site welding costs because of reduced thickness.

The main drawbacks tied to the use of API X100 concern safety aspects, which may also of course have far reaching economic implications. As far as R&D is concerned major areas of concern appear to be:

- the risk of mechanical damage calling for the need to achieve a balance between reduced thickness of the metal and higher internal pressure;
- fracture initiation control begging the question of the applicability of present assessment criteria and testing;
- ductile fracture propagation control (are crack arresters needed?);

– resistance to environmental fracture (is Hydrogen Stress Cracking a problem?).

4. DEEPWATER OFFSHORE PIPELINES

There are two key technological issues posed by deepwater offshore development: firstly the development of adequate technology for laying, maintaining and operating sealines and secondly the question of underwater storage facilities in order to optimise the export rate of gas.

These technological issues relate to well completions, gathering lines, flowlines and to sealines.

Considering the choice of steel pipe products in the case of gathering and flowlines the options are:

- C-microalloyed steel (plus inhibitors);
- Duplex stainless;
- Clad pipes, i.e. (C-Mn) backing pipe plus internal layer by stainless or inconel alloys.

It is therefore a question of selecting the best option for each case, based on a life cycle cost analysis.

In relation to deepwater sealines the only viable option is the use of C-microalloyed steel, with no intermaterial competition involved. The pipes that are required need to have a very high thickness/diameter ratio. Moreover competition in order to be resistant to sour crude gas the steel needs to be of extra-high quality which means:

- a very low level of impurities (S,P);
- inclusion shape control;
- a high degree of internal soundness to minimise risks of macro and micro segregation;
- the use of advanced termomechanical controlled processes and facilities.

THE USE OF STAINLESS STEEL FOR PASSENGER RAILCARS

P. DECLÉTY and L. FASSINA
Consultants to the Nickel Development Institute (NiDI)

INTRODUCTION

In the latter half of the 20th century, many changes have taken place in the approach to the construction of passenger railcars. These changes have exploited new engineering technology and advanced materials, resulting in cars which are stronger, lighter, safer and easier to maintain.

In this paper, we shall consider why stainless steels have enjoyed increasing popularity worldwide for all types of railcar. Reference will be made to the properties of the most suitable grades of stainless steel and to the advantages of their use for the fabrication of railcar body shells. Examples will be given from several countries of the proven performances of stainless steel railcars, some of which have been in service for decades.

1 DEFINITION AND CLASSIFICATION OF STAINLESS STEELS

Stainless steel is an iron–carbon alloy, that means a steel containing a minimum of 11% of chromium which gives it corrosion resistance properties in most environments. The higher the chromium content, the more resistant the steel becomes to corrosive environments. Addition of nickel is of basic importance because it serves to improve corrosion resistance. Molybdenum is sometimes added to enhance corrosion resistance in a chloride environment. Other alloys such as titanium, niobium, nitrogen, are included in order to improve certain properties such as weldability, formability, machinability, strength or hardness.

As a result, a wide range of stainless steel grades is available which can be classified in three categories according to their metallurgical structure:

- Martensitic stainless steels: these are the chromium stainless steels that are hardenable by heat treatment. They are used mainly in bar form for mechanical parts.
- Ferritic stainless steels: these straight chromium stainless steels are not hardened by heat treatment or appreciably by cold working. The grades more currently used in railcar construction contain about 12% chromium with small additions of other alloying elements for improving weldability or mechanical properties.
- Austenitic stainless steels: these are chromium nickel stainless steels that are hardenable by cold working. They are non magnetic in the annealed

Table 1. Nickel-containing stainless steels for railcar fabrication.

	Typical analysis (%)			
Type	Carbon	Chromium	Nickel	Others
304	0.07 max.	17.5	8.5	
304L	0.03 max	17.5	9.0	
301	0.10	18.0	7.5	
301LN	0.03 max	17.5	7.0	N = 0.15

condition. This group includes chromium–nickel–molybdenum stainless steels as well as chromium–nickel–manganese stainless steel.

Table 1 gives the chemical analysis of the nickel-containing grades which are most currently used in sheet or strip form for the fabrication of railcar body shells. Type 301LN stainless steel is the high strength grade which is now in general use for lightweight railcars.

To these three main categories must be added a fourth class, the DUPLEX stainless steels, which have a mixed austenitic/ferritic structure. They have a higher chromium content and a lower nickel content than austenitic stainless steel, with the addition of nitrogen and, eventually, molybdenum and copper. They benefit from improved corrosion resistance in specific environments and from high strength properties. Their use is considered for some elements of railcar underframing.

2 STAINLESS STEEL GRADES FOR RAILCARS

Two points have to be considered for the choice of stainless steel to use for railcar construcion.

2.1 Corrosion Resistance

When exposed to the environment usually encountered by railway rolling stock in service, stainless steels offer a good corrosion resistance. Ferritic, straight chromium steel can not be used for visible parts if not painted because of insufficient corrosion resistance affecting external appearance. Austenitic stainless steels which have an improved corrosion resistance due to the addition of nickel can be used in railcar construction without painting.

For stainless steel railcars, corrosion resistance means:

- long service life
- no repair for corrosion damage
- no painting
- easy maintenance

These performances are attested by the experience of railcars which have remained in service for sometimes more than 30 years in several countries all over the world.

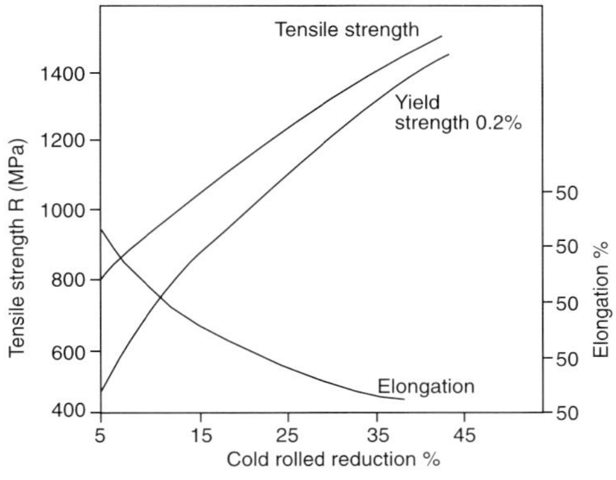

Fig. 1 Type 301 LN stainless steel effect of cold reduction on mechanical properties.

2.2 Mechanical Characteristics

Stainless steel features mechanical characteristics of fundamental interest for railcar construction:

High strength. As annealed, stainless steels have higher mechanical characteristics than carbon steel. Nickel containing austenitic stainless steels have the capacity to be strengthened by cold rolling (see Fig. 1) and high levels of tensile and yield strength can be reached, especially with AISI Type 301 stainless steel. Today the austenitic stainless steel of general use for railcar construction is type 301LN stainless steel, which is a 301 Type grade with extra low carbon content for immunity to intergranular corrosion and addition of nitrogen for improved strength. Typical analysis of 301LN is given in Table 1. Mechanical characteristics at different levels of temper-rolling are indicated in Table 2.

Elevated temperature properties. At elevated temperature, stainless steel retains high mechanical characteristics. This is illustrated by Fig. 2. In addition, the low thermal conductivity of stainless steel tends to retard the spread of heat.

Comparative fire tests, conducted by Darchem Engineering Ltd,[1] clearly demonstrated the superiority of stainless steel compared with carbon steel and

Table 2. Mechanical characteristics of Type 301 LN stainless steel.

		Ultimate tensile strength Minimum MPa	0.2% yield strength Minimum MPa	Elongation Minimum %
Low tensile (2B finish)	LT	700	340	45
Special tensile	ST	760	415	30
Medium tensile	MT	830	550	25
High tensile	HT	1,000	750	20

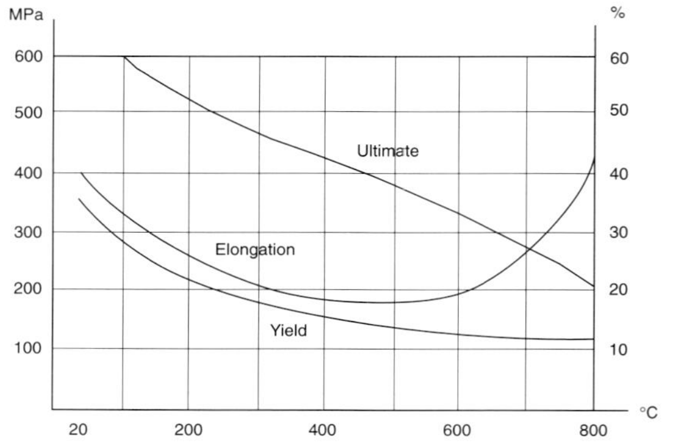

Fig. 2 Mechanical properties of 301 LN stainless steel at elevated temperature.

aluminium. For example, in a test where cable ladders were directly heated by flames to a temperature of 1000–1050 °C, the aluminium ladder suffered total structural failure after 26 seconds, whereas the stainless steel ladder survived 45 minutes without any damage to its integrity. Heat radiation resistance tests, conduction through fittings and conduction through walls showed a similar superiority in performance by stainless steel.

Fatigue resistance. Austenitic stainless steels have a very good resistance to repeated mechanical stresses. Fatigue limits of the 301LN grade are much higher than those of carbon steel and aluminium as shown in Table 3.

Energy absorption properties. The ability of austenitic nickel containing stainless steel to absorb energy is 2,5 times greater than that of carbon steel. Figure 3 shows a comparison between energy absorption properties of carbon steel, aluminium and stainless steel.

Table 3. Fatigue resistance of Type 301LN stainless steel compared with steel and aluminium. R = −1.

	Type 301 LN				
	LT	ST	HT	Steel	Aluminium
0.2% yield strength MPa	380	530	955	430	260 min.
Ultimate tensile strength MPa	720	920	1,120	540	285 min.
Elongation %	50	39	21	30	12 min.
Fatigue limit MPa					
Base metal	365	475	540	240	97
Welded joint	230	215	–	140	–

Source: Ugine SA.

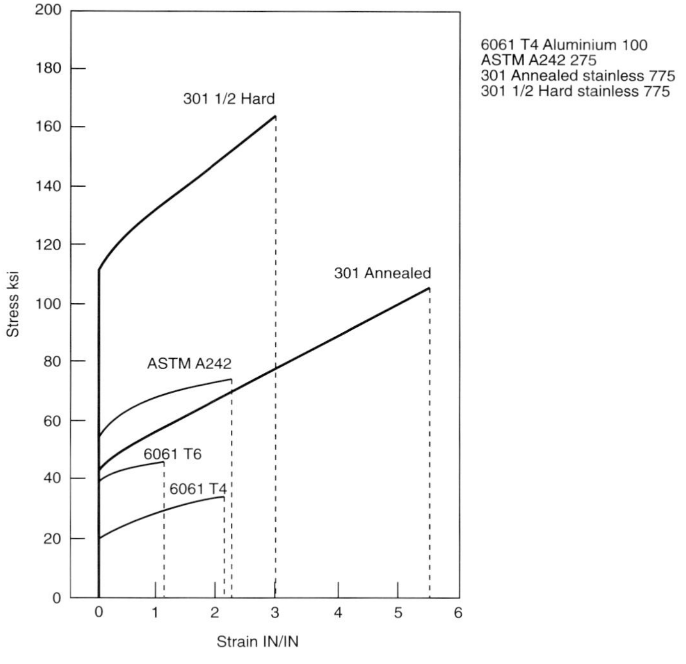

Fig. 3 Plastic energy absorption.

3 ADVANTAGES OF USING STAINLESS STEEL

The advantages of stainless steel for railcars can be summarised under three headings.

Lightweight Construction

Energy saving is one of the major preoccupations of railcar operators. Railcar weight reduction not only saves operating power costs but also brings additional advantages by reducing brake and track wear. Thanks to the high strength properties of 301LN stainless steel, it is possible to design stainless steel carbody structures which are weight-competitive with aluminium structures.

Table 4 gives the result of a weight comparison between an old style carbon steel car and cars of the same size made by using stainless steel to various extents. The table shows that it is not sufficient to use stainless steel for outer cladding or even some structural parts but that, in order to achieve optimum weight saving, it is necessary to extend its use to underframing and to call on advanced engineering and manufacturing technologies.

Figure 4 gives an example of the body structure of a lightweight stainless steel car designed according to the above considerations.

To succeed in manufacturing lightweight stainless steel railcars at competitive costs, it is necessary to consider the three following essential points:

Table 4. Comparison of carbody weights (kg).

Car type and description	Steel weight	Stainless steel weight	Total weight
Carbon steel car	10,000	0	10,000
Skin stainless steel car (outer sidings only in stainless steel)	9,200	800	10,000
Semi stainless steel car (sidings, roof and structural parts except for the underframing)	4,500	5,000	9,500
All stainless steel car (sidings, roof and all structural parts except bolsters of the underframing)	2,900	5,100	8,000
Lightweight stainless steel car (designed by using advanced engineerig and manufacturing technologies)	1,900	4,400	6,300

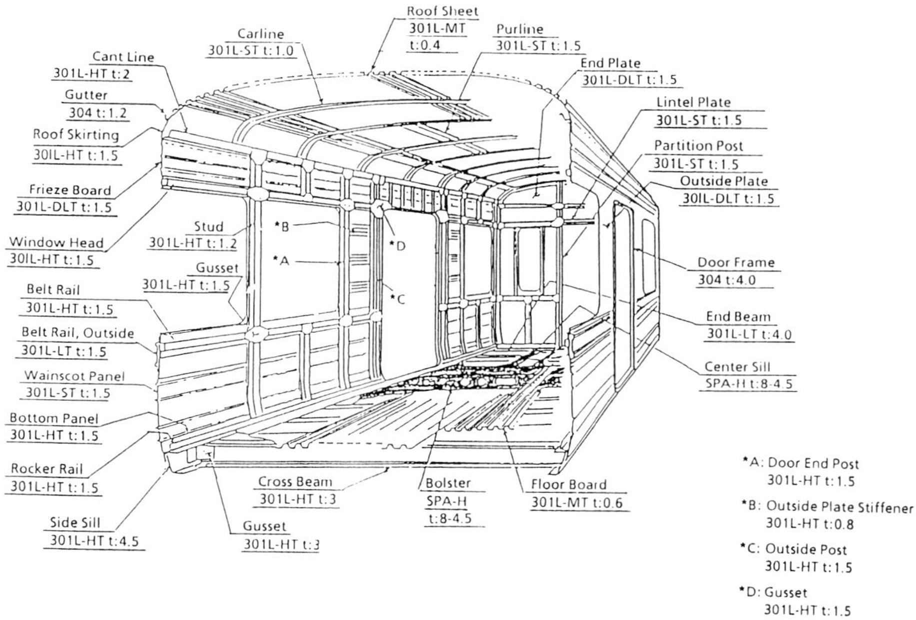

Fig. 4 Main stainless steel carbody materials.

- To use, for each element of the carbody structure, the minimum sheet thickness compatible with the stresses applied. This can be achieved by an accurate three-dimensional stress analysis using computerised finite element method.
- To use welding methods which bring together corrosion resistance, high strength and competitive assembling cost.

Spot welding is the optimum method to meet these requirements, when work hardened light-gauge material is used as in railcar construction. Use of automatic spot welding machines, often fitted with multiple electrodes, allows an accourate control of welding parameters which is essential, among others, to minimise the heat affected zone around the welded point. Use of chromium–copper electrodes help to make spot welds less noticeable.

With the extra low carbon grade 301LN stainless steel, TIG or MIG arc welding can be used advantageously for the heavier underframing elements. It may contribute to a better rigidity of the structure and help save weight by avoiding the overlapping of sheets, as required by spot welding.

- To reduce to a minimum all other manufacturing costs and to take account of eventual aesthetic requirements.

  Prefabrication of standardised elements, use of thinner special shaped sections, automation of welding operations and limitation of their number should also be mentioned among the methods to reduce the manufacturing cost.

  The aesthetics of stainless steel railcars are mainly dependant upon the design of the outside cladding. Fully corrugated panels are commonly used because the corrugations allow the use of light-gauge material. Panels with a reduced number of corrugations stiffened by lightweight shaped-sections bring a new look together with a satisfying flatness. It is also technically possible to realise flat sides.

  One of the decisive advantages of using stainless steel is that it is not essential to apply a protective paint layer. Many possibilities exist nevertheless for diversifying the external appearance of stainless steel railcars. These diversifications may be derived from the design of the sides, from the matching of various surface finishes or from the addition of coloured strips or areas obtained by painting, transfer or self-adhesive surfaces.

3.2 Safety

In case of collision. Passenger safety in a railway collision depends to a large extent on the maintenance of carbody structural integrity. If sound methods of construction are employed, the inherent high strength of stainless steel enables this rigidity to be achieved.

The great ability of stainless steel to absorb energy when deformed, enhances the effectiveness of the deformation zones at the front and rear of the railcar, which act as shock absorbers, helping to minimize injury and save life.

This has been confirmed, in particular, when a stainless steel train crashed at a speed of 80 km/h in Australia. 'The collision caused only minor injuries and the railcar has been put into service again after cutting away the damaged front section and welding on a new one including a crush zone. Other parts of the car were undamaged and, to illustrate the high car body stiffness, even the windows were unbroken'.

In case of fire. Stainless steel retains good mechanical characteristics at elevated temperature. In case of fire, the structural integrity of a stainless steel railcar remains intact much longer than that of a carbon steel car and very much longer than that of an aluminium structure. Also important is the low thermal conductivity of stainless steel which tends to retard the spread of heat damage to structural components. Stainless steel fire resistance is considered of prime importance for passenger safety by many railway companies, in particular subway companies, and is an essential reason to choose stainless steel in preference to any other material.

3.3 Life Cycle Cost

It is assumed today that a cost comparison must not be limited to just a comparison between initial costs but should also take into account all the cost elements incurred during the expected life of the equipment.

A complete life cycle cost analysis includes three items:

- Initial cost and related expenses
- Operating costs
- Maintenance and repair costs

Initial cost. Stainless steel has a significantly higher intrinsic cost than carbon steel. But only a minor part of the initial cost of a railcar is represented by the material chosen for the body shell. The cost differential compared with carbon steel is often partially offset by the lower labour cost necessary for the fabrication of a stainless steel railcar. As a result, the total initial cost of a stainless steel car is rarely more than 2% higher than that of a carbon steel car.

Operating cost. The operating cost is directly related to the overall weight of the railcar. Weight saving means a power consumption saving and is particularly noticeable when a train has to make frequent stops followed by rapid acceleration as is the case of commuter and subway trains. For a commuter car running 15 000 km/year, the power cost saving has been estimated at between 600 and 700 US$ per year and per 1000 kg of weight saved, depending on the cost of energy. Additional operating cost savings result from reduced brake and track wear.

Maintenance and repair costs. The cost of maintenance represents the most significant factor in determining the total life cycle cost of a railcar. It includes not only the necessary periodic washing but also the repairs that have to be done throughout the life of the railcar.

Available experience dating back more than 50 years for stainless steel car bodies proves that repairs of corrosion damage are not to be expected for a stainless steel car during its life cycle, usually assumed to be 30 to 35 years for rolling stock. In case of collision damage, repairs of stainless steel damaged body parts can be undertaken without special difficulty: stainless steel sheet material is currently available and welding is easy to work out.

Despite improvements in protective paint and new design of some body parts in order to avoid corrosion traps, repairs of corrosion damage are still unavoidable on carbon steel cars, after a service life varying between 10 to 12 years.

Aluminium car bodies must be painted as protection against pitting corrosion which occurs on uncoated aluminium surfaces and repainting is necessary every ten years. If repairs of body parts become necessary, in case of collision damage for instance, these repairs are costly due to the problems of supplying extruded sections and due to welding difficulties.

A study was carried out in Japan[2] which compared 20-year maintenance costs for stainless steel commuter cars with similar cars in carbon steel and aluminium. The result of this study, which took into account the effects of interest rates and inflation, are shown in Table 5 and illustrated by Fig. 5. Although the

Table 5. Railcar maintenance expenses (Japanese data in real value terms) (US$).

Year	Carbon steel car	Aluminium car	Stainless steel car
2	1,234	752	196
4	1,319	788	213
6	1,413	827	233
8	1,514	868	254
10	1,624	912	278
12	28,556*	6,629*	830*
14	1,877	1,012	333
16	2,019	1,065	364
18	2,176	1,124	399
20	77,534**	9,767**	1,673**
Total	119,266	23,744	4,773
* of which			
– Repainting	11,514	1,308	196
– Body repairs	10,467	0	0
** of which			
– Repainting	21,651	2,803	773
– Body repairs	46,695	0	0

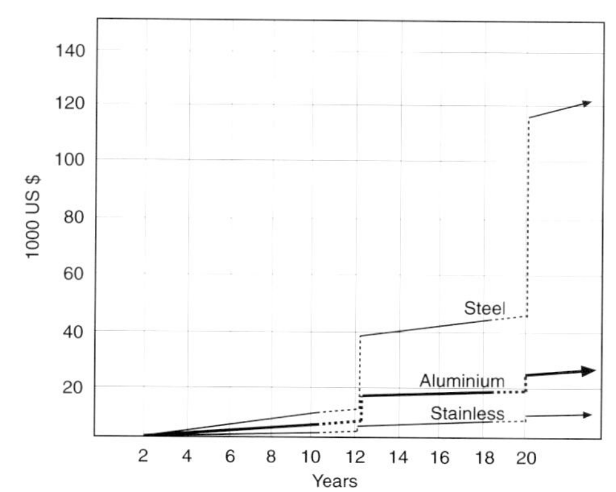

Fig. 5 Cumulative maintenance costs.

cost of maintaining a stainless steel car is marginally lower during the first ten years of service, it is after that period that the real benefit begins to show, when repairs of corrosion damage and repainting are required for the carbon steel car, and when it is also necessary to repaint the aluminium car.

Total life cycle cost. Several life cycle cost comparisons have been undertaken. In particular, the Swedish choice of material to use for the fabrication of the high speed train X2000; the result showed that the stainless steel alternative was less costly than aluminium and carbon steel ones.[3]

A computer program for IBM-compatible personal computers has been recently developed by the European Stainless Steel Development and Information Group EUROINOX.[4] It allows a comparison, in present value terms, of all related future costs that may be incurred as a consequence of an initial investment decision, in the event of using three different materials.

From the results of these comparisons and from the experience of stainless steel railcars maintained in service all over the world often for more than 40 years, the following conclusions can be stated:

- *When comparing carbon steel and stainless steel railcars*, it is clear that the heavy repairs requested on a carbon steel car after ten or twelve years of service easily offset the slightly higher initial cost of a stainless steel car. This is even truer if energy savings resulting from the lighter weight of the stainless steel car are taken into account.
- *When comparing aluminium and stainless steel railcars*, the essential factor to consider is again the maintenance and repair costs and, given that initial costs are similar, the advantage falls in favour of the stainless steel car, because the aluminium car requires repainting after ten to twelve years of service. The higher cost of repair of an aluminium car in case of collision must also be taken into account.

4 THE USE OF STAINLESS STEEL RAILCARS IN THE WORLD

Stainless steel has been used in railcar fabrication for a long time: in the United States, France and in some other countries, stainless steel railcars were put in to service a few years before World War II. Today, the use of stainless steel is growing steadily all over the world.

The development of stainless steel railcars is to be analysed according to the type of rolling stock:

- Commuter cars, subway cars and light rail trains
- Long distance cars and high speed trains

Commuter Cars, Subway Cars and Light Rail Trains

Due to frequent stops and the need for rapid acceleration, weight saving is a major consideration for these vehicles, and the material choice is usually limited to stainless steel and aluminium. While these two materials offer similar weight (and therefore energy-saving potential), there are additional benefits from using stainless steel. They are:

- durability;
- low maintenance cost (and therefore life cycle cost);
- safety in case of collision;
- safety in case of fire;
- easy removal of graffiti.

Tables 6, 7 and 8 give a list of stainless steel commuter and subway cars in service in various parts of the world. It is interesting to point out the following:

Table 6. Stainless steel railcars world development: commuter cars and subway cars in Europe.

Country	Commuter cars	Subway cars
France	All one-deck cars Regional transport cars	
Germany	S-Bahn (only roof & sidings) Berlin S-Bahn (chromium S.S.)	Hamburg subway
Greece		Athens Pyraeus metro
Portugal	All commuter cars	Lisbon subway
Eurotunnel	All shuttle cars	

Table 7. Stainless steel railcars world development: commuter cars and subway cars in America and Africa.

Country		Commuter cars	Subway cars
America	USA	St. Steel cars Dominant – New York (Metro north) – San Francisco – New York (LIRR) – Baltimore – Boston – Detroit etc ...	St. Steel cars Dominant – Atlanta – Baltimore – Chicago – Miami – New York (NYCTA) – New York (PATH) – Philadelphia – Los Angeles etc ...
	Canada	– Montreal	
	Brasil		– Sao Paulo – Recife – Rio de Janeiro – Brasilia
Africa	Egypt		– Cairo subway
	South Africa	– National Railways cars	

Table 8. Stainless steel railcars world development: commuter cars and subway cars in Asia and Oceania.

	Country	Commuter cars	Subway cars
Asia	Hong Kong		Hong Kong subway
	Japan	Most commuter cars operated by JR and other companies	Tokyo subway
	Korea		Seoul–Pusan Taeju–Inchon
	Malaysia	Kuala Lumpur	
	Taiwan	Taipeh commuter cars	
Oceania	Australia	Stainless steel dominant Victoria, NSW, Brisbane, West Australian etc....	

- in the United States, stainless steel is commonly used for commuter and subway services. One of the decisive reasons to choose stainless steel for subway cars is fire resistance;
- in Japan, stainless steel now accounts for no less than 60% of all commuter cars constructed;
- in other East Asian countries (Hong Kong, Korea, Taiwan), use of stainless steel railcars is in rapid expansion;
- in Europe, the shuttle cars used by the Eurotunnel link under the English Channel were built of 100% stainless steel. For these unusually large railcars, only stainless steel was able to meet the requirements of corrosion resistance, stiffness and fire resistance.

Cars for a Light Rail Train (LRT) small sized units need a lighter frame than other railcars. Aluminium is in a better position to compete with stainless steel as far as weight is concerned. But the weight difference is small and the maintenance cost is in favour of stainless steel. The same considerations apply to street cars (tramways).

4.2 Long Distance Cars and High Speed Trains

For long distance cars, weight saving is not as important as for commuter cars but the light weight of a stainless steel car, compared to a carbon steel car, remains an advantage. It is clear that a life cycle cost comparison is in favour of stainless steel cars because repairs of corrosion damage and repainting are still unavoidable after 10 years of service for carbon steel cars. The use of stainless steel adds safety advantages of particular importance for long distance cars in case of collision or fire. Stainless steel has been used for long distance cars since its first introduction in the railcar field. Many examples of long service life can be found in the United States, Canada, France, Portugal, South Africa and

Table 9. Stainless steel railcars world development: long distance cars and high speed trains in Europe.

Country	Long distance cars	High speed trains
France	Trans European Express	Stainless steel use under study
Germany	Intercity cars (roof & sidings only)	Stainless steel use under study
Portugal	All Intercity cars	
Sweden		X 2000 Sprintor

Table 10. Stainless steel railcars world development: long distance cars and high speed trains in America, Africa, Asia and Australia.

	Country	Long distance cars	High speed trains
America	USA	Stainless steel dominant (Amtrack, Caltrans, . . .)	
	Canada	VIA train	
Africa	Algeria	SNTF Intercity cars	
	South Africa	Intercity cars	
	Saudi Arabia	Intercity cars	
Asia	Japan	Tilting trains	
	Malaysia	Eastern and Oriental	
	Thailand	Express	
Oceania	Australia	All Intercity cars	

Australia. Tables 9 and 10 list some of the stainless steel long distance cars actually in service across the world.

The same considerations apply to high speed trains as far as life cycle cost and safety are concerned. Weight saving also becomes an essential point with the very high speeds that are recently being considered. The required increase of power induces an increase in the engine's weight, so that a lightweight body structure becomes necessary to maintain the maximum axle load authorised by the track. Here, there are also special benefits in using a material which does not require painting, as painted surfaces are quickly damaged by the impact of small stones and other particles hitting a train travelling at speeds of 400 km/hr or higher. An important development of High Speed Trains is forecast for the future in the world: stainless steel construction will certainly take part in this development.

CONCLUSION

In many countries around the world, railcar operators have become increasingly aware of the advantages of choosing stainless steel as the principal material of construction. Due to recent advances in stainless steel metallurgy, the designer/builder may select a grade which offers the ideal combination of high strength, weldability and corrosion resistance that is essential in this application. In addition, the use of stainless steel for railcar construction appears to be an advantageous economic choice as long as life cycle costs are considered; it is a modern concept, supported nevertheless by an experience dating back more than 60 years.

REFERENCES

1. *Stainless Steel for Durability, Fire Resistance and Safety*, a publication of the Nickel Development Institute, NiDI Technical Series No. 10041.
   Stainless Steel, The Effective Solution, a 20-minute video sponsored by Avesta Sheffield, the International Chromium Development Association, the Nickel Development Institute and Ugine SA.
2. *Development of Stainless Steel Railroad Cars in Japan*, a Nickel Development Institute Review Series No. 13006.
3. *Life Cycle Cost, LCC, A new approach to material selection*, Avesta AB, Information Series Engineering and Economy.
4. *Life Cycle Costing Programme* – EUROINOX, Postfach, 308 CH 8053, Zurich, Switzerland.

GUIDELINES OF PRODUCT AND PROCESS INNOVATION IN SHIPBUILDING

M. BREDA

Fincantieri, Coutieri Navali Italiani S.p.A., Methods Service – Technical Department, Via Genova, I–34100 Trieste, Italy

The development of the marine transport market requires shipbuilders to achieve a higher degree of competitiveness by an innovative approach to product definition and construction. Reducing costs, increasing the quality of the product and reducing engineering and manufacturing lead-time are all essential.

WHY IS THE MARKET BECOMING MORE COMPETITIVE?

After a long period of slack market demand, the rise of cruise passenger transport has meant an improvement in the market trend and particularly for Fincantieri. This has been coupled with a renewed growth in conventional shipbuilding (i.e. tankers, bulk-carriers, container ships etc.).

THE DYNAMICS OF INNOVATION

Today's latest cruise vessels are significantly different from the old passenger ships. Whereas the latter were used as transport means on transoceanic routes, today's cruise vessels are designed for holiday-makers. This affects the design, the engineering and manufacturing of the ship, the planning of new materials and systems, the product life-cycle, maintenance, refitting and so on.

WEIGHT REDUCTION AND SHIP STABILITY

To build a large cruise vessel demands the weight be reduced drastically in order to comply with very stringent stability conditions. Thinking of the weight balance between the hull and the added fixtures (components) we have to consider that the average ratio is 1:1 for a cruise vessel, whereas for a conventional ship it is 3:1. How to design under such severe weight constraints? The solution in our opinion lies in designing:

- lay-out arrangements (machinery accommodation, public areas etc.) of optimal size;
- an optimal integration of the fixtures within the hull structure;
- paying particular attention to the choice of materials, especially for the

added components which could be built in alternative materials to steel (i.e. polymeric matrix composites for pipes, funnels, cabin walls and ceilings, spryed thermal and fireproof insulations etc.);
- adopting innovative engineering solutions for hydraulic, mechanical, electric and HVAC plants such as:
  - press fitted piping systems;
  - prefabricated galvanised steel pipes (std. DIN 19530);
  - prefabricated air ducts and more efficient air distribution and treatment machines (local fan-coils etc.);
  - optical fibres and digital controls for automation.

NEW MATERIALS AND NEW ENGINEERING SOLUTIONS

The main applications and technical solutions adopted by Fincantieri, are explained in greater detail. At first we have to consider that there are two main alternative solutions in design planning to weight reduction and compliance with stability criteria.

The first is to build the upper decks of the ship using light alloys. This, however, is not as satisfactory as it may seem for in fact it implies:

- higher costs of materials;
- hull structures built and designed other than in steel;
- manufacturing and operating processes more expensive than steel-making;
- insulation in order to comply with very severe fire protection rules (see SOLAS – IMO);
- evaluating and solving a number of fatigue, ageing and corrosion problems;
- eventual irregularites due to noise and vibrations.

The second is Fincantieri's approach to redesign and to build the hull structure optimising the integration between the hull and the added components and achieving the required weight reduction on the latter. Instead of using steel and copper alloys pipe components, manufactured by in-house workshops, we prefer to use:

- GRP pipes complying with IMO rules (low flame spread and fire resistant) which allow us to achieve both an easier installation and an easier maintainance as well as to avoid corrosion troubles;
- prefabricated galvanised steel pipes internally coated with polymeric resins or press fitted stainless steel pipes;
- wall and ceiling panels for cabins and public areas completely made of incombustible composite matrix which:
  - have a better external lining;
  - are specifically designed to support designated loads;
  - have a better noise insulation;

- bring about a significant weight reduction;
- are more easily assembled due to kit assembly composition.

The polymeric composite materials are also used for structural and non-structural applications on the hull where you need a flexible configuration, moulding facilities or the like.

Another important field of innovative approaches are engineering applications for:

- energy production and distribution;
- HVAC distribution;
- hydraulic controls;
- mooring systems;
- passengers comfort, i.e. active noise and vibration control of exhaust gases and air conditioning machines

However steel remains the most common material in shipbuilding insofar as it is also affected by an innovation process involving:

- the wider use of high tensile steel products to counterbalance the reduction in the hull weight as well as the hull's new design configuration;
- an improvement in coating protection;
- the development of new methodologies and operation procedures designed to cut, to weld and to assemble the hull components.

Other points to be considered in relation to engineering and production methods involve cost reduction and delivery times. The use of advanced CAD/CAE systems is a successful approach because by allowing for 'virtual modelling' it enables one to achieve:

- time reduction in engineering activities making them practically contemporary to the construction work;
- a good integration between components manufacture and the hull structuring;
- a higher quality of technical documentation.

In relation to the construction activities in the yard it is possible to pursue the aim of:

- higher standards of manufacturing accuracy;
- on-shore prefabrication and components' assembling, which allow a reduction in transformation costs and in production lead-time.

Moreover the production environment is characterised by important developments due to the introduction of a 'Flexible Manufacturing Technology' named FASP – Flexible Automation in Shipbuilding Prefabrication – using robots and N.C. machineries operating within an unattended environment.

COOPERATING WITH OTHER PRODUCERS

The processes mentioned above are best developed through a joint effort of a number of producers' associations and bodies i.e. Shipbuilders, Classification Societies, Material and Engineering Suppliers. The main tools for involving them are:

- partnerships;
- co-design and co-engineering methodologies;
- value analysis tools;
- technical cooperation inside the International Committee for Rules and Regulations.

CONCLUSION

The above summary of innovative guidelines is little more than a brief report of all the available solutions. The solutions that need to be adopted to advance to the rank of 'competitive manufacturer' can be achieved through careful creative effort. This is the only way to advance an industrial policy by means of actual technological and systems innovations.

MOTOR VEHICLES: MATERIALS AND PROCESSES

C. V. FOLONARI
FIAT Research Centre, Turin, Italy

INTRODUCTION: PROCESSES OF CHANGE IN THE MOTOR VEHICLE INDUSTRY

The process of change in the motor vehicle industry is mainly evolutionary in nature. This is reflected in current trends in the vehicle industry. Present indications, based on 'industrial experience' recorded in the main car producing centres, confirm there is a general trend towards the use of new technologies for small and medium scale car production. Components of the new technology include the use of composite materials, a switch to bonded materials, adoption of light alloys and the introduction of mixed and mechanical joint technology in place of welding.

Innovations of this sort are always linked to substantial changes in overall vehicle design. An illustration is the switch from monocoque bodies to space-frame construction. Or, alternatively, existing design approaches find new uses. Perturbations in this pattern of gradual evolution may arise for all sorts of reasons. For example, a reduction in the cost of materials such as carbon fibre would modify construction. Other foreseeable incentives to change include solutions to the problem of recycling composite material, or pollution legislation.

Here we review future trends in the use of the main materials used in vehicle manufacture and forecast their likely application in car body shells.

CONFLICTING DEMANDS AFFECTING MOTOR VEHICLE DEVELOPMENT

Conflicting demands affecting motor vehicle development can be illustrated with the help of a simple diagram. This shows the factors which need to be taken into account when designing a vehicle.

MATERIAL AND PROCESS REQUIREMENTS

These design considerations are affected by pressures to reduce the cost of car manufacture, improve product quality and take account of the potential for recycling at the end of the vehicle's life. More specifically, the manufacturing process has to take account of the following factors:

- The need to reduce overall vehicle cost both in materials and assembly
- Function integration

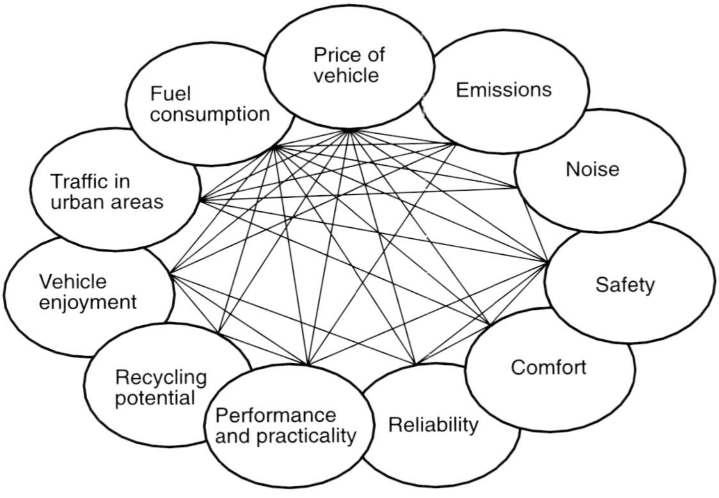

Fig. 1 Conflicting demands affecting motor vehicle development.

- New production technologies
- Elimination of components having no added value
- Simplification of manufacturing processes
- Quality and reliablitiy
- Recycling potential

THE FUTURE USE OF FERROUS ALLOYS

As a result of these cost and quality pressures, the share of each material used in a typical vehicle is likely to change as follows.

The share of ferrous material in vehicle body weight is likely to decline from 66 per cent to 55 per cent over fifteen years. But this also conceals an overall

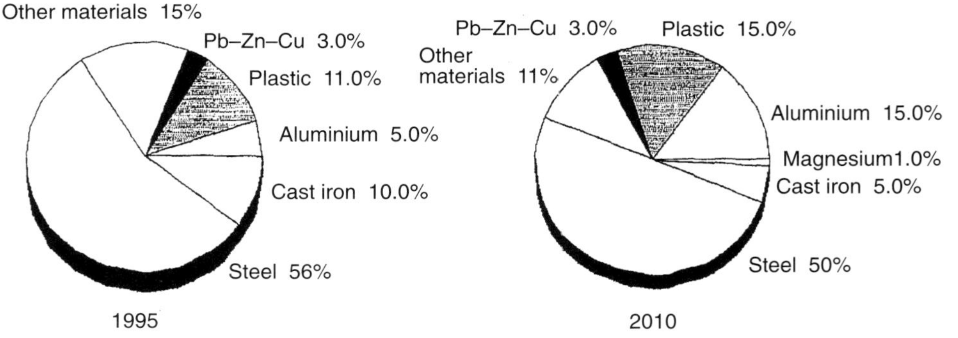

Fig. 2 Percentages of each material used in the vehicles.

drop in vehicle body weight of some 10 per cent. So steel and cast iron are likely to account for a smaller proportion of a lighter vehicle.

Weight reduction will be achieved in two ways. Overall project optimisation will result in a lighter design. In addition, use of continuous joints between components will make bodies lighter. Bonding technology and laser welding will replace spot welding of overlapping joints, for example.

THE FUTURE USE OF ALUMINIUM ALLOYS

Growing future use of aluminium alloys is illustrated in figure 3.

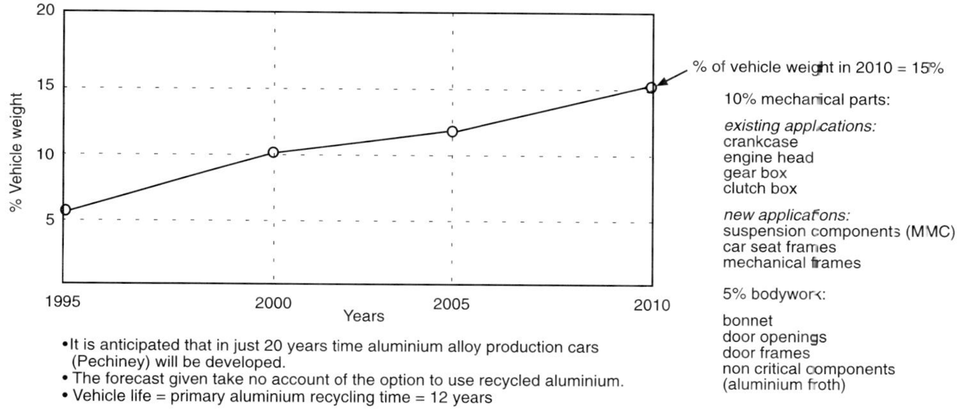

Fig. 3 Anticipated future use of aluminium alloys.

Out of the total 15 per cent of overall vehicle weight accounted for by aluminium alloys in the year 2010, some 10 per cent will be in the form of mechanical components and 5 per cent as part of the vehicle bodywork.

Mechanical Parts Formed from Aluminium Alloys

existing applications
- crankcase
- engine head
- gear box casing
- clutch box

new applications
- suspension components
- car seat frames
- mechanical frames

Bodywork Formed from Aluminium Alloys

- bonnet
- door openings
- door frames
- non-critical components

It is anticipated by Pechiney, the French aluminium producer, that in just 20 years time complete aluminium alloy production cars will be developed. These forecasts take no account of the option to use recycled aluminium components. This implies a recycling time of 12 years, since the stock of aluminium parts will turn at the end of the life of a typical vehicle – some 12 years.

THE FUTURE USE OF PLASTIC MATERIALS

Further development in the use of plastics material depends upon four factors:

- the potential of plastic for recycling
- the cost of plastic material
- likely production volumes
- the recycling time.

The anticipated future use of plastic materials as a proportion of vehicle weight is shown in figure 4.

By now, all possible uses for plastic materials in *non*-structural components of vehicles have been exhaustively investigated. Their use is strictly linked to their recycling potential. For instance, the option to use mechanically specialised

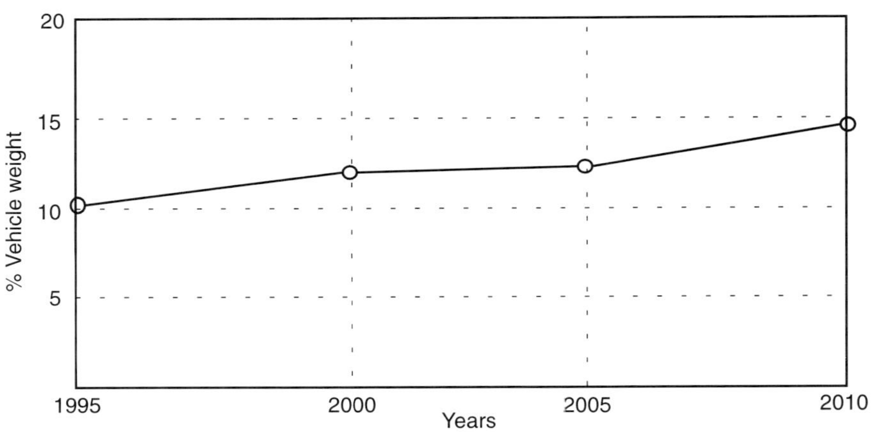

- All possible uses for plastic materials in non structural components are exhaustively investigated.
- Their use is strictly linked to their recycling potential.
- The option to use highly mechanically specialised composites (fibre toughened) is limited by recycling problems which are quite independent of the type of matrix in question.

Fig. 4 Anticpated future use of plastic materials.

composite materials toughened with fibres is limited by recycling problems. The difficulties of recycling composite materials are quite independent of the type of matrix in question. They all pose problems.

An estimated future use of plastic materials of only 15 per cent of total vehicle weight is conservative. A second scenario can be suggested where plastic materials are increasingly employed in structural applications until they form up to 60 per cent of the vehicle's weight. This optimistic forecasts implies:

- a drop in the price of carbon fibre to $4/kg
- a solution to the problem of repairing minor damage.

THE FUTURE USE OF MAGNESIUM ALLOYS

World production of magnesium is 300,000 tonnes, of which 20% is used by the motor vehicle industry, amounting to some 60,000 tonnes. In addition, there is 50,000 tonnes of Israeli production dedicated solely to Volkswagen, giving a total usage of 110,000 tonnes by the motor industry. The use of primary magnesium is required to assure and guarantee good mechanical characteristics for the resulting magnesium alloy components. More widespread use will be limited by the number and capacity of recycling plants. Viewed overall, the percentage proportion of magnesium used in vehicles is unlikely to exceed 1%.

FORECASTS RELATING TO THE MAIN TECHNOLOGIES

The future of the main vehicle maufacturing technologies can be summed up as follows.

thumbs down:	Welding Function integration
thumbs up:	Bonding New transformation technologies
thumbs down:	Mechanical manufacture Greater use of light alloys
thumbs up:	Net shape

Net Shape

Net shape manufacture, or near net shape manufacture, of components has considerable advantages in terms of cost reduction. It permits rapid manufacture and requires less machining and fabrication. However, developments are still needed in pouring processes for cast components.

Essentially, two different approaches could be adopted to the production of net shape parts. The first is the *direct* approach. The evolution of rapid prototyping into rapid manufacturing will enable highly complex shaped components to be produced for small production runs by means of layered manufacturing methods. The *indirect* approach is to rely on net shape components being supplied from the foundry from near net shape casts.

In turn, net shape production can be exploited by *rapid tooling*. Direct net shape technologies will enable highly complex, low cost dies to be manufactured for both sheet pressings and injection moulding. As a result, innovative press technologies and rapid tooling will be more extensively applied in vehicle manufacture.

FORECASTS FOR THE MAIN APPLICATIONS

Bodyshells

Here multi-material designs will be able to deliver the greatest weight savings (up to 50%). It is therefore likely that materials such as aluminium, steel and composites will be integrated into the bodyshell in a variety of different applications. Space frames will be limited to use in 'niche' vehicles. Steel space frames are only likely to be employed for production runs of up to 100,000 vehicles a year.

Bodyshells are set to benefit from new joining technologies. Mechanical joint techniques and bonding ought to apply to approximately 30% of a vehicle bodyshell, thereby reducing the number of weld points by at least a half. Door frames can be rolled and folded from aluminium components rather than fabricated. Again, hydroformed steel tubes are likely to be used for scaffolding components, and hydroformed components for steel space frames. Mixed structural bonding will need to be adopted, for instance to join aluminium to steel or metals to thermoplastics.

The Weight Reducing Steel Space Frame

The *U*ltra *L*ight *S*teel *A*uto *B*ody project (USLAB), in which thirty-two steel manufacturers are participating, is aimed at illustrating the feasibility of developing a steel bodyshell having similar or superior characteristics to current steel bodyshells, but weighing less than aluminium shells. The first stage of the project concentrated on developing a new design concept and on redesigning key structural elements.

The resulting steel bodyshell has been developed and verified against FEM calculations. It weighs only 205 kg, which is 66 kg less than the average for the nine reference bodyshells: *Honda Accord, Accura Legend, BMW 5, Mercedes 190E, LS400, Mazda 929, Ford Taurus/Mondeo, Toyota Cressida* and *Chevrolet Lumina* (table). Briefly, flexural and torsional rigidity have been increased on average by 74%, with the weight being reduced by 35% and costs by approximately 15%. Technologies used to achieve this include: multi-thickness metal sheets, bonding and laser welding, and hydroformed tubes. Two-thirds of the materials used are categorised as being high endurance (35 BH and above).

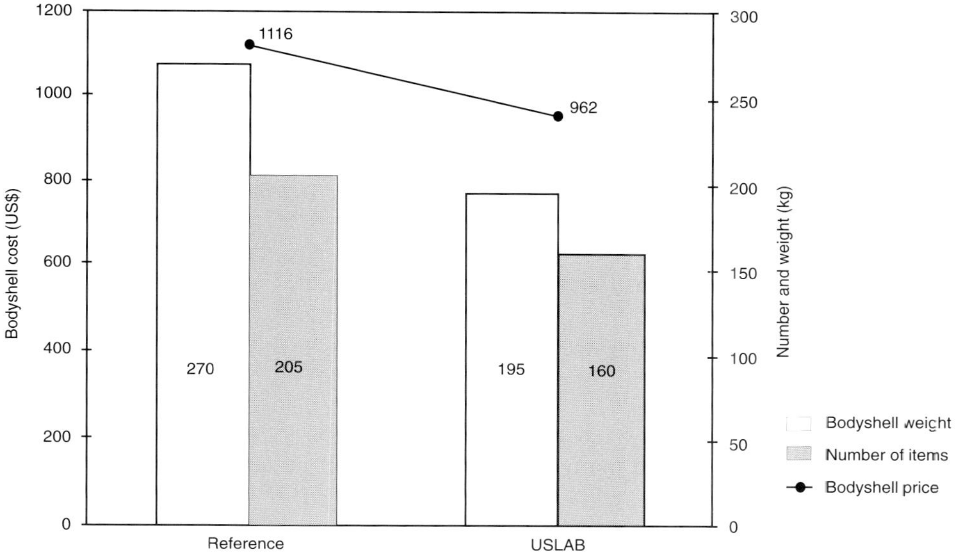

Fig. 5 Comparison between a steel space frame (USLAB) and a conventional bodyshell.

USLAB will continue to develop the prototype within the next two years and take steps to 'sell' the new desigm concept to motor manufacturers.

Chassis

We can summarise the main technical developments in chassis design and manufacture as follows. There will be extensive use of aluminium (both extrusions and castings) in mechanical frames, suspension components and brake systems. Hydroformed tubes are likely to be used for steel frames.

Production Technologies

In the assembly area there will be greater use of mechanical connectors (such as bolts) or laser welding of aluminium frames to replace conventional spot welding of steel bodyshells. Laser or water-jet cutting will be used for finishing hydroformed components. The wider use of precision castings will be accompanied by high speed finishing operations (600–900 m/s) using diamond tipped tools. Manufacturing operations will be conducted on highly abrasive materials. In general, car production will be characterised by rapid manufacturing for small production runs.

Section IV
Conclusion

SUMMING UP: PROSPECTS FOR THE STEEL INDUSTRY INTO THE NEW MILLENNIUM

WILLIAM HOGAN, S.J.
Industrial Economics Research Institute, Fordham University, East Fordham Road, Bronx, New York 10458, USA

The conference covered a multitude of subjects on a global basis. To present conclusions, one must do so on a global basis. Therefore, I will discuss some aspects of the present world situation and make some observations on the future.

Crude steel produced on a worldwide basis in 1950 was slightly less than 200 million tonnes. In 1974, it had more than tripled to 703 million tonnes, an increase of 350%. In 1995, output had reached 748 million tonnes, a really small growth of 7% over 21 years.

In terms of the first period, 1950 to 1974, the worldwide increase in output of products made from steel was tremendous. One example is the growth in worldwide automobile production from 10 million units in 1950 to 35 million in 1974. Further, a capital goods boom as well as a significant growth in infrastructure was reflected in steel demand. In addition crude oil production rose from 4 billion barrels in 1950 to over 20 billion in 1974. On the production side, Japan increased from 5 million tonnes of crude steel to 117 million during this period.

In contrast to the global growth which took place between 1950 and 1974, there was a distinct decline in the industrialised countries of the world for the next 20-year period up to the present. For example, in the United States, crude steel production fell from 136 million tonnes in 1973 to 91 million in 1994. In Japan, crude steel production declined from 117 million tonnes to 98 million in the same period. Germany witnessed a drop from 53 million tonnes in 1974 to 41 million, in spite of the fact that the latter figure included both eastern and western Germany. In the former Soviet Union, due to its dissolution, crude steel production fell from 160 million tonnes to less than 80 million.

It is interesting to note that the decline in steel production in the past 20 years took place principally in the industrialised countries. In contrast, lesser developed countries, such as China, India, Brazil, South Korea and Taiwan, showed a remarkable increase over the same period, thus balancing out the decline in the industrialised countries, so that worldwide production showed a slight increase. For example, China increased its output between 1980 and 1995 from 37 million tonnes of crude steel to 93 million tonnes. India increased production from 7 million tonnes in 1974 to 20 million in 1995. South Korea rose from 2 million tonnes in 1974 to 37 million in 1995. Taiwan increased from less than

600 000 tonnes to 11.6 million in the same period. Other countries in southeastern Asia, such as Malaysia, Indonesia and Thailand, showed a distinct percentage growth; in some instances well over 100% between 1987 and 1995. In 1974, Brazil produced 7.5 million tonnes of crude steel which increased to 25 million tonnes in 1995.

The decline in the production of crude steel in the industrialised countries was the result of a number of factors, among which the installation of continuous casting was probably the most significant, since yield from crude steel to finished product rose from approximately 70% to well over 85% in these countries, thus less crude steel was required for an equivalent tonnage of shipments. The steel industry itself contributed to this decline through the production of lighter steels.

Capital investment changed to a great extent from large buildings and machinery requiring heavy amounts of steel to investment involving computer installations requiring less steel. Further, the automobile industry, which is a principal market for steel in the industrialised countries, reduced the size of its product. In the United States, this accounted for a reduction of 10 million tonnes of steel shipments during this period. The lesser developed countries, on the other hand, witnessed a boom in capital investment as well as the development of more consumer goods requiring steel, such as automobiles and appliances.

So much for the past, we now look to the future in terms of growth to the year 2000, which represents a projection of five years that can be reasonably forecast.

ABOUT THE FUTURE

The growth of the steel industry will not be in the industrialised world. Industrialised countries, including the United States, Japan and Western Europe, have very modest plans for the next 10 years. In the United States, increases in capacity will come through the installation of minimills, particularly those entering the sheet business. In Japan, there are very modest plans for increases in capacity. Most of the major companies have no plans. In Western Europe, the same applies; perhaps a few minimills but not many of these.

The growth of the industry will be centred in the developing countries, principally in Asian countries, such as China, India, Taiwan, South Korea, Malaysia, Indonesia and Thailand. Significant tonnages will be added in China and India as well as in South Korea. The smaller countries will have large percentage increases; however, tonnages will not be that significant. For example, a country with a million and a half tonnes in 1995 may well increase by 50%, thus adding three-fourths of a million tonnes by the year 2000.

South Korea, China and India all have ambitious plans. These will involve not only increases in steel, but also in iron ore to support the steel. The iron will be from iron ore or pretreatments of iron ore before it is reduced to iron. There will also be electric furnaces in these areas. This depends to a great extent on the availability of scrap and substitutes as well as the availability and cost of electric power. Miniature blast furnaces will be constructed in India as a means of providing iron and avoiding the very large capital costs of a large blast furnace.

Because of the limitations of time, I will confine my discussion of individual countries to four: India, China, South Korea and Taiwan.

India

A task force appointed by the Ministry of Steel in 1992 drew up a 20-year plan for the growth of the Indian steel industry. The action plan had projected a crude steel demand of 37 million tonnes by the year 2012. This is to be provided by 11 million tonnes from electric furnaces and 26 million from others which include basic oxygen and a small portion from the open hearth. This requires an addition of some 17 million tonnes from the crude steel production figure of 20 million tonnes in 1995.

One asks the question 'What is the possibility of achieving this growth?' In examining announced plans we find that a number of integrated plants are to be built and others are to be modernised and expanded. Two integrated plants, one by the Tata organisation and the other by Indian Seamless Tube, are planned and will be located in the state of Orissa. The Tata plant is ultimately to be 10 million tonnes; the first phase of which will amount to 2.5 million tonnes and will be built by the year 2000. The second phase will be 5 million tonnes and the final phase 2.5 million tonnes. Tata has also undertaken a significant modernisation plan for its existing equipment which will be completed in 1998. Indian Seamless Tube's mill will be an integrated plant also located in Orissa with 1.2 million tonnes.

Another expansion programme has been undertaken by the Steel Authority of India at its Bokaro and Bhilai plants. The Bokaro expansion will be carried out in two phases. The first phase will increase crude steel output by 500 000 tonnes to 4.6 million, and the second will expand it to 7.5 million tonnes. The Bhilai plant will increase its capacity to 6 million tonnes with the addition of 1 million tonnes in the next five years and another million tonnes thereafter.

The minimill segment of the industry in India will add significant capacity. At one time, there were over 100 mills in operation. However, because of the lack of scrap, many of these were shut down. Presently, Mukand, the country's largest electric-furnace steelmaker, is planning to build 1.5 million tonnes of capacity. Further, there is a group of 14 companies all with projects to build steel capacity which will amount to about 20 million tonnes.

In addition to expanded steel capacity, a number of mini blast furnaces capable of producing between 75 000 and 300 000 tonnes of pig iron annually have been built and at least six are under construction. Mini furnaces are constructed at a cost of some US$15–17 million, thus saving the investment required for a large blast furnace of over US$200 million.

In terms of direct reduction of iron, a number of plants are under construction or consideration throughout the country. Some of these are coal-based and others are gas-based. Currently, there are about 5 million tonnes of direct-reduced-iron capacity in the country with an additional 1 million tonnes planned.

China

The steel industry in China has grown remarkably in the past 15 years. The increase in tonnage has been staggering. In 1980, China produced 37 million tonnes of crude steel. By 1995, the figure had risen to 93 million tonnes, a rate of growth which is outstanding in the history of the steel industry on a world basis.

Currently, China is the third largest steel producer in the world and has plans to expand to over 100 million tonnes before the turn of the century. If this is achieved, there is a very strong probability that China will surpass Japan in crude steel output and become the world's leading steel producer by the year 2000. The central factor in the plans for expansion was three 10-million-tonne plants, which would add 30 million tonnes to China's capacity after the year 2000. The plants were to be built in stages of approximately 3 million tonnes each, so that a significant portion of their capacity would be functioning by the turn of the century. However, a significant change has been announced by the Ministry of Metallurgical Industry.

Their statement is 'It is time for us to change our concept from the pursuit of quantity to quality, since steel demand will not grow very fast in the years to come.' This statement indicates that China will indefinitely postpone the construction of new large integrated steel plants.

There are a number of other projects in addition to the three 10-million-tonne plants that will most probably be pursued to their completion, particularly those that have already been started. The movement from emphasis on quantity to quality would indicate that such items as continuous casting, by which China now processes 47% of its crude steel output, will be increased, hopefully to about 70% by the turn of the century.

Further, the amount of crude steel made by the open-hearth process in 1992 was approximately 14 million tonnes. Much of this is considered obsolete and should be replaced. China also has plans to build plants with thin-slab casters to make sheets.

In addition, a number of joint ventures have been entered into with outside countries, such as South Korea and Japan. It is relatively certain that these will be completed, and thus, the steel production in China will grow in the next few years. Presently, the steel industry in China has three 8-million-tonne companies which rank respectively with the steelmakers of the world. They are in 15th, 17th and 18th position. There are a number of other integrated companies with somewhat less capacity and a large number of smaller producers.

By the year 2000, there is a strong possibility that Chinese crude steel production will be in excess of 100 million tonnes, despite the indefinite postponement of the three 10-million-tonne plants. Other construction projects will add a significant tonnage to the capacity of the steel industry of China by the turn of the century.

South Korea

The steel industry in South Korea was virtually non-existent in 1968, when the Pohang Iron and Steel Company (POSCO) was formed. It built its first plant at Pohang and operated the first phase in 1974. At that time, it produced in excess of 1 million tonnes. Three more phases were added to the plant which brought its capacity up to 9.4 million tonnes. In 1985, construction was started on a second integrated plant at Kwangyang. This proceeded through four phases with the final tonnage at 11.4 million tonnes in 1992. Over a 20-year period, investment in the two plants exceeded US$12 billion for a capacity of approximately 21 million tonnes. In 1995, POSCO produced 23 million tonnes which moved it into second position among the steel companies of the world, outranked only by Nippon Steel Company of Japan. Total crude steel production for the country was 37 million tonnes.

Future plans involve an increase in steelmaking capacity for POSCO of some 8 million tonnes, 6 million of which will be located in South Korea and 2 million in joint ventures in other countries of southeastern Asia. POSCO's current expansion programme involves a large Corex unit at the Pohang works, which is the second in the world and is much larger than the currently operating unit in South Africa. There was also a plan to install two Corex units of the same size, i.e. 700,000 tonnes a year, at the Kwangyang plant. However, this program has been cancelled and has been replaced by a blast furnace capable of producing 3 million tonnes of iron.

At the Kwangyang facility, a complete minimill will be installed with two DC electric furnaces and two thin-slab casters, as well as a four-high, strip mill with five stands. There is also consideration being given to two more thin-slab casters for this facility. As all of these are completed, it is more than possible that POSCO will be the largest steel company in the world.

In addition to POSCO's tonnage, which will be 28 million tonnes in Korea and 2 million tonnes outside of Korea, there are 17 minimills located in the country with a capacity of over 12 million tonnes. These minimills are expanding their capacities considerably and adding Corex units. The plan is to add a minimum of 10 million tonnes to the minimill segment, thus bringing South Korea's total production to over 50 million tonnes in the next few years.

South Korea has modern facilities and, unlike China and India, its continuous-casting rate is over 95% of crude steel produced. There is no question about the addition of this steelmaking, since the financing is available, and some of it is under construction. Of the three countries, India, China and South Korea, South Korea is the most certain to finish its expansion plans.

Taiwan

There are a number of other countries in Asia that have definite plans to add steelmaking capacity through the next five years. One of these is Taiwan which grew in steel output from less than 500,000 tonnes in 1969 to 11.6 million in

1995. In spite of this increase in production, the country imported over 14 million tonnes in 1994.

There are significant plans for the future in terms of steel production. China steel is adding 2.4 million tonnes to boost its capacity to 8 million tonnes. Other Steel companies have plans for the construction of integrated steel plants, one to the extent of 5 million tonnes and another with a capacity of 6.5 million tonnes. Other finishing facilities include a plant for structural steel, as well as cold-rolling capacity for sheets. Total expansion for steel capacity through the next few years adds up to 14 million tonnes which, added to its current capacity, will provide a total capacity of 25 million tonnes by the turn of the century.

MARKET

The market for steel in the Asian countries should grow appreciably in the next five years as the automobile industry increases its output. In the period from 1990 to 1994, automotive production in Asia, exclusive of Japan, increased from approximately 2 million vehicles to over 5 million. Currently, most of the major automobile producers in the world have already built plants in Asia within the last few years and are planning to increase their presence there. General Motors is participating in a US$1.5 billion plant in China and Ford in a US$700 million plant in India. European companies such as VW are also involved. As a consequence, it is conceivable that vehicle production in Asia, exclusive of Japan, will increase by more than 50% by the year 2000, resulting in an increased market for steel. This will require an improvement in the quality of the steel in some of the Asian countries. Thus, in China and India, additional investment will be needed to improve not only the amount of continuous casting, but also some of the rolling and finishing facilities to meet the specifications required by the automotive industry for passenger cars.

OTHER ASIA

Other countries in Asia, including Malaysia, Indonesia and Thailand, have relatively small steelmaking capacities, ranging from 1 million tonnes to 4 million. These countries have plans to increase the tonnage they produce mainly through joint ventures with other countries such as South Korea and Japan.

CONCLUSION

To sum up, the growth of the steel industry will be in Asia during the next five years. In 1995, Asia, excluding the output of Japan, produced 175 million tonnes of crude steel. Without question, this will be increased by 30–40 million tonnes in the next five years, raising the figure beyond 200 million tonnes. The industrialised countries have very limited plans, and in some cases no plans for increasing steel capacity. Thus, the world will look to Asia for growth in the steel industry in the century ahead. This has implications not only for the steel industry, but for the suppliers of steel mill equipment.

The general theme of this conference, which has treated a wide variety of subjects, can be characterised by the word 'change'. In almost every session, this has been evident including the sessions on technology which have reviewed some of the radical changes that have taken place in the past decade, the changes in environmental regulations which have become more stringent, and the changes from government ownership to private ownership which have been pronounced in many countries the world over.

In the future, more changes can be expected as research in the development of technology continues and is, if anything, accentuated. The fact is that environmental regulations will undoubtedly become stricter and that government ownership will decline as many of the steel companies are transferred into private ownership throughout a number of countries.

INDEX